ANNALS *of* THE NEW YORK ACADEMY OF SCIENCES

T0188666

EDITOR-IN-CHIEF
Douglas Braaten

ASSOCIATE EDITOR
Rebecca E. Cooney

PROJECT MANAGER
Steven E. Bohall

EDITORIAL ADMINISTRATOR
Daniel J. Becker

Artwork and design by Ash Ayman Shairzay

The New York Academy of Sciences
7 World Trade Center
250 Greenwich Street, 40th Floor
New York, NY 10007-2157

annals@nyas.org
www.nyas.org/annals

**The New York
Academy of Sciences**

Published by Blackwell Publishing
On behalf of the New York Academy of Sciences

Boston, Massachusetts
2012

ANNALS *of* THE NEW YORK ACADEMY OF SCIENCES

VOLUME
1253

ISSUE
Glycobiology of the Immune Response

ISSUE EDITORS
Gabriel Rabinovich,[a] Brian Cobb,[b] and Yvette van Kooyk[c]

[a]Institute of Biology and Experimental Medicine, Buenos Aires, [b]Case Western Reserve University, and [c]VU University Medical Center, Amsterdam

TABLE OF CONTENTS

1   Glycobiology of immune responses
Gabriel A. Rabinovich, Yvette van Kooyk, and Brian A. Cobb

16   Multifarious roles of sialic acids in immunity
Ajit Varki and Pascal Gagneux

37   Siglecs as sensors of self in innate and adaptive immune responses
James C. Paulson, Matthew S. Macauley, and Norihito Kawasaki

49   Interleukin-2, Interleukin-7, T cell-mediated autoimmunity, and N-glycosylation
Ani Grigorian, Haik Mkhikian, and Michael Demetriou

58   T cells modulate glycans on CD43 and CD45 during development and activation, signal regulation, and survival
Mary C. Clark and Linda G. Baum

68   Interplay between carbohydrate and lipid in recognition of glycolipid antigens by natural killer T cells
Bo Pei, Jose Luis Vela, Dirk Zajonc, and Mitchell Kronenberg

80   Galectins in acute and chronic inflammation
Fu-Tong Liu, Ri-Yao Yang, and Daniel K. Hsu

92   Mechanisms underlying *in vivo* polysaccharide-specific immunoglobulin responses to intact extracellular bacteria
Clifford M. Snapper

102   CD33-related siglecs as potential modulators of inflammatory responses
Paul R. Crocker, Sarah J. McMillan, and Hannah E. Richards

112 Sulfated glycans control lymphocyte homing
Hiroto Kawashima and Minoru Fukuda

122 Acute phase glycoproteins: bystanders or participants in carcinogenesis?
Eugene Dempsey and Pauline M. Rudd

133 Glycans, galectins, and HIV-1 infection
Sachiko Sato, Michel Ouellet, Christian St-Pierre, and Michel J. Tremblay

149 An evolutionary perspective on C-type lectins in infection and immunity
Linda M. van den Berg, Sonja I. Gringhuis, and Teunis B.H. Geijtenbeek

159 Integrated approach toward the discovery of glyco-biomarkers of inflammation-related diseases
Takashi Angata, Reiko Fujinawa, Ayako Kurimoto, Kazuki Nakajima, Masaki Kato, Shinji Takamatsu, Hiroaki Korekane, Cong-Xiao Gao, Kazuaki Ohtsubo, Shinobu Kitazume, and Naoyuki Taniguchi

170 Novel roles for the IgG Fc glycan
Robert M. Anthony, Fredrik Wermeling, and Jeffrey V. Ravetch

181 The effect of galectins on leukocyte trafficking in inflammation: sweet or sour?
Dianne Cooper, Asif J. Iqbal, Beatrice R. Gittens, Carmela Cervone, and Mauro Perretti

193 Engineering cellular trafficking via glycosyltransferase-programmed stereosubstitution
Robert Sackstein

201 The expanding role of α2-3 sialylation for leukocyte trafficking *in vivo*
Markus Sperandio

206 Beyond glycoproteins as galectin counterreceptors: tumor-effector T cell growth control via ganglioside GM1
Robert W. Ledeen, Gusheng Wu, Sabine André, David Bleich, Guillemette Huet, Herbert Kaltner, Jürgen Kopitz, and Hans-Joachim Gabius

Online only

Polarization of host immune responses by helminth-expressed glycans
Donald Harn Jr., Smanla Tundup, and Leena Srivastava

Carbohydrate-recognition in the immune system: contributions of NGL-based microarrays to ligand discovery
Ten Feizi

Diversity in recognition of glycans by F-type lectins and galectins: molecular, structural, and biophysical aspects
Gerardo R. Vasta, Hafiz Ahmed, Mario A. Bianchet, José A. Fernández-Robledo, and L. Mario Amzel

Ann. N.Y. Acad. Sci. ISSN 0077-8923

ANNALS OF THE NEW YORK ACADEMY OF SCIENCES
Issue: *Glycobiology of the Immune Response*

Glycobiology of immune responses

Gabriel A. Rabinovich,[1,2] Yvette van Kooyk,[3] and Brian A. Cobb[4]

[1]Laboratorio de Inmunopatología, Instituto de Biología y Medicina Experimental (IBYME), Consejo Nacional de Investigaciones Científicas y Técnicas (CONICET), [2]Laboratorio de Glicómica Funcional, Departamento de Química Biológica, Facultad de Ciencias Exactas y Naturales, Universidad de Buenos Aires, C1428 Ciudad de Buenos Aires, Argentina. [3]Department of Molecular Cell Biology and Immunology, VU University Medical Centre, Amsterdam, the Netherlands. [4]Department of Pathology, Case Western Reserve University School of Medicine, Cleveland, Ohio

Addresses for correspondence: Gabriel A. Rabinovich, gabyrabi@gmail.com; Yvette van Kooyk, Y.vanKooyk@vumc.nl; or Brian A. Cobb., brian.cobb@case.edu

Unlike their protein "roommates" and their nucleic acid "cousins," carbohydrates remain an enigmatic arm of biology. The central reason for the difficulty in fully understanding how carbohydrate structure and biological function are tied is the nontemplate nature of their synthesis and the resulting heterogeneity. The goal of this collection of expert reviews is to highlight what is known about how carbohydrates and their binding partners—the microbial (non-self), tumor (altered-self), and host (self)—cooperate within the immune system, while also identifying areas of opportunity to those willing to take up the challenge of understanding more about how carbohydrates influence immune responses. In the end, these reviews will serve as specific examples of how carbohydrates are as integral to biology as are proteins, nucleic acids, and lipids. Here, we attempt to summarize general concepts on glycans and glycan-binding proteins (mainly C-type lectins, siglecs, and galectins) and their contributions to the biology of immune responses in physiologic and pathologic settings.

Keywords: glycobiology; glycoimmunology; glycans; lectins; C-type lectins; siglecs; galectins

Glycobiology—an overview

Calling nucleic acids "cousins" of carbohydrates is an accurate representation of the fact that both DNA and RNA are essentially *polysaccharides* composed of phosphate-linked polyribose cores. Without carbohydrates, nucleic acids would not form the linear scaffolding required for their function. Likewise, proteins and carbohydrates can be termed "roommates" because, in mammals, carbohydrates are nearly always associated with either proteins or lipids. With some notable exceptions (e.g., β_2-microglobulin and the galectin family), all secreted proteins and all cell surface proteins and lipids contain significant amounts of covalently attached glycans. In some cases the molecular mass of a glycoprotein can be as much as 50% carbohydrate or more,[1] with the carbohydrates falling into two categories, N-linked or O-linked (Fig. 1), depending on the nature of the linkage to the underlying protein backbone.[2,3] Molecular pathways required

for N- and O-linked carbohydrate pathways, or at least primitive versions of them, are also present in the microbial world, with the most remarkable example being *Campylobacter* species that can add both O- and N-linked carbohydrates to their proteins (e.g., flagellin).[4] Importantly in humans, loss of genes in these pathways (conditions known as *congenital disorders of glycosylation*) is often embryonically lethal; but when individuals survive, the pathologies present are typically quite devastating.[5–7] Remarkably, the mammalian glycome repertoire is estimated to include thousands of glycan structures generated by the concerted action of an endogenous group of glycosyltransferases and glycosidases.[8]

The general physiologic roles of carbohydrates and their binding partners are highly diverse and include critical functions in cell–cell communication and adhesion,[9] protein structural stability,[10] membrane structure,[11] and cellular signaling.[12] Indeed, while carbohydrates play as varied roles in

doi: 10.1111/j.1749-6632.2012.06492.x

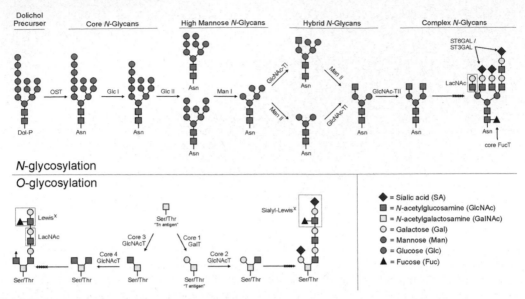

Figure 1. Schematic of the N- and O-linked glycosylation pathway in mammals. (Top) The asparagine (N)-linked pathway begins in the endoplasmic reticulum (ER), where OST (oligosaccharyltransferase) moves the core *N*-glycan from the dolichol precursor to an asparagine residue with the N-X-S/T consensus sequence. This structure is trimmed by glucosidases (Glc I and Glc II) within the ER, which assist in the folding quality control system mediated by the calnexin/calreticulin pathway. Once released from this quality control, the nascent glycoprotein traffics to the Golgi apparatus where further trimming occurs initially, which is then followed by the creation of significant diversity through addition of other saccharides by a variety of transferases ("T" in the abbreviations) in a nontemplate-driven process. Another key addition are 2,3-linked and 2,6-linked terminal sialic acids, which are critical for a number of biological functions. Finally, the LacNAc disaccharide unit (*N*-acetyllactosamine) is a key recognition site for a number of glycan-binding molecules, including some of the galectin family. (Bottom) The serine/threonine (O)-linked glycosylation pathway is distinct from the N-linked pathway in a number of ways. The nature of the linkage and the enzymes involved are separate, and the resulting structures are quite divergent, although some similarities exist, such as the presence of the LacNAc unit. The O-linked glycans are broken down into core subgroups and carry names like "Tn antigen" and "Sialyl-Lewis^x" that are common in the literature. These have been labeled for reference.

biology as proteins do, they have been problematic to study due to the inherent differences between the complex synthetic pathways for carbohydrates and the template-driven synthesis of nucleic acids and proteins, and this has unfortunately led many investigators to ignore the "complicating factor" that carbohydrates can seem to represent. One particularly stark example is the understanding of protein structure and function, much of which depends on the decades of crystallography that have created a cache of structural information with which we interpret key functional data. Protein crystallography depends on well-ordered crystals forming. But the typical mammalian pattern of glycosylation often precludes the formation of high-quality crystals. As a result, the approaches to determining the three-dimensional structure of a glycoprotein include recombinant overexpression in *E. coli* to prevent the

addition of glycans completely[13] and expression in Sf9 insect cells,[14] which produces proteins that carry only limited mannose structures that resemble the hybrid *N*-glycan structures shown in Figure 1 and are more homogeneous than the mammalian counterparts.[15] At the same time, a wealth of data have shown that fully complex-type glycans characteristic of mammalian glycoproteins are required for native function. One can immediately recognize the problem with assigning a single structure or function—or a list of binding partners—to a glycoprotein when the protein used in the experiments lacks the native glycosylation patterns found *in vivo*. As of 2011, the reality is that our structural knowledge of how carbohydrates—integral components of proteins—(possibly as many as 80–90% of all human proteins are glycosylated) may affect function remains in its infancy.

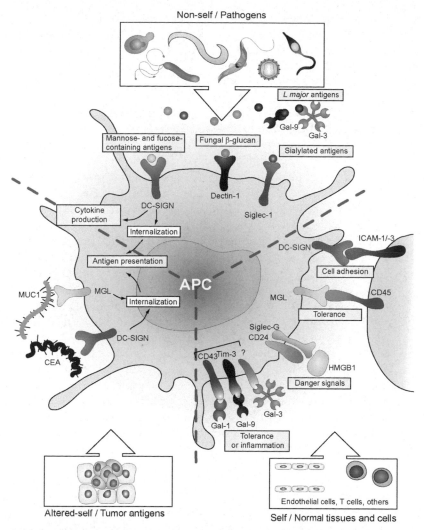

Figure 2. Lectin-glycan interactions in innate immunity: Discrimination of non-self, altered-self, and self by antigen-presenting cells (APCs). GBPs (e.g., DC-SIGN, Dectin-1, siglec-1, galectin-3 (Gal-3), galectin-9 (Gal-9)) can function as pattern recognition receptors (PRR) through the recognition of non-self glycans exposed on different pathogens, including viruses, bacteria, yeasts, and parasites. In particular those CLRs with glycan specificity for Lewis and mannose glycans (such as DC-SIGN) have been shown to bind a multitude of pathogens such as HIV-1 and other viruses, bacteria such as *Mycobacterium tuberculosis* and *Helicobacer pylori*, helminths such as *Schistosoma mansoni*, and yeasts such as *Candida albicans*. On the other hand, Gal-3 and Gal-9 play key roles as soluble PPRs by discriminating *Leishmania* species. Also, CLRs (e.g., MGL, DC-SIGN) can detect changes in glycosylation of certain tumor-associated antigens, such as the carcinoma embryonic antigen (CEA) and MUC1 occurring during onco-transformation and tumor progression. These changes include increased expression of the Lewis blood group family of antigens, particularly Le^x, and Le^y, that are often associated with poor prognosis of the tumor. This recognition allows antigen internalization, presentation to CD4$^+$ T cells, cross-presentation to CD8$^+$ T cells, and potentiation of antitumor immunity, although in some cases these interactions can also lead to inhibition of T cell responses. In addition, GBPs may play important roles in self-recognition in a variety of cellular processes, including (among others) cell adhesion (DC-SIGN-ICAM1/3 interactions), T cell signaling (MGL interactions with GalNAc-expressing CD45 glycoforms), discrimination of danger signals (interactions among siglec-G, CD24, and the HMGB1 alarmin), and modulation of immunogenic or tolerogenic APC programs (e.g., interactions between galectins and APC glycoproteins, such as Tim-3 and CD43).

Another point worth making is the issue of heterogeneity, the central complicating aspect of glycobiology. Complex-type *N*-glycans, for example, are hallmarks of mammalian glycoproteins, and as one moves down the evolutionary ladder protein glycosylation becomes simpler despite the genetic similarities among organisms. As such, it seems entirely reasonable to suggest that the inherent heterogeneity associated with these pathways is integral to their function in biology. The reductionist in each of us wants our experiments to be as clean as possible by limiting the variables, but perhaps it is the very molecular heterogeneity that we try to avoid in experiments that is the key to unlocking our understanding life and the differences among species and differences associated with disease. If correct, it is impossible to study systems biology in the absence of understanding and investigating the contributions of glycobiology and the underlying carbohydrate heterogeneity to biological processes.

Although one can paint a relatively bleak picture of our ability to dissect the global complexities of carbohydrate biological function, the last decade has seen some remarkable progress in terms of the creation of tools to determine glycan structure and diversity, as well as in the collaborative efforts to pool expertise in tackling the "glycomics problem." Under the leadership of James Paulson at the Scripps Research Institute, the Consortium for Functional Glycomics (CFG) was created and funded by the National Institute of General Medical Sciences. Although the ten years of funding is now complete, the CFG was responsible for the creation and development of numerous tools to assist investigators with understanding the role of carbohydrates in a variety of biological systems. One particularly important resource is the glycan-binding array in which hundreds of carbohydrates representing a wide variety of structures were covalently attached to a chip for binding studies.[16] Another key resource created by the CFG is a gene chip including genes that encode all carbohydrate-associated proteins and pathway enzymes, which enables studies on the regulation of glycosylation and other mechanisms as a function of disease state.[17] The collective efforts represented by the CFG and other worldwide efforts, such as EuroCarb and the Japanese Consortium for Glycobiology and Glycotechnology, the world of glycomics has begun to open to the nonspecial-

ist and will likely lead to rapid progress in understanding the multifaceted roles of carbohydrates in physiology.

Glycans and the immune system—an introduction

The focus of this issue of *Annals of the New York Academy of Sciences* is one part of the overall picture of carbohydrate structure and function: the immune system. The immune response is an ideal system in which to begin painting a larger picture of carbohydrates and their role in biology, because of the importance of glycans and their binding partners in nearly every aspect of immunology (Fig. 2).

As with all cells, essentially all surface-localized immune receptors are glycoproteins, including pattern recognition receptors (PRRs) such as toll-like (TLR) and NOD-like receptors, the class I and class II major histocompatibility complex proteins (MHC class I and MHC class II), chemokine receptors, cytokine receptors, T and B cell coreceptors, and the T and B cell receptors themselves. The role of the glycans on these molecules is as varied as the molecules themselves. For example, prevention of complex-type *N*-glycan synthesis on T cells substantially increases T cell receptor signaling, making T cell activation much less stringent.[18] In addition, the loss of complex-type *N*-glycans on MHC II significantly impedes the ability to present bacterial polysaccharides to T cells,[19] whereas mutagenesis of TLR molecules to remove sites of glycosylation directly influences their signaling.[20,21]

Other components of the immune system are the various glycan-binding proteins (GBPs) or lectins, including the C-type lectins, siglecs, and galectins. Through binding to lectins and sterically modulating molecular interactions, mammalian glycans participate in a diversity of cellular mechanisms that contribute to innate and adaptive immune responses.[8] The theme with these molecules is that they are carbohydrate-binding molecules, and they also tend to be glycoproteins, with the exception of the galectin family. One well-known family member is DC-SIGN, a C-type lectin receptor (CLR) found on the surface of dendritic cells (DCs) that serves to bind a variety of mannose- and fucose-containing ligands, including glycoproteins of the envelope of human immunodeficiency virus (HIV), and signals into the cell for activation.[22] Another example is CD22, a member of the siglec (sialic acid–binding Ig-like lectin) family and a crucial part of the B cell receptor complex that assists with antigen uptake and B cell activation.[23,24] Finally, the galectins are thought to be able to associate with host glycans, such that the cell surface can be organized into receptor lattices for optimal receptor spacing and signaling.[11,25]

Secreted glycoproteins in the immune system include essentially all cytokines and chemokines, many of the complement components, and antibodies. One recent and important discovery is that the *N*-glycans attached to the Fc portion of IgG molecules modulate antibody activity,[26,27] for example, antibodies' terminal sialic acids send an inhibitory signal to responding immune cells, whereas those missing terminal sialic acids send activating signals. Another pathway to note is the lectin pathway for complement activation, which requires the mannose-binding lectin and ficolins, rather than the classical components such as C1q.[28]

Two other areas of note are lymphocyte development and leukocyte homing. The Notch family of molecules relies upon glycan changes, especially the addition of fucose residues, to mediate T and B cell differentiation into the various known lineages from common progenitors.[29] For homing, migrating cells rely upon the selectin family of adhesion molecules on circulating leukocytes; the selectins bind to specific carbohydrates expressed on the surface of endothelial cells near sites of inflammation and infection.[30,31] This enables the cells to home to sites where they are needed.

Turning to microbes, a majority of their immunogens are glycans or glycan conjugates, and nearly all bacteria and viruses are coated with a thick glycocalyx (sugar coat). For HIV, its often-cited gp120 coat protein is glycosylated to the point that most of the underlying protein is inaccessible to antibodies.[32] Some bacteria are completely encapsulated with polysaccharides,[33] whereas others have thick coats of endotoxin and other glycolipids. Gram-positive bacteria also have externally exposed peptidoglycans. From these groups, essentially all of the so-called *pathogen-associated molecular patterns* can be found (e.g., endotoxin/LPS, viral coats, muramyl dipeptide, and bacterial capsules). An important point is that many pathogens have evolved to make host-like carbohydrates at their cell surface to mimic the host and evade immune detection.[34] This fact alone shows the importance of carbohydrate immunogens, because if the immune system did not recognize carbohydrates specifically, there would be no need to be tolerant of self-carbohydrates, and pathogens would not be able to hide through mimicry.

Although the diversity of glycans and GBPs, coupled with the diversity of pathways involved in shaping and directing the immune response, is as broad as immunology itself, this issue touches on several key topics under active investigation around the world. These areas are described below and further elaborated on in the papers collected in this special issue.[a]

GBPs in self, non-self, and altered-self recognition

Several GBPs are expressed on cells of the immune system and/or released to the extracellular milieu where they play essential roles in the control of innate and adaptive immunity.[35] These include the families of CLRs, siglecs, and galectins that mainly function through recognition of defined carbohydrate structures decorating glycoproteins or glycolipids.[35] All these GBPs carry one or more carbohydrate recognition domains (CRD) that coordinate the interaction with a specific glycan. CLRs are calcium-dependent carbohydrate binding proteins, where specific amino acids in the CRD coordinate the carbohydrate ligand and a Ca^{2+} ion.[36] C-type lectins that contain an Glu–Pro–Asn (EPN) amino acid motif, such as DC-SIGN, mannose receptor (MR), Langerin, and L-SIGN/DC-SIGNR,[37] have specificity for mannose- and/or fucose-terminated glycans. In contrast, galactose-specific C-type lectins like macrophage galactose lectin (MGL) and DC-ASGPR,[38,39] which contain the Gln–Pro–Asp (QPD) sequence in the CRD, recognize galactose-terminated or *N*-acetylgalactosamine (GalNAc)-terminated glycan structures. The type II subfamily of CLRs, of which 17 members have been cloned in humans, is mainly restricted to antigen-presenting cells (APCs) such as macrophages and DCs, whereas some of them have also been identified on NK cells or endothelial cells.[40] Most

---

[a]Rabinovich, G., Cobb, B., van Kooyk, Y., eds. 2012. Glycobiology of the Immune Response. Special issue, *Ann. N.Y. Acad. Sci.* **1253**.

C-type lectins have internalization motifs in their cytoplasmic tail and play a role in the uptake of glycosylated antigens that they recognize.[41,42] In particular, CLRs are involved in antigen presentation, as they take up and route antigens to the MHC class II loading compartments to present the antigen in the context of MHC class II molecules to prime antigen-specific CD4[+] T cell responses.[43] Several C-type lectins expressed on DCs have also been reported to shuttle antigen intracellularly to the MHC class I loading compartment through a yet uncharacterized cross-presentation route, thereby promoting antigen-specific CD8[+] T cell responses.[44–47] CLRs can have inhibitory or activating domains in the cytoplasmic tail, leading to signaling that either negatively or positively influences immune responses.[37,48] This illustrates that amino acid sequences of CLRs can have profound regulatory functions in antigen processing and presentation for instructing T cells and altering signaling cascades.

Mice also express C-type lectins, but certain members of the mouse CLR family differ substantially from human CLRs, making the direct comparison of mouse data to a human setting difficult. For example, DC-SIGN has various homologues in the mice, such as SIGNR1, SIGNR3, and mDC-SIGN.[49] In total, seven DC-SIGN homologues have been identified, but none of them has the exact APC-restricted expression or glycan specificity as human DC-SIGN. Also, for human MGL, which exhibits GalNAc specificity, two mouse homologues have been described, MGL1 and MGL2, with glycan specificity of Le[X] and GalNAc, respectively.[50,51] On the other hand, some CLRs can be quite homologous between human and mice, such as Langerin, on Langerhans cells in the skin, and Dectin-1, which recognizes β-glucans on yeasts.[52]

GBPs have been shown to be involved in pathogen binding and therefore function as PRRs through the recognition of glycans exposed on pathogens, ranging from viruses and bacteria to yeasts and parasites.[23] In particular those receptors with glycan specificity for Lewis and mannose glycans (such as DC-SIGN) have been shown to bind a multitude of pathogens, such as HIV-1 and other viruses, bacteria such as *Mycobacterium tuberculosis* and *Helicobacter pylori*, helminths such as *Schistosoma mansoni*, and yeasts such as *Candida albicans*.[23,48,53,54] Glycan structures on these pathogens—high mannose,

Lewis X or Lewis Y, or LDNF—are recognized by DC-SIGN. Pathogen-binding CLRs are also internalization receptors, by which recognition of pathogens results in the uptake of the pathogen by APCs.[55] Because CLRs play essential roles in the recognition of pathogens, they are formally considered PRRs, similar to TLRs; however, the function of CLRs is distinct from TLRs because the latter are not thought to internalize antigens for processing and presentation by MHC class I or II molecules. Surprisingly, many reports also illustrate the different signaling function that CLRs play by interfering with TLRs or Fc receptor (FcR) triggering. In particular, CLRs may signal via Syk- or Raf-1–dependent pathways upon interactions with specific glycans.[48,56,57] CLRs are equipped with signaling motifs in their cytoplasmic tails and/or associate with signaling complexes, although some CLRs (e.g., DC-SIGN) may only induce gene expression in the presence of other PRRs. Upon recognition of fungal β-glucan structures, the well-described CLR Dectin-1 triggers signaling through the Syk tyrosine kinase and activation of ERK- and JNK-mediated cascades, which, in turn, activate the transcription factor nuclear factor kappa B (NF-κB).[48,52,58] These signaling events are dependent on the adaptor CARD9 and are independent of TLR signaling. Other CLRs, such as DC-SIGN, interact with mycobacterial ManLAM, and thereby promote the production of proinflammatory cytokines, such as IL-6 and IL-12, via a Raf-1–mediated signaling pathway. In contrast, fucose interaction with DC-SIGN initiates Raf-1–independent signaling pathways, resulting in the induction of anti-inflammatory cytokines, such as IL-10, and inhibition of proinflammatory cytokines, such as IL-6 and IL-12.[57] This illustrates the notion that the glycan signature of pathogens can trigger different signaling processes, all through the same CLR.

Most GBPs are not pathogen-restricted, as they often interact with self-glycoproteins that have exposed mannose/fucose or GalNAc glycan structures.[59] These glycans can be found on proteins or lipids, and the multivalency of glycans can often increase the avidity to GBPs.[60] Tumor antigens commonly show an altered glycan composition that is reflected by a shorter glycosylation backbone or by the multivalent exposure of glycan epitopes on repeated sequences. This is well illustrated by the mucin repeated sequences in the tumor antigen MUC-1, which shows increased binding to MGL, or by the binding of tumor-specific carcinoembryonic antigen (CEA) to DC-SIGN.[61,62] Most often, GBPs can recognize single glycan structures; but in a few cases the protein backbone is also crucial for specific glycans to be recognized by CLRs. As an example, DC-SIGN does not recognize sialylated glycans, yet sialylated IgGFc is recognized by DC-SIGN and has been shown to contribute to the anti-inflammatory activity of intravenous immunoglobulins (IVIGs), together with the inhibitory FcR FcγRIIB, in promoting the production of the Th2-type cytokine IL-33 and expansion of IL-4$^+$ basophils.[26]

Pathogens can also display specific glycans, as exemplified by helminth glycans such as LDNF (GalNAcβ1–4(Fucα1–3)GlcNAc-R) and LDN (GalNAcβ1–4GlcNAc-R) that can interact with GBPs with high affinity.[53] Other pathogens, such as HIV-1, display high-mannose structures that are not particularly pathogen-specific but show high-affinity interactions for certain GBPs, such as the C-type lectins DC-SIGN, MR, langerin, and dendritic cell immunoreceptor (DCIR).[63–65] CLRs such as DC-SIGN, MR, or MGL are well-known to recognize endogenous glycoproteins, and thereby function as adhesion molecules, signaling receptors, or antigen uptake mediators.[66] DC-SIGN has been demonstrated to interact with ICAM-2 and ICAM-3 to mediate cell adhesion and rolling of DCs due to expression of Lewis Y on endothelial cells. DC-SIGN has also been reported to interact with Lewis X (Fig. 1) and Y glycans present on Mac-1 and CEACAM-1 on neutrophils, thereby allowing regulated adhesion between neutrophils and DCs.[67] The presence of GalNAc on CD45 expressed by effector memory CD4$^+$ and CD8$^+$ T cells has been shown to interact with the C-type lectin MGL, found mainly on tolerogenic APCs, and induce apoptosis and proliferation of these cells—demonstrating a homeostatic regulatory role of glycans exposed on CD45.[68] Furthermore, C-type lectins such as MR, Langerin, and DC-SIGN also recognize tissue antigens, for example serum hydrolases, tissue plasminogen activator, type I collagen, and Fc-IgG. Because MR, Langerin, and DC-SIGN have internalization capacity, their uptake serves to control homeostatic surveillance by APCs, such as DCs and macrophages.[35]

Similar to CLRs, siglecs are cell surface GBPs capable of conveying regulatory signals that positively or negatively control immune responses.

Although most siglecs are known to be negative regulators of cell signaling, because they contain immune receptor tyrosine-based inhibitory motifs (ITIMs), others have been shown to promote immune cell activation.[69] Structurally, siglecs belong to the immunoglobulin superfamily and can be divided into two subsets: the less-related group (25–35% identity), which includes sialoadhesin (Sn: siglec-1), CD22 (siglec-2), and myelin-associated glycoprotein (MAG; siglec-4), and the rapidly-evolving group of CD33-related siglecs that share high sequence similarity (50–90%) (siglec-3, -5, -7, -8, -9, -10, -11, and -14 in humans and siglec-E, -F, -G, and -H in mice).[70] The latter GBPs are defined by their well-known carbohydrate specificity for sialic acid–containing glycans, and for their ability to discriminate among different linkages (α2–3, α2–6, and α2–8).[71] These glycan structures can be found on immune cells, epithelial cells, and tumor cells. Interestingly, microbes may capture sialic acid from host cells, which prevents their immune recognition as foreign and enables infection of host target cells through specific recognition of siglecs.[71] This is clearly exemplified by *Campylobacter jejuni*, *Neisseria meningitides*, and group B streptococci, which interact with CD33-related siglecs.[72] Unique to this lectin family, siglecs can mediate *cis* and *trans* interactions with sialylated glycans. Although with *cis* interactions the siglec is often masked by low-affinity ligands on neighboring receptors, these linkages do not prevent *trans* interactions with other cell types.[69]

Although some siglecs have very restricted localization patterns, others are more widely expressed within hematopoietic cells. For example, Sn is preferentially expressed on macrophages, CD22 on B cells, and siglec-8 on eosinophils.[69] Interestingly, siglec-9 and siglec-E are selectively expressed on human and mouse myeloid–derived DCs, respectively, while siglec-5 and siglec-H are expressed on human and mouse plasmacytoid DCs, respectively.[69] For most CD33-related siglecs, ligand engagement results in tyrosine phosphorylation of ITIMs by Src family tyrosine kinases and recruitment of Src homology 2 (SH2)–containing phosphatases (SHPs), such as SHP-1 and SHP-2, which control cellular activation by attenuating tyrosine phosphorylation.[73] Recent work has identified siglec-10 in humans and siglec-G in mice as binding partners that recognize CD24 and discriminate between pathogen-associated and danger-associated molecular partners.[74] Moreover, like CLRs, siglec-H has been identified as a specific endocytic receptor on plasmacytoid DCs that can take up viruses and other pathogens for delivery to intracellular TLRs and, ultimately, for induction of antiviral immunity.[75] In contrast to other members of the family, siglec-H lacks tyrosine-based signaling motifs and associates with the adaptor DAP-12 to exert its function.[75] Thus, selective targeting of individual siglecs may serve to differentially shape immune cell signaling.

In contrast to C-type lectins and siglecs, which are mostly cell surface–associated receptors, galectins belong to a category of soluble immunomodulatory proteins that may act either intracellularly, by modulating signaling pathways, or extracellularly, as bioactive regulatory mediators.[76] Galectins bind to N-acetyllactosamine (Galβ1,3GlcNAc or Galβ1,4GlcNAc), a common disaccharide found on many N- and O-linked glycans present in different cellular counter-receptors and matricellular proteins[25] (Fig. 1). Although galectins do not have the signal sequence required for the classical secretion pathway, most of them are secreted via a nonclassical export route that is still poorly understood.[77] Traditionally, galectins are classified into *prototype* galectins (galectin-1, -2, -5, -7, -10, -11, -13, -14, -15), which have one CRD that can dimerize, *tandem repeat* galectins (galectin-4, -6, -8, -9 and -12), which contain two homologous CRDs in tandem in a single polypeptide chain, and the single member galectin-3, which is unique in that it contains a CRD connected to a non-lectin N-terminal region that is responsible for lectin oligomerization.[78] Although some galectins have a wide tissue distribution (e.g., galectin-1 and -3), others have preferential localization, for example, galectin-7 in skin, galectin-12 in adipose tissue, and galectin-10 in eosinophils and regulatory T cells.[77] Interestingly, galectin-3 and galectin-9 play key roles as PRRs by discriminating *Leishmania* species to stimulate appropriate innate immune responses.[78]

Different factors can regulate the function of galectins, including (1) their dynamic expression during the development and resolution of innate and adaptive immune responses; (2) the programmed remodeling of cell surface glycans generated by the concerted action of glycosyltransferases; (3) their dimerization or oligomerization

status; and (4) the stability of these proteins in oxidative or reducing microenvironments.[77] Through their ability to recognize specific glycan structures, galectins are potential soluble mediators involved in immune cell communication, signaling, host–microbial interactions, T helper cell homeostasis, preservation of fetomaternal tolerance, and suppression of autoimmune pathology. In addition, galectins may contribute to the creation of immunosuppressive microenvironments at sites of tumor growth and metastasis.[25]

Interestingly, accruing evidence from a number of laboratories indicates that N- and O-glycosylation can modulate interactions of receptors and ligands themselves, thereby altering receptor endocytosis, clustering, and signaling. Lectins, particularly galectins, have been shown to play critical roles in these cellular processes through binding and cross-linking specific glycans attached to a preferred set of counter-receptors, thereby forming supramolecular interactions or lattices that fine-tune the dynamics of receptor-ligand interactions.[12,79] This includes modulation of cytokine receptor signaling, T cell receptor (TCR)–mediated activation at sites of immunological synapse, and B cell receptor (BCR)–integrin signaling during B cell maturation.[12] However, galectins can also bind their ligands in a carbohydrate-independent manner, which is often the case for galectins localized in the intracellular compartment, where ligand binding occurs predominantly via protein–protein interactions.[25]

Although the *in vivo* roles of endogenous galectins are just now emerging, recent studies have demonstrated distinctive immunological phenotypes in mice lacking individual members of the family. For example, mice lacking the gene for either galectin-1 or galectin-9 show increased Th1 and Th17 responses and augmented inflammatory responses,[80–82] whereas mice deficient in galectin-3 exhibit reduced inflammation in mouse models of multiple sclerosis and arthritis.[83,84] Thus, in spite of protein sequence similarities, overlapping distribution, and functional redundancy, individual members of the galectin family may play distinct roles in the control of immune cell homeostasis. Moreover, different lectin families may overlap, synergize, and/or counter-regulate the effect of each other through recognition of a similar set of glycan ligands. Given their multifunctional properties and versatility, GBPs, including CLRs, siglecs, and galectins, are increasingly recognized as molecular targets for innovative drug discovery in immune-mediated disorders.

Glycobiology of autoimmunity and chronic inflammation

In spite of compelling evidence of the role of glycosylation in pathogen recognition, our awareness of the impact of lectin–glycan interactions in the control of immune tolerance, autoimmunity, and chronic inflammation is relatively new. These emerging data are based on pioneering observations demonstrating programmed remodeling of cell surface glycans in the transition from normal to inflamed tissues.[85] These changes affect emigration and trafficking of immune cells to sites of inflammation, as demonstrated for selectins that bind to sialylated and fucosylated epitopes (e.g., sLe[x]) often found in sulphated form on glycans present on most leukocytes and endothelial cells.[86] Thus, inhibition of selectins and glycans during leukocyte–endothelium interactions has been proposed as an attractive target for anti-inflammatory therapy.

However, glycosylation can control inflammation and autoimmunity through modulation of other homeostatic checkpoints. In fact, in several inflammatory autoimmune diseases, such as systemic lupus erythematosus and rheumatoid arthritis, effector T cells express altered cell surface glycans, especially those exposing terminal GalNAc and Gal-β(1–4)-GlcNAc structures.[87] For example, MGL and galectin interactions with these structures thereby downregulate TCR signaling and alter CD45 phosphatase activity.[18,39,68] In addition, glycosylation can alter the threshold of TCR activation, which can have profound effects on T cell function during adaptive immunity and autoimmune diseases. The GnT5 glycosyltransferase, encoded by the mannoside acetylglucosaminyl transferase 5 (*Mgat5*) gene, initiates formation of the β1–6-N-glycan-branch structure on various glycoproteins, including the TCR. This N-glycan branch typically includes N-acetyllactosamine (LacNAc; Fig. 1) disaccharide repeats that serve as ligands for galectins. T cells lacking GnT5 have a decreased threshold for T cell activation, and GnT5-deficient mice have increased delayed-type hypersensitivity responses and increased susceptibility to autoimmune diseases, including ex-

perimental autoimmune encephalomyelitis (EAE), glomerulonephritis, and immune complex deposition.[18] Interestingly, treatment with high doses of GlcNAc was shown to increase GnT5-mediated N-glycan branching and inhibited TCR activation in autoimmune models of EAE and diabetes.[88]

Although these studies highlight the relevance of glycosylation in the control of adaptive immunity and the development of autoimmune diseases, other studies have been instrumental to understanding the contribution of innate immune mechanisms to autoimmune and inflammatory settings. Deficiency in α-mannosidase-II (αM-II), which results in diminished complex-type N-glycan branching, has been shown to induce an autoimmune disease in mice similar to human systemic lupus erythematosus.[89] This effect was independent of the adaptive immune system and was linked to chronic activation of innate immune components. The authors proposed a mechanism by which the exposure of cryptic N-glycans under αM-II deficiency (i.e., mannose-dependent ligands) results in recognition by endogenous lectins (including MR) and aberrant activation of innate immune responses in the absence of infection, thus leading to the development of lupus-like autoimmune disease.[89] Thus, selective targeting of glycan-modifying enzymes, early or late during N-complex glycan branching, can result in aberrant activation of innate or adaptive immune responses.

However, it is not only N-glycans that influence the development and resolution of autoimmune response. In patients with IgA nephropathy, for example, O-glycans attached to IgA are truncated to expose GlcNAc, which correlates with antibody deposition in the inflamed kidney;[90] this suggests that differential glycosylation of immunoglobulins critically regulates their biological functions. A striking example of this is the presence of α2,6 sialic acid in the Fc region of IgG, which determines the anti-inflammatory function of IVIGs in several autoimmune diseases.[26]

Antibodies specific for glycans have been found in different settings. For example, because humans lack the enzyme that converts N-acetylneuraminic acid (NeuAc) to N-glycolylneuraminic acid (NeuGc), incorporation of NeuGc in the diet can lead to the generation of antibodies against NeuGc.[91] In addition, antibodies specific for particular glycans have also been found in the human Tn syndrome, which is characterized by thrombocytopenia and leukocytopenia and caused by a mutation in the *COSMC* gene encoding a chaperone required for β1,3 galactosyltransferase (T-synthase) activity required for *O*-glycan biosynthesis.[92]

The biological roles for the different repertoires of cell surface glycans found on different immune cell types are just emerging. For example, α2,6-linked sialic acid is selectively upregulated on the surface of Th2 but not Th1 and Th17 effector cells, and this difference determines the susceptibility of these T cell subsets to the GBP galectin-1.[80] Accordingly, galectin-1–deficient (*Lgals1*$^{-/-}$) mice develop more severe autoimmune manifestations during EAE due to selective expansion of antigen-specific Th1 and Th17 cells and to increased immunogenicity of DCs.[80,93] Therapeutic administration of recombinant galectin-1 has been shown to restore tolerance and suppress chronic inflammation in several experimental models of autoimmunity (arthritis, uveitis, EAE, hepatitis, diabetes, and inflammatory bowel disease); to prevent fetal loss; and to attenuate graft-versus-host disease by suppressing Th1 and Th17 responses, by expanding regulatory T cell populations, and by fueling tolerogenic DC circuits (reviewed in Ref. 76). These effects were also observed following administration of the tandem-repeat galectin-9, which suppresses harmful autoimmune responses through its ability to bind the Tim-3 glycoreceptor.[81]

C-type lectins can also play important roles in the modulation of inflammatory and autoimmune diseases. *In vivo* targeting of a mannosylated encephalitogenic peptide has been shown to inhibit the onset of EAE, most likely by targeting the mannose receptor on immature DCs.[94] Similarly, oral administration of mannose-enriched antigens can induce oral tolerance and favor the generation of IL-10–producing regulatory T (Tr1) cells through mechanisms involving the CLR SIGNR1 expressed on lamina propria DCs.[95] Moreover, siglec-2 (CD22) delivers inhibitory signals to B cells, and siglec-G/CD22 double knockout mice spontaneously develop B cell–dependent autoimmunity characterized by the presence of anti-DNA and antinuclear autoantibodies.[96] These data suggest that different GBPs may amplify or silence tolerogenic circuits within DC, T cell, or B cell compartments.

Glycobiology of tumor immunity

Changes in the glycosylation status of certain tumor-associated antigens (TAAs) are often accompanied by onco-transformation. Such TAAs include the CEA and MUC1.[62,63] Both antigens are expressed on cells of normal colonic mucosa and the epithelium. Changes in glycosylation include increased expression of the Lewis blood group family of antigens, particularly Le^x and Le^y, that are often associated with poor prognosis of the tumor. How these posttranslational modifications contribute to tumor cell dissemination and disease severity is not fully understood. However, recent findings demonstrate that DCs recognize these aberrantly glycosylated structures on CEA or MUC1 through CLRs, such as DC-SIGN and MGL, respectively, while these CLRs do not interact with normal CEA or MUC1 from colon tissue.[97]

Several studies have demonstrated that specific targeting of antigens to CLRs can result in more robust antitumor responses. An elegant study demonstrated that *in vivo* targeting of CLRs in the mouse can induce antigen-specific autoimmune responses when the antigen is coupled to a DEC-205–specific antibody.[98] In contrast, when the antigen coupled to DEC-205-Ab is simultaneously combined with a strong DC activator, such as a TLR agonist, a strong antigen-specific immune activation is induced that results in potentiation of antitumor responses and, ultimately, eradication of the tumor. Moreover, other CLRs have also been used for targeting purposes resulting in strong induction of immunity.[99–101]

CLEC9A is a CLR predominantly expressed on a subset of DCs in human and mice that functions to promote phagocytosis of dead cells for MHC class I–dependent cross-presentation and stimulation of CD8$^+$ T cell responses. The precise ligand recognized by CLEC9A on dead cells has not yet been identified, and it is under debate whether CLEC9A ultimately recognizes glycan structures or protein frameworks. The targeting of antigen to CLEC9A, using antibody–antigen conjugates, has led to the efficient endocytosis of antigen and to enhanced CD4$^+$ and CD8$^+$ T cell proliferation and more robust antibody responses. Furthermore, targeting of CLEC9A in cancer cells in mouse models has been shown to heighten T cell–mediated tumor rejection.[47,102] In addition, another strategy to target antigen to CLRs on DCs involves using natural ligands (glycan structures) to modify antigen.[103] The use of glycans for *in vivo* DC-targeting purposes has advantages over CLR-specific antibodies because glycans, which are nonimmunogenic and of self-origin, can be produced synthetically to large scale using simple chemistry.[104] In contrast, the production of humanized antibodies is expensive, and such antibodies can be immunogenic in patients. Furthermore, antibodies themselves are highly glycosylated, a property that is currently being capitalized on to increase or decrease their therapeutic potential *in vivo*.

Modification of the melanoma antigen gp100 with high mannose has been demonstrated to enhance both gp100-specific $CD4^+$ and $CD8^+$ T cell responses, because high-mannose structures target DC-SIGN on DCs and this leads to enhanced endosomal-dependent antigen presentation.[55] Ovalbumin (OVA) modified with fucose-containing Lewis[b] structures not only improved targeting of OVA to DC-SIGN, as shown by enhanced antigen uptake by DCs from DC-SIGN transgenic mice, but also increased cross-presentation to OVA-specific $CD8^+$ and $CD4^+$ T cells.[44] Moreover, two other MR-specific ligands, sulfo Lewis[A] or GlcNAc, significantly enhanced targeting of OVA to the MR when they were ligated to OVA, thereby increasing cross-presentation of OVA to OT-I (OVA-specific) T cells *in vitro*.[100] Interestingly, in both studies using the natural ligands to target antigen to CLR, cross-presentation was increased in the absence of DC maturation signals. Improvement in $CD4^+$ and $CD8^+$ T cell priming has been demonstrated not only by modification of antigens with glycans; glycosylated nanoparticles have also been designed that specifically target DCs to improve antigen-specific immune responses. In addition, CLRs are currently being used to enhance delivery of several vaccines to DCs, as well as to improve $CD4^+$ and $CD8^+$ T cell priming, due to the high efficacy of CLRs to promote antigen processing in these APCs.

Emerging evidence indicates that tumors may exploit lectin–glycan interactions to evade immune responses.[105] Galectin-1 secreted by tumor cells contributes to the immunosuppressive potential of different tumor types, including melanoma,[106,107] Hodgkin's lymphoma,[108] lung carcinoma,[109,110] pancreatic carcinoma,[111] and neuroblastoma,[112] by selectively modulating T cell and DC com-

partments. The mechanisms underlying this immunoregulatory effect involve a bias toward a Th2-dominant cytokine profile and activation of tolerogenic circuits mediated by IL-27–producing DCs and IL-10–producing type 1 T regulatory (Tr1) cells.[93,110] Moreover, overexpression of galectin-9 results in increased frequency of $CD11b^+Ly-6G^+$ granulocytic myeloid suppressor cells and inhibition of antitumor responses,[113] while the presence of galectin-3 appears to control the anergic state of T cells.[114] Thus, selective inhibition of individual members of the galectin family may contribute to circumventing immunosuppression at sites of tumor growth by targeting distinct immunoevasive programs.

Recent work has demonstrated a novel glycosylation-dependent mechanism of tumor-immune escape. Bladder tumor cells overexpressing the core-2 β-1, 6-N-acetylglucosaminyltransferase (GCNT1 encoded by *C2GnT*), which is responsible for elongating core-2 *O*-glycans (i.e., *O*-glycans containing a GlcNAc branch connected to GalNAc; Fig. 1), are highly metastatic due to their ability to evade NK cell immunity. The recent data suggest that interactions between poly-LacNAc residues present on core-2 *O*-glycans of tumor-associated MHC class I–related chain A (MICA) and galectin-3 reduced the affinity of MICA for the activating NK receptor NKG2D, thereby impairing NK cell activation and antitumor activity.[115] Similarly, expression of the ganglioside GD3 by tumor cells can also alter NK cell cytotoxicity via siglec-7–dependent mechanisms.[116] Moreover, mucins derived from human cancer patients can control the immunogenicity of DCs through mechanisms involving siglec-9.[117]

Collectively, these data indicate that, in addition to differential gene and protein expression profiles, the lectin repertoire, as well as the cellular glycosylation signatures prevailing in the tumor microenvironment, may provide distinctive cellular codes that influence immune cell fate and cancer progression.

Conclusions

In this issue of *Annals of the New York Academy of Sciences*, reviews are presented on the multifaceted and often unexpected roles played by carbohydrates within the immune system. These include how sialic acids control immunity; the impact of glycans and glycosylation on cancer, autoimmunity, and lymphocyte development; and the role of carbohydrates on host and microbe during infection. This brief introduction highlights some of the best-characterized areas of glycan-focused immunology research, with the goal of illustrating the fact that carbohydrates are as integral to immune pathways as are proteins and lipids. The field of glycoimmunology is just beginning to find solid grounding through improved and more accessible carbohydrate-specific analytical tools. We hope the reviews presented in this special issue will encourage the immunology community to take notice of myriad unanswered questions on how carbohydrates influence immune responses. Deep investigation of these questions will surely lead to a greater understanding of the function of the immune system, and eventually of biology as a whole.

Acknowledgments

Work in G.A.R.'s lab is supported by Grants from the National Agency for the Promotion of Science and Technology of Argentina (PICT 2010–870), the University of Buenos Aires, the Mizutani Foundation for Glycoscience (Japan), the National Multiple Sclerosis Society (U.S.A.), Fundación Sales, and Prostate Action (U.K.). Work in Y.v.K.'s lab is supported by grants from the Dutch Cancer Foundation (KWF), the Dutch Society for MS research, the Netherlands Organisation for Scientific Research (NWO), Agentschap NL, and the VUmc Institute for Cancer and Immunology. Work in B.A.C.'s lab is supported by Grants from the National Institute of Allergy and Infectious Diseases (AI079756), the National Institute of General Medical Sciences (GM082916), the National Institutes of Health, Office of the Director (OD004225), the American Asthma Foundation, and the Chronic Granulomatous Disorder Research Trust. We thank M.A. Toscano for help in figure illustration.

Conflicts of interest

The authors declare no conflicts of interest.

References

1. Levine, M.J., M.S. Reddy, L.A. Tabak, *et al.* 1987. Structural aspects of salivary glycoproteins. *J. Dent. Res.* **66:** 436–441.
2. Brockhausen, I., H. Schachter & P. Stanley. 2009. *O*-GalNAc Glycans. In *Essentials of Glycobiology*. A. Varki, R.D. Cummings, J.D. Esko, H.H. Freeze, P. Stanley, C.R. Bertozzi, G.W. Hart & M.E. Etzler, Eds.: 115–127. Cold Spring Harbor: Cold Spring Harbor Laboratory Press.
3. Stanley, P., H. Schachter & N. Taniguchi. 2009. *N*-Glycans. In *Essentials of Glycobiology*. A. Varki, R.D. Cummings, J.D. Esko, H.H. Freeze, P. Stanley, C.R. Bertozzi, G.W. Hart & M.E. Etzler, Eds.: 101–114. Cold Spring Harbor: Cold Spring Harbor Laboratory Press.
4. Nothaft, H. & C.M. Szymanski. 2010. Protein glycosylation in bacteria: sweeter than ever. *Nat. Rev. Microbiol.* **8:** 765–778.
5. Theodore, M. & E. Morava. 2011. Congenital disorders of glycosylation: sweet news. *Curr. Opin. Pediatr.* **23:** 581–587.
6. Van Geet, C., J. Jaeken, K. Freson, *et al.* 2001. Congenital disorders of glycosylation type Ia and IIa are associated with different primary haemostatic complications. *J. Inherit. Metab Dis.* **24:** 477–492.
7. Wang, Y., J. Tan, M. Sutton-Smith, *et al.* 2001. Modeling human congenital disorder of glycosylation type IIa in the mouse: conservation of asparagine-linked glycan-dependent functions in mammalian physiology and insights into disease pathogenesis. *Glycobiology* **11:** 1051–1070.
8. Ohtsubo, K. & J.D. Marth. 2006. Glycosylation in cellular mechanisms of health and disease. *Cell* **126:** 855–866.
9. Takahashi, M., Y. Kuroki, K. Ohtsubo & N. Taniguchi. 2009. Core fucose and bisecting GlcNAc, the direct modifiers of the *N*-glycan core: their functions and target proteins. *Carbohydr. Res.* **344:** 1387–1390.
10. Oberg, F., J. Sjohamn, G. Fischer, *et al.* 2011. Glycosylation increases the thermostability of human aquaporin 10 protein. *J. Biol. Chem.* **286:** 31915–31923.
11. Garner, O.B. & L.G. Baum. 2008. Galectin-glycan lattices regulate cell-surface glycoprotein organization and signalling. *Biochem. Soc. Trans.* **36:** 1472–1477.
12. Boscher, C., J.W. Dennis & I.R. Nabi. 2011. Glycosylation, galectins and cellular signaling. *Curr. Opin. Cell. Biol.* **23:** 383–392.
13. Li, Y., H. Li, N. Dimasi, *et al.* 2001. Crystal structure of a superantigen bound to the high-affinity, zinc-dependent site on MHC class II. *Immunity* **14:** 93–104.
14. Dai, S., G.A. Murphy, F. Crawford, *et al.* 2010. Crystal structure of HLA-DP2 and implications for chronic beryllium disease. *Proc. Natl. Acad. Sci. USA* **107:** 7425–7430.
15. Harrison, R.L. & D.L. Jarvis. 2006. Protein *N*-glycosylation in the baculovirus-insect cell expression system and engineering of insect cells to produce "mammalianized" recombinant glycoproteins. *Adv. Virus Res.* **68:** 159–191.
16. Blixt, O., S. Head, T. Mondala, *et al.* 2004. Printed covalent glycan array for ligand profiling of diverse glycan binding proteins. *Proc. Natl. Acad. Sci. USA* **101:** 17033–17038.
17. Comelli, E.M., S.R. Head, T. Gilmartin, *et al.* 2006. A focused microarray approach to functional glycomics: transcriptional regulation of the glycome. *Glycobiology* **16:** 117–131.
18. Demetriou, M., M. Granovsky, S. Quaggin & J.W. Dennis. 2001. Negative regulation of T-cell activation and autoimmunity by Mgat5 *N*-glycosylation. *Nature* **409:** 733–739.
19. Ryan, S.O., J.A. Bonomo, F. Zhao & B.A. Cobb. 2011. MHCII glycosylation modulates Bacteroides fragilis carbohydrate antigen presentation. *J. Exp. Med.* **208:** 1041–1053.
20. Amith, S.R., P. Jayanth, S. Franchuk, *et al.* 2009. Dependence of pathogen molecule-induced toll-like receptor activation

and cell function on Neu1 sialidase. *Glycoconj. J.* **26:** 1197–1212.

21. Amith, S.R., P. Jayanth, S. Franchuk, *et al.* 2010. Neu1 desialylation of sialyl alpha-2,3-linked beta-galactosyl residues of TOLL-like receptor 4 is essential for receptor activation and cellular signaling. *Cell Signal.* **22:** 314–324.

22. van Kooyk, Y. & T.B. Geijtenbeek. 2003. DC-SIGN: escape mechanism for pathogens. *Nat. Rev. Immunol.* **3:** 697–709.

23. Andersson, K.B., K.E. Draves, D.M. Magaletti, *et al.* 1996. Characterization of the expression and gene promoter of CD22 in murine B cells. *Eur. J. Immunol.* **26:** 3170–3178.

24. O'Reilly, M.K., H. Tian & J.C. Paulson. 2011. CD22 is a recycling receptor that can shuttle cargo between the cell surface and endosomal compartments of B cells. *J. Immunol.* **186:** 1554–1563.

25. Liu, F.T. & G.A. Rabinovich. 2010. Galectins: regulators of acute and chronic inflammation. *Ann. N. Y. Acad. Sci.* **1183:** 158–182.

26. Anthony, R.M., T. Kobayashi, F. Wermeling & J.V. Ravetch. 2011. Intravenous gammaglobulin suppresses inflammation through a novel TH2 pathway. *Nature* **475:** 110–113.

27. Kaneko, Y., F. Nimmerjahn & J.V. Ravetch. 2006. Antiinflammatory activity of immunoglobulin G resulting from Fc sialylation. *Science* **313:** 670–673.

28. Matsushita, M. 2010. Ficolins: complement-activating lectins involved in innate immunity. *J. Innate. Immun.* **2:** 24–32.

29. Stanley, P. & T. Okajima. 2010. Roles of glycosylation in Notch signaling. *Curr. Top. Dev. Biol.* **92:** 131–164.

30. Lowe, J.B. 2003. Glycan-dependent leukocyte adhesion and recruitment in inflammation. *Curr. Opin. Cell Biol.* **15:** 531–538.

31. Mitoma, J., X. Bao, B. Petryanik, *et al.* 2007. Critical functions of N-glycans in L-selectin-mediated lymphocyte homing and recruitment. *Nat. Immunol.* **8:** 409–418.

32. Pashov, A., S. Garimalla, B. Monzavi-Karbassi & T. Kieber-Emmons. 2009. Carbohydrate targets in HIV vaccine research: lessons from failures. *Immunotherapy* **1:** 777–794.

33. Krinos, C.M., M.J. Coyne, K.G. Weinacht, *et al.* 2001. Extensive surface diversity of a commensal microorganism by multiple DNA inversions. *Nature* **414:** 555–558.

34. Carlin, A.F., A.L. Lewis, A. Varki & V. Nizet. 2007. Group B streptococcal capsular sialic acids interact with siglecs (immunoglobulin-like lectins) on human leukocytes. *J. Bacteriol.* **189:** 1231–1237.

35. van Kooyk, Y. & G.A. Rabinovich. 2008. Protein-glycan interactions in the control of innate and adaptive immune responses. *Nat. Immunol.* **9:** 593–601.

36. Drickamer, K. 1999. C-type lectin-like domains. *Curr. Opin. Struct. Biol.* **9:** 585–590.

37. Figdor, C.G., Y. van Kooyk & G.J. Adema. 2002. C-type lectin receptors on dendritic cells and Langerhans cells. *Nat. Rev. Immunol.* **2:** 77–84.

38. Kawasaki, T., M. Li, Y. Kozutsumi & I. Yamashina. 1986. Isolation and characterization of a receptor lectin specific for galactose/N-acetylgalactosamine from macrophages. *Carbohydr. Res.* **151:** 197–206.

39. van Vliet, S.J., E. Saeland & Y. van Kooyk. 2008. Sweet prefer-

ences of MGL:carbohydrate specificity and function. *Trends Immunol.* **29:** 83–90.

40. Zelensky, A.N. & J.E. Gready. 2005. The C-type lectin-like domain superfamily. *FEBS J.* **272:** 6179–6217.

41. Engering, A., T.B. Geijtenbeek, S.J. van Vliet, *et al.* 2002.The dendritic cell-specific adhesion receptor DC-SIGN internalizes antigen for presentation to T cells. *J. Immunol.* **168:** 2118–26.

42. Unger, W.W.J. & Y. van Kooyk. 2011. Dressed for success; C-type lectin receptors for the delivery of glyco-vaccines to dendritic cells. *Curr. Opin Immunol.* **23:** 131–137.

43. Birkholz, K., M. Schwenkert, C. Kellner, *et al.* 2010. Targeting of DEC-205 on human dendritic cells results in efficient MHC class II-restricted antigen presentation. *Blood* **116:** 2277–2285.

44. Singh, S.K., J. Stephani, M. Schaefer, *et al.* 2009a.Targeting of glycan modified OVA to murine DC-SIGN transgenic dendritic cells enhances MHC class I and II presentation. *Mol. Immunol.* **47:** 164–174.

45. Idoyaga, J., C. Cheong, K. Suda, *et al.* 2008. Langerin/CD207 receptor on dendritic cells mediates efficiënt antigen presentation of non MHC I and II products in vivo. *J. Immunol.* **180:** 3647–3650.

46. Bozzacco, L., C. Trumpfheller, F.P. Siegal, *et al.* 2007. DEC-205 receptor on dendritic cells mediates presentation of HIV gag protein to CD8+ T cells in a spectrum of human MHC I haplotypes. *Proc. Natl. Acad. Sci. USA* **104:** 1289–1294.

47. Sancho, D., Mourao-Sa, D., Joffre, O.P., *et al.* 2008. Tumor therapy in mice via antigen targeting to a novel DC restricted C-type lectin. *J. Clin. Invest.* **118:** 2098–2110.

48. Osorio, F. & C. Reis e Sousa. 2011. Myeloid C-type lectin receptors in pathogen recognition and host defense. *Immunity* **34:** 651–664.

49. Park, C.G., K. Takahara, E. Umemoto, *et al.* 2001. Five mouse homologues of the human dendritic cell C-type lectin, DC-SIGN. *Int. Immunol.* **13:** 1283–1290.

50. Singh, S.K., I. Streng-Ouwehand, M. Litjens, *et al.* 2009b. Characterization of murine MGL 1 and MGL 2 C-type lectins: Distinct glycan specificities and tumor binding properties. *Mol. immunol.* **46:** 1240–1249.

51. Denda-Nagai, K., S. Aida, K. Saba, *et al.* 2010. Distribution and function of macrophage galactose-type C-type lectin 2 (MGL2/CD301b): efficient uptake and presentation of glycosylated antigens by dendritic cells. *J. Biol. Chem.* **285:** 19193–19204.

52. Brown, G.D. 2006. Dectin-1: a signaling non-TLR pattern-recognition receptor. *Nat. Rev. Immunol.* **6:** 33–43.

53. Van Die, I. & R.D. Cummings. 2010. Glycan mimmickry by parasitic helminths: a strategy for modulating the host immune response? *Glycobiology* **20:** 2–12.

54. Gow, N.A.R., F.L. van de Veerdonk, A.J.P. Brown & M.G. Netea. 2012. Candida albicans morphogenesis and host defense: discriminating invasion from colonization. *Nat. Rev. Microbiol.* **10:** 112–122.

55. Aarnoudse, C.A., M. Bax, M. Sánchez-Hernández, *et al.* 2008. Glycan modification of the tumor antigen gp100 targets DC-SIGN to enhance dendritic cell induced antigen presentation to T cells. *Int. J. Cancer* **122:** 839–46.

56. Gringhuis, S.I., J. van Dunnen, M. Litjens, *et al.* 2007. C-type lectin DC-SIGN modulates Toll-like receptor signaling via Raf-1 kinase-dependent acetylation of transcription factor NF-kappaB. *Immunity* **26:** 605–616.

57. Gringhuis, S.I., J. den Dunnen, M. Litjens, *et al.* 2009. Carbohydrate-specific signalling through the DC-SIGN signalosome tailors immunity to *Mycobacterium tuberculosis*, HIV-1 and *Helicobacter pylori*. *Nat. Immunol.* **10:** 1081–1088.

58. Geijtenbeek, T.B. & S.I. Gringhuis. 2009. Signalling through C-type lectin receptors: shaping immune responses. *Nat. Rev. Immunol.* **9:** 465–479.

59. Geijtenbeek, T.B.H., S.J. van Vliet, A. Engering, *et al.* 2003. Self- and non-self recognition by C-type lectins on dendritic cells. *Ann. Rev. Immunol.* **22:** 33–54.

60. Dam, T.K. & C.F. Brewer. 2010. Lectins as pattern recognition molecules: the effects of epitope density in innate immunity. *Glycobiology* **20:** 270–279.

61. van Gisbergen, K.P.J.M., C.A. Aarnoudse, G.A. Meijer, *et al.* 2005a. Dendritic cells recognize tumor-specific glycosylation of carcinoembryonic antigen on colorectal cancer cells through dendritic cell-specific intercellular adhesion molecule-3-grabbing nonintegrin. *Cancer Res.* **65:** 5935–5943.

62. Saeland, E., S.J. van Vliet, M. Bäckström, *et al.* 2007. The C-type lectin MGL expressed by dendritic cells detects glycan changes on MUC1 in colon carcinoma. *Cancer Immunol. Immunother.* **56:** 1225–1236.

63. Geijtenbeek, T.B.H., D.S. Kwon, R. Torensma, *et al.* 2000. DC-SIGN, a dendritic cell specific HIV-1 binding protein that enhances trans-infection of T cells. *Cell* **100:** 587–597.

64. de Witte, L., A. Nabatov, M. Prion, *et al.* 2007. Langerin is a natural barrier to HIV-1 transmission by Langerhans cells. *Nat. Med.* **13:** 367–371.

65. Lambert, A.A., C. Gilbert, M. Richard, *et al.* 2008. The C-type lectin surface receptor DCIR acts as a new attachment factor for HIV-1 in dendritic cells and contributes to trans- and cis-infection pathways. *Blood* **112:** 1299–1307.

66. Garcia-Vallejo, J.J. & Y. van Kooyk. 2009. Endogenous ligands for C-type lectin receptors: the true regulators of immune homeostatis. *Immunol. Rev.* **230:** 22–37.

67. van Gisbergen, K.P.J.M., M. Sanchez-Hernandez, T.B. Geijtenbeek, & Y. van Kooyk. 2005b. Neutrophils mediate immune modulation of dendritic cells through glycosylation-dependent interactions between Mac-1 and DC-SIGN. *J. Exp. Med.* **201:** 1281–1292.

68. van Vliet, S.J., S.I. Gringhuis, T.B. Geijtenbeek & Y. van Kooyk. 2006. Regulation of effector T cells by antigen-presenting cells via interaction of the C-type lectin MGL with CD45. *Nat. Immunol.* **24:** 1200–1208.

69. Crocker, P.R., J.C. Paulson & A. Varki. 2007. Siglecs and their roles in the immune system. *Nat. Rev. Immunol.* **7:** 255–266.

70. O'Reilly, M.K. & J.C. Paulson. 2009. Siglecs as targets for therapy in immune-cell-mediated disease. *Trends Pharmacol. Sci.* **30:** 240–248.

71. Angata, T. & A. Varki. 2002. Chemical diversity in the sialic acids and related alpha-keto acids: an evolutionary perspective. *Chem. Rev.* **102:** 439–469.

72. Avril, T., E.R. Wagner, H.J. Willison & P.R. Crocker. 2006. Sialic acid-binding immunoglobulin-like lectin 7 mediates selective recognition of sialylated glycans expressed on Campylobacter jejuni lipooligosaccharides. *Infect. Immun.* **74:** 4133–4141.

73. Paul, S.P., L.S. Taylor, E.K. Stansbury & D.W. McVicar. 2000. Myeloid specific human CD33 is an inhibitory receptor with differential ITIM function in recruiting the phosphatases SHP-1 and SHP-2. *Blood* **96:** 483–490.

74. Liu, Y., G.Y. Chen & P. Zheng. 2009. CD24-Siglec G/10 discriminates danger- from pathogen-associated molecular patterns. *Trends Immunol.* **30:** 557–561.

75. Blasius, A.L. & M. Colonna. 2006. Sampling and signaling in plasmacytoid dendritic cells: the potential roles of Siglec-H. *Trends Immunol.* **27:** 255–260.

76. Rabinovich, G.A. & M.A. Toscano. 2009. Turning 'sweet' on immunity: galectin-glycan interactions in immune tolerance and inflammation. *Nat. Rev. Immunol.* **9:** 338–352.

77. Di Lella, S., V. Sundblad, J.P. Cerliani, *et al.* 2011. When galectins recognize glycans: from biochemistry to physiology and back again. *Biochemistry* **50:** 7842–7857.

78. Sato, S., C. St-Pierre, P. Bhaumik & J. Nieminen. 2009. Galectins in innate immunity: dual functions of host soluble beta-galactoside-binding lectins as damage-associated molecular patterns (DAMPs) and as receptors for pathogen-associated molecular patterns (PAMPs). *Immunol. Rev.* **230:** 172–187.

79. Brewer, C.F., M.C. Miceli & L.G. Baum. 2002. Clusters, bundles, arrays and lattices: novel mechanisms for lectin-saccharide-mediated cellular interactions. *Curr. Opin. Struct. Biol.* **12:** 616–623.

80. Toscano, M.A., G.A. Bianco, J.M. Ilarregui, *et al.* 2007. Differential glycosylation of TH1, TH2 and TH-17 effector cells selectively regulates susceptibility to cell death. *Nat. Immunol.* **8:** 825–834.

81. Zhu, C., A.C. Anderson, A. Schubart, *et al.* 2005. The Tim-3 ligand galectin-9 negatively regulates T helper type 1 immunity. *Nat. Immunol.* **6:** 1245–1252.

82. Cooper, D., J.M. Ilarregui, S.A. Pesoa, *et al.* 2010. Multiple functional targets of the immunoregulatory activity of galectin-1: control of immune cell trafficking, dendritic cell physiology, and T-cell fate. *Methods Enzymol.* **480:** 199–244.

83. Jiang, H.R., Z. Al Rasebi, E. Mensah-Brown, *et al.* 2009. Galectin-3 deficiency reduces the severity of experimental autoimmune encephalomyelitis. *J. Immunol.* **182:** 1167–1173.

84. Forsman, H., U. Islander, E. Andréasson, *et al.* 2011. Galectin 3 aggravates joint inflammation and destruction in antigen-induced arthritis. *Arthritis Rheum.* **63:** 445–454.

85. Dube, D.H. & C.R. Bertozzi. 2005. Glycans in cancer and inflammation—potential for therapeutics and diagnostics. *Nat. Rev. Drug Discov.* **4:** 477–488.

86. Sperandio, M., C.A. Gleissner & K. Ley. 2009. Glycosylation in immune cell trafficking. *Immunol. Rev.* **230:** 97–113.

87. Buzás, E.I., B. György, M. Pásztói, *et al.* 2006. Carbohydrate recognition systems in autoimmunity. *Autoimmunity* **39:** 691–704.

88. Grigorian, A., L. Araujo, N.N. Naidu, *et al.* 2011. N-acetylglucosamine inhibits T-helper 1 (Th1)/T-helper 17 (Th17) cell responses and treats experimental autoimmune encephalomyelitis. *J. Biol. Chem.* **286:** 40133–40141.

89. Green, R.S., E.L. Stone, M. Tenno, *et al.* 2007. Mammalian *N*-glycan branching protects against innate immune self-recognition and inflammation in autoimmune disease pathogenesis. *Immunity* **27**: 308–320.

90. Hiki, Y., H. Odani, M. Takahashi, *et al.* 2001. Mass spectrometry proves under-*O*-glycosylation of glomerular IgA1 in IgA nephropathy. *Kidney Int.* **59**: 1077–1085.

91. Padler-Karavani, V., H. Yu, H. Cao, *et al.* 2008. Diversity in specificity, abundance, and composition of anti-Neu5Gc antibodies in normal humans: potential implications for disease. *Glycobiology* **18**: 818–830.

92. Ju, T. & R.D. Cummings. 2005. Protein glycosylation: chaperone mutation in Tn syndrome. *Nature* **437**: 1252.

93. Ilarregui, J.M., D.O. Croci, G.A. Bianco, *et al.* 2009. Tolerogenic signals delivered by dendritic cells to T cells through a galectin-1-driven immunoregulatory circuit involving interleukin 27 and interleukin 10. *Nat. Immunol.* **10**: 981–991.

94. Kel, J., J. Oldenampsen, M. Luca, *et al.* 2007. Soluble mannosylated myelin peptide inhibits the encephalitogenicity of autoreactive T cells during experimental autoimmune encephalomyelitis. *Am. J. Pathol.* **170**: 272–280.

95. Zhou, Y., H. Kawasaki, S.C. Hsu, *et al.* 2010. Oral tolerance to food-induced systemic anaphylaxis mediated by the C-type lectin SIGNR1. *Nat. Med.* **16**: 1128–1133.

96. Jellusova, J., U. Wellmann, K. Amann, *et al.* 2010. CD22 x Siglec-G double-deficient mice have massively increased B1 cell numbers and develop systemic autoimmunity. *J. Immunol.* **184**: 3618–3627.

97. Saeland, E., A.I. Belo, S. Mongera, *et al.* 2011. Differential glycosylation of MUC1 and CEACAM5 between normal mucosa and tumour tissue of colon cancer patients. *Int. J. Cancer.* Aug 5. doi: 10.1002/ijc.26354. [Epub ahead of print].

98. Bozzacco, L., C. Trumpfheller, Y. Huang, *et al.* 2010. HIV gag protein is efficiently cross-presented when targeted with an antibody towards the DEC-205 receptor in Flt3 ligand-mobilized murine DC. *Eur. J. Immunol.* **40**: 36–46.

99. Klechevsky, E., A.L. Flamar, Y. Cao, *et al.* 2010. Cross-priming CD8+ T cells by targeting antigens to human dendritic cells through DCIR. *Blood* **116**: 1685–1697.

100. Singh, S.K., I. Streng-Ouwehand, M. Litjens, *et al.* 2011. Design of neo-glycoconjugates that target the Mannose Receptor and enhance TLR independent cross-presentation and Th1 polarization. *Eur. J. Immunol.* **41**: 916–25.

101. Burgdorf, S., V. Lukacs-Kornek & C. Kurtc. 2006. The mannose receptor mediates uptake of soluble but not of cell-associated antigen for cross-presentation. *J. Immunol.* **176**: 6770–6776.

102. Caminischi, I., A.I. Proietto, F. Ahmet, *et al.* 2008. The dendritic cell subtype-restricted C-type lectin Clec9A is a target for vaccine enhancement. *Blood* **112**: 3264–3273.

103. Sánchez-Navarro, M. & J. Rojo. 2010. Targeting DC-SIGN with carbohydrate multivalent systems. *Drug News Perspect* **23**: 557–572.

104. Streng-Ouwehand, I., W.W.J. Unger & Y. van Kooyk. 2011.

105. Salatino, M. & G.A. Rabinovich. 2011. Fine-tuning antitumor responses through the control of galectin-glycan interactions: an overview. *Methods Mol. Biol.* **677**: 355–374.

106. Rubinstein, N., M. Alvarez, N.W. Zwirner, *et al.* 2004. Targeted inhibition of galectin-1 gene expression in tumor cells results in heightened T cell-mediated rejection; A potential mechanism of tumor-immune privilege. *Cancer Cell* **5**: 241–251.

107. Cedeno-Laurent, F., M.J. Opperman, S.R. Barthel, *et al.* 2012. Metabolic inhibition of galectin-1-binding carbohydrates accentuates antitumor immunity. *J. Invest. Dermatol.* **132**: 410–420.

108. Juszczynski, P., J. Ouyang, S. Monti, *et al.* 2007. The AP1-dependent secretion of galectin-1 by Reed Sternberg cells fosters immune privilege in classical Hodgkin lymphoma. *Proc. Natl. Acad. Sci. USA* **104**: 13134–13139.

109. Banh, A., J. Zhang, H. Cao, *et al.* 2011. Tumor galectin-1 mediates tumor growth and metastasis through regulation of T-cell apoptosis. *Cancer Res.* **71**: 4423–4431.

110. Kuo, P.L., J.Y. Hung, S.K. Huang, *et al.* 2011. Lung cancer-derived galectin-1 mediates dendritic cell anergy through inhibitor of DNA binding 3/IL-10 signaling pathway. *J. Immunol.* **186**: 1521–1530.

111. Tang, D., Z. Yuan, X. Xue, *et al.* 2011. High expression of galectin-1 in pancreatic stellate cells plays a role in the development and maintenance of an immunosuppressive microenvironment in pancreatic cancer. *Int. J. Cancer.* doi: 10.1002/ijc.26290. [Epub ahead of print].

112. Soldati, R., E. Berger, A.C. Zenclussen, *et al.* 2011. Neuroblastoma triggers an immunoevasive program involving galectin-1-dependent modulation of T cell and dendritic cell compartments. *Int. J. Cancer.* doi: 10.1002/ijc.26498. [Epub ahead of print].

113. Dardalhon, V., A.C. Anderson, J. Karman, *et al.* 2010. Tim-3/galectin-9 pathway: regulation of Th1 immunity through promotion of CD11b+Ly-6G+ myeloid cells. *J. Immunol.* **185**: 1383–1392.

114. Demotte, N., G. Wieërs, P. Van Der Smissen, *et al.* 2010. A galectin-3 ligand corrects the impaired function of human CD4 and CD8 tumor-infiltrating lymphocytes and favors tumor rejection in mice. *Cancer Res.* **70**: 7476–748.

115. Tsuboi, S., M. Sutoh, S. Hatakeyama, *et al.* 2011. A novel strategy for evasion of NK cell immunity by tumours expressing core2 *O*-glycans. *EMBO J.* **30**: 3173–3185.

116. Nicoll, G., T. Avril, K. Lock, *et al.* 2003. Ganglioside GD3 expression on target cells can modulate NK cell cytotoxicity via siglec-7-dependent and -independent mechanisms. *Eur. J. Immunol.* **33**: 1642–1648.

117. Ohta, M., A. Ishida, M. Toda, *et al.* 2010. Immunomodulation of monocyte-derived dendritic cells through ligation of tumor-produced mucins to Siglec-9. *Biochem. Biophys. Res. Commun.* **402**: 663–669.

Ann. N.Y. Acad. Sci. ISSN 0077-8923

ANNALS OF THE NEW YORK ACADEMY OF SCIENCES
Issue: *Glycobiology of the Immune Response*

Multifarious roles of sialic acids in immunity

Ajit Varki and Pascal Gagneux

Glycobiology Research and Training Center, Departments of Medicine, and Cellular and Molecular Medicine, University of California at San Diego, La Jolla, California

Address for correspondence: Ajit Varki or Pascal Gagneux, Glycobiology Research and Training Center, Departments of Medicine, and Cellular and Molecular Medicine, UC San Diego, La Jolla, CA 92093-0687, a1varki@ucsd.edu or pgagneux@ucsd.edu

Sialic acids are a diverse family of monosaccharides widely expressed on all cell surfaces of vertebrates and so-called "higher" invertebrates, and on certain bacteria that interact with vertebrates. This overview surveys examples of biological roles of sialic acids in immunity, with emphasis on an evolutionary perspective. Given the breadth of the subject, the treatment of individual topics is brief. Subjects discussed include biophysical effects regulation of factor H; modulation of leukocyte trafficking via selectins; Siglecs in immune cell activation; sialic acids as ligands for microbes; impact of microbial and endogenous sialidases on immune cell responses; pathogen molecular mimicry of host sialic acids; Siglec recognition of sialylated pathogens; bacteriophage recognition of microbial sialic acids; polysialic acid modulation of immune cells; sialic acids as pathogen decoys or biological masks; modulation of immunity by sialic acid *O*-acetylation; sialic acids as antigens and xeno-autoantigens; antisialoglycan antibodies in reproductive incompatibility; and sialic-acid–based blood groups.

Keywords: sialic acids; immunity; evolution; selectins; Siglecs; sialidases

Sialic acids (Sias) are unusual sugars with a shared nine-carbon backbone that are widely expressed on the surfaces of all cells in all animals of the deuterostome lineage (vertebrates and so-called "higher" invertebrates), and also in certain pathogenic or symbiotic bacteria that associate with them (Refs. 1–7; Fig. 1A). Given their remarkable diversity in structure, glycosidic linkage, and underlying glycan chains, as well their exposed location, it is not surprising that Sias have numerous roles in many aspects of immunity (by "immunity," we here mean immunology, as well as aspects of microbiology that are relevant to symbiosis and pathogenesis). Details regarding the occurrence, biosynthesis, structural diversity, cellular expression patterns, rapid evolution, and species variations of sialic acids have been extensively reviewed elsewhere[1-7] and will not be repeated here. This overview surveys the multifarious roles of sialic acids in selected aspects of immunity. Given the vast breadth of the subject under consideration, the treatment of the selected topics is necessarily brief, and references to the primary literature are not comprehensive. The emphasis is

also on topics with which the authors are more familiar.

Biophysical effects of sialic acids

Given their ubiquitous presence and abundance at the surface of all cell types (including those of the immune system), Sias have major biophysical effects.[8,9] The typical cell displays tens of millions of Sia molecules, and it is estimated that the local concentrations on the cell surface glycocalyx can approach 100 mM.[10] Sialic acids thus provide a large component of negative charge repulsion between cells, which could alter the biophysical properties of cellular interactions (Fig. 1B). Many earlier studies removed Sias from immune cell surfaces using sialidases and showed marked changes in behavior of such cells.[11] However, such studies are often confusing, because wholesale removal of cell surface Sias has many potentially pleiotropic effects. First, removal reduces the net charge and hydrophilicity of the cell surface. Second, it can reduce the charge repulsion between adjacent cell surface molecules. Third, it eliminates ligands for

doi: 10.1111/j.1749-6632.2012.06517.x

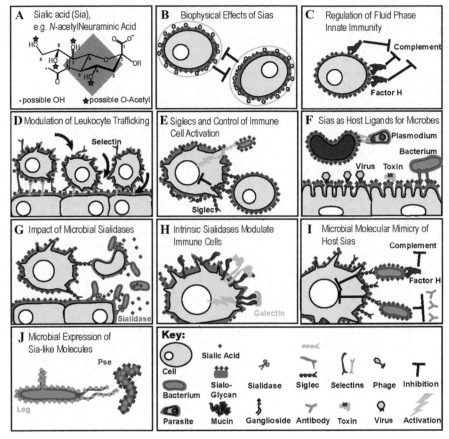

Figure 1. Examples of roles of sialic acids in immunity. Sialic acids are shown as pink diamonds. See the text for details.
(A) Neu5Ac, the most common sialic acid in mammals. These acidic sugars share a nine-carbon backbone and can be modified in many ways. (B) The high density of terminal sialic acids on the glycocalyx of vertebrate cells imparts negative charge and hydrophilicity to cell surfaces, altering biophysical properties. (C) Factor H binds cell surface Sias, protecting cell surfaces from the alternative complement pathway. (D) Intrinsic Sia-binding molecules such as selectins on endothelia, leukocytes, and platelets initiate leukocyte rolling on endothelial surfaces, a key initial step for leukocyte extravasation. (E) Intrinsic Sia-binding Siglec molecules on immune cells detect sialylated ligands and can inhibit immune cell activation. There are also activatory Siglecs. (F) Host Sias are frequently exploited as attachment sites ("receptors") by pathogens including protozoa, viruses, bacteria, and toxins. (G) Microbial sialidases can help pathogens to expose underlying glycan-binding sites, to avoid sialylated decoys (see below), and/or provide Sias as food sources. The loss of SAMPs from cells may then be used by host immune cells to react to pathogens, and/or to clear away desialylated cells or glycoproteins. (H) Endogenous sialidases such as Neu1 can modulate immune cell function by modulating receptor clustering, possibly by exposing underlying galactose residues and facilitating galectin-mediated cross-linking of surface molecules. (I) Microbial mimicry of host Sias allows manipulation of host immune response by engaging inhibitory Siglecs, inhibiting complement via factor H binding, and reducing the opportunity of the host to form antibodies. (J) Microbial synthesis of Sia-like molecules, such as legionaminic acid and pseudaminic acid stabilizes fimbriae.

endogenous receptors like Siglecs and selectins (see later). Fourth, sialic acid removal exposes underlying glycans (mostly galactose residues), which can be recognized by other endogenous receptors, such as galectins and the galactose-binding proteins of macrophages. Finally, there is potential for sialidase treatment to enhance cell surface interactions and lattices of galectins with the uncapped N-glycans of various surface receptors (see later).

Thus, more subtle alterations in cell surface Sias are needed to investigate their specific functions. In this regard, the use of mild periodate oxidation to eliminate only the C9 and C8 side chain carbon atoms of Sias is a remarkably specific manipulation, which leaves the rest of the sialic acid molecule and its negative charge intact, and is not known to affect other surface structures.[12] However, periodate oxidation also generates a C7-aldehyde on the sialic acid side chain, which can potentially react with lysine residues on adjacent proteins, cross-linking cell surface glycoproteins. This may help explain some reported dramatic effects of periodate oxidation on lymphocytes.[13] A less risky option would be to modify the type of sialic acid on the surface. One approach is to feed unnatural (bio-orthogonal) precursors to generate unnatural Sias.[14–16] If high levels of incorporation can be achieved on the cell surface, the altered biophysical properties can then be further studied. However, Sia recognition phenomena could also be altered. In this regard, it would also be worth asking whether the single oxygen atom difference between the common Sia (Neu5Ac) and the nonhuman one (Neu5Gc)[17] can alter immune cell behavior.

Meanwhile, the high densities of Sias found on surface polysaccharides of various pathogenic bacteria (see later) can also markedly alter the biophysical properties of these organisms.[3–7] But again, simply eliminating these Sias by genetic or enzymatic means can have secondary effects, due to exposure of underlying glycans. Overall, while it is clear that the biophysical properties of Sias modulate many cellular and microbial interactions in the immune system, these effects are not easy to study, because the experimental approaches may perturb the very thing that is being explored in a pleiotropic fashion. This is an important and challenging area for future studies.

Sialic acid regulation of fluid phase innate immunity

Classic studies showed a role for Sias in regulating the alternative pathway of the complement activation.[18,19] The mechanism involves the major serum protein factor H, which recognizes Sias as "self," gets recruited to native cell surfaces, and so helps to downregulate the constant "tick-over" of the complement pathway on all surfaces (Refs. 18–21; Fig. 1C). Details of this mechanism have been elucidated, including accelerated dissociation of the C3bBb C3 convertase and acting as a cofactor for factor I–mediated cleavage of C3b.[21] While Sias may thus act as "self-associated molecular patterns" (SAMPs)[22] for recognition by factor H, this does not fully explain the relative specificity of factor H for sialoglycan structural variants. There are also complexities involving the type of glycosidic linkage of Sias to the underlying glycan (its presen-

tation in space), which may alter recognition.[23,24] Furthermore, studies have shown that this factor H self-recognition is mediated by certain anion-binding sites, which can also recognize sulfated glycosaminoglycans as "self." The factor H domains involved in these recognition phenomena are largely domains 19–20 (Refs. 25–28), and this mechanism is also hijacked by bacteria that express Sias on their surface polysaccharides[21,23] (see later). Interestingly, mutations in some of these domains were found by genome-wide association studies to correlate with increased risk of complement-mediated inflammatory processes, such as hemolytic-uremic syndrome, membranoproliferative glomerulonephritis,[29] and age-dependent macular degeneration.[27,30] These experiments of nature provide evidence of the functional significance of Sias as SAMPs for recognition by factor H.

Variation in sialic acid side chain O-acetylation can also affect factor H binding. Classic studies suggested that the amount of sialic acid on red blood cells on erythrocytes of different strains of mice might restrict the extent of control of the alternate complement pathway activation.[31] It was then shown that the difference was not the amount of sialic acid, but in the extent of sialic acid side-chain O-acetylation, that is, such modified Sias are not good targets for factor H binding.[32] These older observations need to be revisited in the light of modern evidence regarding the mechanisms that control Sia O-acetylation,[1,33] as well as genetic and genomic sequences of these strains. Also of note is that ficolins (circulating soluble activators of the lectin pathway of complement activation) can recognize sialic acids, particularly on the surfaces of sialylated bacteria.[34–36] This appears to be a host response to molecular mimicry by bacteria (see later).

Modulation of leukocyte trafficking via sialylated selectin ligands

Until the 1980s, factor H was the only known intrinsic vertebrate sialic acid–binding protein. A classic study[37] then noted that pretreatment of lymph node sections with a sialidase abolished the interaction of lymphocytes with the high-endothelial venules, which normally provide exit sites for lymphocytes. These and other observations eventually led to recognition of the selectin family of cell adhesion molecules and their role in leukocyte trafficking (Refs. 38–42; Fig. 1D). Different isoforms of these endogenous lectins were found expressed on leukocytes (L-selectin), platelets (P-selectin), and endothelium (P- and E-selectin). It then became apparent that Sias in the glycan sequence Siaα2–3Galβ1–3/4(Fucα1–3/4)GlcNAcβ1-R (Sialyl Lewis X/A) are critical components of the natural ligands for these selectins.[38,39,43–50] In some instances, Sialyl Lewis X/A motifs combine with other features such as sulfation on the Gal or GlcNAc residues (L-selectin ligands), and/or sulfation of adjacent tyrosine residues (P-selectin) and contribute toward specific recognition sites on specific proteins, particularly mucin-like glycoproteins.[51] A particularly striking example was the elucidation of a defined amino-terminal sulfoglycopeptide motif on P-selectin glycoprotein ligand-1 (PSGL-1), which serves as a specific high affinity ligand of P-selectin.[52]

Whereas Sias form a critical part of most ligands for selectins, recognition does not seem to be affected by other structural details of the Sias themselves, except that the α2–3 linkage to the underlying galactose residue is critical. In keeping with this, some selectin ligands can function with a sulfate ester at the three position of galactose, instead of a sialic acid.[53] Overall, it appears that Sias are primarily acting as conveyers of a necessary negative charge for selectin interactions, and the details of Sia diversity do not matter. In keeping with this, some 6-O-sulfated glycosaminoglycans, such as heparan sulfate, can act as alternate selectin ligands.[54,55] However, if one oxidizes the side chain of sialic acid with periodate and generates an aldehyde, this reactive group can be cross-linked into the binding pocket of the selectin via a covalent interaction.[56] Overall it is evident that α2–3 linkage-specific sialyltransferases play a key role in generating selectin ligands, along with α1–3/4 fucosyltransferases, and GlcNAc sulfotransferases and/or tyrosine sulfotransferases.[57]

Siglecs in the control of immune cell activation

In the mid-1980s, some macrophage types were found to form rosettes with sheep erythrocytes *in vitro*, and that binding could be abolished by sialidase pretreatment of the erythrocytes.[58] This sialic acid–dependent receptor was purified and shown to be a very large protein called sialoadhesin, which was then demonstrated to bind sialic acid–containing ligands *in vitro*.[59] However given the size of the sialoadhesin molecule and the era in which this work occurred, cloning proved difficult. Meanwhile, expression cloning of the presumed ligand for a B cell "adhesion molecule" called CD22 had surprisingly yielded a sialyltransferase.[60] In fact, it turned out that CD22 was a sialic acid–binding lectin, with recombinant soluble CD22 shown to bind Sias through its extracellular domain, and not to the sialyltransferase identified through expression cloning (the transferase is not at the cell surface but rather generates sialylated ligands of CD22 in the Golgi).[61] Moreover, recognition by CD22 was specific for the α2–6 linkage, with no binding to α2–3–linked Sia.[62] Additional studies defined the highly conserved preference of CD22 for this linkage and characterized the interactions further.[63] Soon thereafter, the cloning of sialoadhesin revealed that its amino-terminal domains had a homology with CD22 and to similar domains of two other previously known proteins, CD33 and myelin-associated glycoprotein (MAG), suggesting that these molecules might belong to a single family of sialic acid–binding proteins.[64] Further studies showed that this was indeed the case, resulting in recognition of a new family of sialic acid–binding proteins.[65] It was initially suggested that these molecules be called *sialoadhesins*.[66–68] However, besides the confusing relationship to the first molecule already with this name, some of these proteins did not seem to mediate cell–cell adhesion. The alternate term suggested was *Siglec*, to stand for sialic acid–recognizing Ig-superfamily lectins, as a subset of I-type lectins (Ig-like lectins).[69] Discussions among those working in the field eventually led to general acceptance of this term, with the founding members sialoadhesin, CD22, CD33, and MAG being designated Siglecs-1–4 (Ref. 70). Studies of expressed sequence tags and mining of genomic sequence data then led to extension of this family of intrinsic vertebrate lectins, which now comprise at least 16 members in primates.[71–79]

Interestingly, a subset of Siglecs seems relatively conserved in mammals (CD22/Siglec-2 and sialoadhesin/Siglec-1), and even among vertebrates (MAG/Siglec-4 and Siglec-15). In contrast, another subset found in one large syntenic cluster (on chromosome 19q in humans) shows the highest amount of variation between species.[71,72] This subfamily was named *CD33-related Siglecs* (or CD33rSiglecs) and shown to have a variety of Sia-binding properties. CD33rSiglecs can be further subdivided into two categories. Most have cytosolic domains containing immune receptor tyrosine-based inhibitory motifs (ITIMs) that can be tyrosine phosphorylated, resulting in recruitment of tyrosine phosphates like SHP-1 and SHP-2.[80–84] This in turn results in dephosphorylation of tyrosine residues on various kinases associated with other receptors, effectively downregulating their functions (Fig. 1E). Thus, these inhibitory CD33rSiglecs likely serve as innate immune detectors for SAMPs, thereby downregulating unwanted inflammation, particularly that occurring in response to tissue damage.[85] Notably several of these CD33rSiglecs also have a second cytosolic ITIM-like motif whose functions are much less clear. Moreover, there is evidence that some of the inhibitory effects of such Siglecs do not require either of these tyrosine-based motifs.[86,87] This suggests that more attention should be paid to the better-conserved extracellular C2-set domains, as mediators of additional and/or complementary functions.

In contrast to the inhibitory Siglecs, a positively charged amino acid in the transmembrane domain of some activating CD33rSiglecs allows them to engage immune receptor tyrosine-based activatory motif (ITAM)-containing adapter molecules like DAP12, which in turn recruits the tyrosine kinase Syk and mediates tyrosine phosphorylation of various receptors and kinases.[88–91] In some instances, CD33rSiglecs with inhibitory and activatory properties have undergone gene conversion events that maintain their amino-terminal identity, suggesting that they may be paired receptors, sending opposite signals on binding of the same ligand(s). In this context, it seems likely that the activatory Siglecs represent an evolutionary response to bacteria that are "hijacking" inhibitory Siglecs (see later).[88,92]

The general subject of Siglecs and their biology and evolution has been extensively discussed elsewhere[73,74,76,77,79,84,93–98] and details will not be repeated here. However, certain features are worthy of special note. First, most Siglecs are typically bound by so-called "*cis* ligands," that is, sialylated glycans on the same cell surface.[99,100] However, another cell surface or a soluble ligand with a high enough density of sialylated ligands can compete

out the *cis* ligands and cause engagement.[76,101,102] Second, the amino-terminal V-set Ig-like domain contains the sialic acid recognition site, including a canonical conserved arginine residue that is critical for interaction with the carboxylate of sialylated ligands.[103–107] Interestingly, this arginine residue can be naturally mutated, affecting one or more Siglecs unique to a given species or taxon.[71,88,107] One possibility is that these events occur randomly because the arginine codon (CGN) is highly mutable. However, there are instances where the arginine appears to mutate and then reappears in one phylogenetic branch, for example, for Siglecs-5 and Siglecs-14 in humans versus great apes.[88] Taken together with the high frequency of such events, it is more likely that these mutations are an evolutionary mechanism for rapidly "retiring" a Siglec, that is, curtailing its interactions with sialic acid–containing ligands without losing the entire molecule, leaving the option to "resurrect" it later. By convention, such Siglecs are referred to by a Roman numeral (e.g., Siglec-XII in humans and Siglec-V in chimpanzees). Both the arginine mutations and the paired receptors mentioned above are likely to be evolutionarily related to the interactions of sialylated microbes with Siglecs (see later).

Sialic acids as host ligands (receptors) for microbes

Given the location and abundance of Sia on cell surfaces, it is not surprising that numerous viruses and some bacteria use host-sialylated structures as targets for binding and recognition (Refs. 108–112; Fig. 1F). The same is also true of several important bacterial toxins.[112] In the case of viruses that bind Sia via a hemagglutinin, most also often express a sialidase (neuraminidase) that cleaves the same receptor.[113] This dualistic recognition and removal of sialic acid is best studied for influenza viruses.[114] The traditional term *neuraminidase* is being replaced by *sialidase*, since neuraminic acid (with a free amino group) is not only vanishingly rare in nature but is also actually resistant to the neuraminidases studied to date.

Unfortunately, given the history of virology, where viruses were originally characterized by their hemagglutinin (H) or neuraminidase (N), by antigenicity/serology, and now by their RNA genotypes, it would be difficult to ask this particular field to change nomenclature, for example, from H1N1 to H1S1.

When it comes to natural sialic acid modifications as pathogen ligands, further subtleties abound. For example, some viruses recognize O-acetyl-Sias and have a receptor-destroying enzyme that removes the O-acetyl group.[115–125] Several eukaryotic pathogens also employ sialic acid recognition as part of interactions with hosts (the falciparum malarial merozoite).[126–129] Meanwhile, a bacterial SubAb toxin selectively recognizes ligands bearing the Neu5Gc sialic acid.[130] Examples of such binding phenomena are numerous and have been reported in detail elsewhere.[108,109,112]

Impact of microbial sialidases on the immune system

We have already mentioned the striking effects on immune cell function of adding exogenous sialidases. Given the marked instability of vertebrate sialidases in extracellular fluids, the only sialidases that could have been used for such studies have been of microbial origin, particularly soluble bacterial sialidases, which are easily found in nature.[113,131] Why would so many bacteria express sialidases? The most obvious answer is that the first structure encountered by them on and around most cell surfaces is likely to be a sialic acid. Thus, without a mechanism to bind to sialic acid (as is the case with a majority of bacteria), it is useful to bacteria remove this negatively charged sugar. This may help in the breakdown of both soluble mucins (sialic acid–rich glycoproteins secreted by epithelia) and cell surface glycoconjugates on the way to cellular entry or interactions (Fig. 1G). Some bacteria, for example, *Haemophilus influenzae* also use the free Sia as a source of energy by "browsing" on host Sia.[132,133] Such free Sia can be broken down to the useful energy sources pyruvate and ManNAc (the latter after it is converted into GlcNAc).[132,133]

Other functions for bacterial sialidases are now becoming apparent. For example, released Sia may be taken up by some bacteria and used to decorate their surfaces (see the section "Modulation of immune cell responses by intrinsic sialidases" on the expression of Sia by certain bacteria using exogenous sources). There is also evidence that free Sia can act as a signal to certain bacteria, for example, *Pneumococcus*,[134] directing them toward biofilm formation and/or colonization. Perhaps free sialic acid is a way

for the bacterium to recognize that it has arrived in a vertebrate environment suitable for colonization. In most of the situations mentioned above, the roles of different sialic acid types and glycosidic linkage types have not been considered. However, in some instances it is clear that modifications of Sias, such as the N-glycolyl group at the five position or O-acetyl groups on the side chain, can limit the action of bacterial sialidases (by "masking the mask").[135] Further studies are needed to understand the significance of this inhibitory effect. Finally, given the role of Sias as a SAMP recognized by molecules such as Siglecs,[22] bacterial desialylation could also perturb natural self-recognition phenomena, perhaps increasing inflammatory responses by exposing desialylated danger-associated molecular patterns (DAMPS).[136,137] This concept requires further study.

Modulation of immune cell responses by intrinsic sialidases

As mentioned earlier, active vertebrate sialidases are not reported in extracellular fluids, as they are unstable. While there are four sialidases in vertebrate cells (Neu1–4), the major one in most cells is Neu-1.[131,138–140] The fact that the *Neu1* gene is located within the major histocompatibility locus, and that it has altered activity in some mouse strains, is of great interest from the immunological perspective.[141,142] Although this enzyme is primarily in the lysosome, it is now known to also exist at the cell surface. In both instances, Neu1 is very unstable unless it is in a complex with two other proteins, beta galactosidase and protective protein/cathepsin A (PPCA).[143] A selective advantage for this instability can be considered. It is reasonable to suggest that a vertebrate organism with Sias covering all its cell surfaces and terminating the glycans on extracellular glycoproteins would not benefit from having constitutive extracellular sialidase activity, risking damage to its SAMPs, and exposing underlying glycans. Following this logic, it may be that maintaining the extracellular fluid in a sialidase-free state also allows exploitation of the sudden appearance of a sialidase as a "danger signal," indicating the presence of a bacterial or viral organism—that is, giving a potential higher fidelity to the "desialylation signal." Regardless of these speculations, it appears that endogenous Neu1 is capable of being translocated to the cell surface and desialylating certain surface molecules, such as TLRs, TCRs, and integrins (Refs. 144–146; Fig. 1H), and modifying signaling[147,148] and phagocytosis.[149] The resulting alteration of receptor functions is poorly understood, and candidate mechanisms include the loss of charge repulsion and/or altered galectin-mediated clustering.[150] Meanwhile, Neu3 is also found on the cell membrane, but has been shown to act specifically on the sialic acids of gangliosides.[151,152]

Microbial molecular mimicry of host sialic acids

As mentioned earlier and extensively documented elsewhere, several bacterial pathogens express Sias on their surfaces (Table 1 of Ref. 153; Fig. 1I). Every possible way in which Sias might be expressed has been exploited, indicating a strong evolutionary selection pressure to achieve this state. Given the apparent restriction of Sias to multicellular animals of the deuterostome linage, these examples were once assumed to be due to co-opting of vertebrate genes. However, in every case examined, bacterial sialic acid biosynthesis appears to represent convergent (parallel) evolution, recruiting and modifying ancient pathways for synthesis of bacterial nonulosonic acids[3,154,155] to instead produce Sias (see later). In combination with independently evolved sialyltransferases that catalyze addition of Sias to the tips of glycan chains, this has enabled remarkable levels of molecular mimcry, involving multiple novel genes that can recreate sialylated glycans essentially identical to those found on host cell surfaces. Based on the discussion mentioned above about the role of Sias and SAMPs,[22] one can see the benefit to the microbe of synthesizing vertebrate host-like sialylated glycans. They could provide the pathogen with suppression of the alternate pathway via factor H,[156] hijack host Siglecs, and inhibit the formation of antibodies against underlying glycan structures. Finally, Sias also serve to block recognition of underlying (nonmammalian) glycans by naturally occurring antibodies circulating in most vertebrates.[157]

Mechanism of microbial expression of sialic acids and sialic acid–like molecules

With regard to the repeated convergent evolution of sialic acid biosynthesis pathways, a picture is now emerging that can explain this remarkable phenomenon. In turns out that ~20% of the first thousand prokaryotic genomes that were sequenced have clusters of genes similar to those involved in the biosynthesis of Sias.[3] In most cases, these organisms are synthesizing a more ancient family of nine-carbon backbone acidic sugars called nonulosonic acids, such as pseudaminic acid (Pse) and legionaminic acid (Leg; Ref. 155; Fig. 1J). As discussed in detail elsewhere, the homology of the genes, metabolic intermediates and steps in the pathway make it very likely that bacteria have co-opted this ancient pathway for nonulosonic acid biosynthesis and simply remodeled it for the production of vertebrate-like Sias.[3] In this regard, we have suggested that the term *Sias* be reserved for the molecules based on Neu or Kdn backbone originally found in deuterostomes and some of their pathogens, and that the term *bacterial Sias* be replaced by the family name *nonulosonic acid* (NulO), which includes the Sias.[3] It remains to be seen whether NulOs, such as pseudaminic acid (Pse) or legionaminic acid (Leg), are also recognized by Siglecs.

Sialoadhesin recognition of sialylated pathogens

If bacteria mimic vertebrate cells by expressing Sias, the immune system must find a way to distinguish sialylated pathogens (and perhaps other pathogens that express nonulosonic acids), even while maintaining tolerance toward self sialic acid structures. While several of the previously mentioned CD33rSiglecs can recognize sialylated pathogens and mediate endocytosis,[158–161] it is unclear whether this is an *in vitro* artifact, related to the unexplained tendency of CD33rSiglecs to undergo endocytosis when cross-linked. However, Siglec-1 (sialoadhesin) has no signaling properties, and instead has the size, length, and structure to carry out this protective phagocytic function (Fig. 2A). The highly conserved specificity of Siglec-1 for α2–3–linked or α2–8–linked Neu5Ac (not Neu5Gc) supports this idea—as these are exactly the types of structures that bacteria express (no microorganism has ever been shown to synthesize Neu5Gc, and α2–6–linked Neu5Ac is rare in bacteria). Sialoadhesin is also expressed on the right cell types (macrophages) and in the right locations (the marginal zones on lymph nodes and filtering areas of the spleen) to carry out the functions described. Conversely, certain viruses

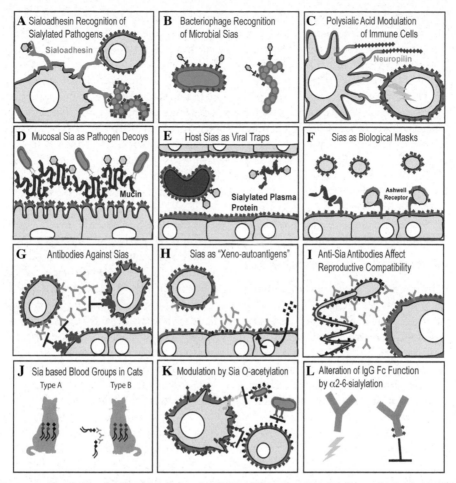

Figure 2. More examples of roles of sialic acids in immunity. Sialic acids are shown as pink diamonds. See key in Figure 1, and the text for details. (A) Siglec-1 (sialoadhesin) expressed on macrophages recognizes Sias in patterns commonly found on microbial pathogens and facilitates phagocytosis. Siglec-1 may also mediate immune cell interactions with one another. Some viruses exploit Siglec-1 binding to gain access to host cells. (B) Certain bacteriophages use Sias on their microbial hosts as "receptors" for invasion. (C) Polysialic acid on immune molecules such as neuropilin on dendritic cells modulates interactions with T cells. (D) Sia-rich secretions on host epithelia can act as decoys for Sia-binding microbes. (E) Sia-covered erythrocytes and Sia-rich plasma proteins can act as "viral traps." (F) Sias act as biological masks by blocking interactions between intrinsic receptors and underlying glycan structures. (G) Sias on potentially antigenic glycoconjugates prevent the formation of antibodies to "cryptoantigens." Less commonly, Sias can be autoantigens. (H) Nonself Sias can be metabolically incorporated from dietary sources and become "xeno-autoantigens," targeted by intrinsic anti-Sia antibodies. (I) Female genital tract reactions to nonself Sia on sperm can lead to reproductive incompatibility. (J) Some mammals, such as cats, have blood groups defined by Sia-containing glycolipids. (K) O-acetylation of Sias can block Sia recognition by intrinsic lectins like Siglecs, and modulate microbial lectin interactions, in a positive or negative fashion. (L) Alpha-2–6 sialylation of IgG-Fc region N-glycans can change the effects of IgG antibodies from activating to inhibitory.

that emerge from host cells with a coating of sialic acids can "hijack" sialoadhesin for use as a mode of entry into macrophages.[162,163] Sialoadhesin also contributes to intrinsic immune functions, and immune changes found in sialoadhesin-deficient mice indicate a role in regulation of the adaptive immune system.[164–167]

Bacteriophage recognition of microbial sialic acids

When bacteria mimic vertebrate cells by expressing similar sialoglycans, they are also likely using variations of this mechanism to escape recognition by bacteriophages that normally recognize underlying bacterial glycans. However, phages evolve even faster than bacteria, and some have evolved to recognize these sialylated bacterial capsules (Refs. 168 and 169; Fig. 2B). Many more such phages probably exist and remain to be discovered. For example, this may help explain why Group B *Streptococcus* has evolved so many different polysaccharide variants, each terminating with the same human-like sialylated trisaccharide.[170] The possibility of using such phages as alternates to antibiotic therapy has also been considered.[171] It is also interesting to note that the cholera toxin ganglioside-binding subunit B is encoded by nonlysogenic phage that was recruited by *Vibrio cholerae* to mediate its own pathogenicity.[172]

Polysialic acid modulation of immune cells

Sias are usually found as single monosaccharide unit at the end of glycan chains. However they can sometimes be linked to each other, generating short or long homopolymers. In the typical form, the polysialic acid consists of α2–8–linked Neu5Ac units. This structure is found on certain proteins in the brain and is known to have major functions in the development, morphogenesis, and function of various neural systems.[173,174] However, polysialic acid has also since been discovered on a few immune cells. These include dendritic cells,[175,176] and some stages of T cell development.[177,178] In these instances, the polysialic acid is attached to specific proteins such as neuropilins, modulating cellular interactions in meaningful ways (Fig. 2C).

Host sialic acids as pathogen decoys

We have already discussed how host Sias can be the binding target (often called "receptor") for a variety of viruses, toxins, and some bacteria. One ubiquitous and simple function of Sias is likely to act as a decoy against such organisms. Thus, for example, a virus or other sialic acid–binding organism that reaches a mucosal surface will first encounter mucins, which are heavily sialylated but mostly secreted glycoproteins. Even membrane-anchored mucins can be shed. These soluble molecules can act as decoys, preventing the organism from reaching its intended target on the cell surface (Refs. 179 and 180; Fig. 2D). This is a testable hypothesis that has yet to be carefully evaluated, though loss of O-linked glycans on mucins has been linked to increased frequency and severity of colitis.[181] It is also

possible that the destruction or bypassing of these decoys represents a function of the sialidases (neuraminidases) found on numerous viruses. However, the only study to date on this subject was done in cell culture,[182] in the absence of the mucins that could be the dominant decoys in the natural state.

Another decoy function might be mediated by the heavily sialylated glycoproteins found in plasma and extracellular fluids.[179,180] Again, sialic acid–binding pathogens would first encounter these heavily sialylated glycoproteins and have to escape from them before approaching the intended target. In the case of viruses, the target cells typically need a nucleus, since the virus needs to take advantage of the cellular machinery for synthesis and replication of its own nucleic acid. Another decoy example might be non-nuclear cells such as erythrocyte, which represents about 50% of the total volume of blood, and could act as "viral traps" (Fig. 2E). A sialic acid–binding virus such as influenza virus that manages to make its way into the bloodstream would immediately encounter this extensive cell surface that it can bind to, but lacks appropriate mechanisms to allow invasion and replication. Ironically then, the hemagglutination reaction that helps define the binding preferences and specificities of a variety of viruses may actually represent the host's attempt to evade the very same virus. Over large scales of evolutionary time one might thus expect the sialome of erythrocytes to evolve to keep up with the rapidly evolving sialic acid–binding specificities of pathogens and those of new pathogens that arrive at various times.[179,180] Meanwhile there is a propensity of malarial parasites to use Sias to invade erythrocytes, inside of which they asexually replicate.[183,184] Taken together, all of the above considerations may explain why there is such extreme inter-species variation in sialomes of erythrocytes, mucins, and plasma glycoproteins. It might also explain the sudden changes in sialylation patterns and levels occurring during inflammation within a given species (e.g., the acute phase reaction).[185]

Sialic acids as biological masks

Schauer originally emphasized the dualistic roles of Sias as binding sites and as biological masks.[186] The first discovered vertebrate glycan-binding protein was the asialo-glycoprotein receptor on hepatocytes (the so-called "Ashwell receptor").[187] As the name suggests, binding to this receptor occurs when one removes Sias from a glycoprotein and exposes the underlying beta-linked galactose residues (Fig. 2F). Since this discovery, additional beta-galactose–binding receptors in macrophages have been discovered. However, in most instances gene knockouts of these proteins failed to uncover a clear-cut natural function in intrinsic systems.[188] On the other hand, when a sialidase of bacterial origin enters the circulation, there can be extensive desialylation of cells and proteins, and these receptors become relevant. This was recently shown as a host mechanism to clear away the excess of platelets that might result in increased coagulopathy that is associated with microbial sepsis.[189] The removal of Sias could also generate "eat me" signals that allow macrophages to recognize and eliminate dying or apoptotic cells.[190] It is important to recognize and differentiate the galactose-binding receptors involved in such phenomena from soluble galectins, which can bind terminal or subterminal galactose residues on cell surfaces.[191,192] Galectins may actually function in the opposite direction, acting to reduce endocytosis of the cell surface proteins by forming lattices, and may thus be more important in regulation of signaling, as shown by others.[193–195] The numerous other functions of galectins in the immune system[192,195] will not be discussed here, except to say that they can be modulated by the presence or absence of terminal Sias, particularly α2–6–linked ones.[196] Also, unlike most galectins that prefer N-acetyllactosamine ligands with non-sialylated terminal beta-Gal residues, galectin-8 and galectin-9 have domains that preferentially recognize α2–3–sialylated N-acetyllactosamines.[197,198]

Antibodies against intrinsic sialic acids

Not surprisingly, it is uncommon to find antibodies against sialic acid–containing glycans, if the sialic acid in question is already intrinsic within the host (Fig. 2G). Presumably, this is because B cells that happen to express a B cell receptor (sIgM) that can recognize sialylated glycans are tolerized and eliminated before they leave the bone marrow. Indeed, as discussed earlier, this might be one of the selective advantages to pathogens that express Sias. In mammals, most of these comments reflect upon the common Neu5Ac sialic acid. On the other hand, it is possible to induce mice to generate monoclonal antibodies that detect Neu5Ac-containing glycans,[199,200] and the addition of an O-acetyl group to

the sialic acid can increase the probability of getting such an antibody.[201] Overall, while host-intrinsic Sias can be generally considered "immunosuppressive" for the host organism, exceptions can be found.

Sialic acids as xenoautoantigens

The above comments do not apply if a particular sialic acid is missing in a species. This appears to be the case both in humans[17] and in the *sauropsid* lineage of animals (birds and reptiles),[202] which appear unable to synthesize the common mammalian sialic acid Neu5Gc from its precursor Neu5Ac. In the case of humans, the basis for this phenotype is a fixed loss-of function mutation of the cytidine monophosphate N-acetylneuraminic acid hydroxylase (*CMAH*) gene,[203,204] which remains intact in our closest evolutionary relatives the chimpanzees.[204] The significance of the independent loss in the sauropsid lineage is unclear, though it does make for convenient source of anti-Neu5Gc antibodies by immunizing chickens, which generate a robust response.[205] Unexplained also is the fact that similar antibodies appear when chickens that become infected with the Marek's disease lymphoma virus.[206]

In the case of humans, more information is available. It appears that Neu5Gc from dietary sources can be metabolically incorporated either into our tissues[207] or into commensal bacteria such as *H. influenzae*, which specialize in taking up low quantities of Sias present in the upper oropharynx.[208] One or both mechanisms appear to be the cause of moderate-to-high levels of anti-Neu5Gc antibodies in humans (Fig. 2H). Current studies suggest that these antibodies may be interacting with the metabolically incorporated Neu5Gc of dietary origin to generate chronic inflammation.[17] This may help explain the propensity of red meat (beef, pork, and lamb, the richest sources of dietary Neu5Gc) to increase the risk of inflammation-associated diseases such as carcinomas, cardiovascular disease, and macular degeneration. These findings also relate to the classic reports of the Hanganutziu-Deicher "heterophile" antibodies, which reacted with animal red blood cells.[209,210]

Antisialyl antibodies can affect reproductive incompatibility

The female reproductive tract has levels of IgG antibodies and complement levels similar to that found in the serum.[211] Thus a sperm that enters the uterus must negotiate this immunological gauntlet before it can reach the ovum and fertilize it, further up in the fallopian tube. There may well be multiple antiglycan systems that can affect sperm, but the one so far documented involves antibodies against Neu5Gc, which can enter the uterine fluid and affect both sperm and embryos that happen to have Neu5Gc on them (Ref. 211; Fig. 2I). In this regard, it is suggested that this could even be a mechanism of speciation in the genus *Homo*, due to the loss of Neu5Gc in ancestors, after it initially became polymorphic.

Sialic-acid–based blood groups in mammals

The above-mentioned considerations about the lack of immunogenicity of Neu5Ac do not apply if the Neu5Ac is attached to a specific polypeptide that is foreign to the host. An example is the MN blood groups in humans, where individual variations in the amino acid sequence of the red cell protein glycophorin result in differential presentation of small *O*-linked sialylated chains at the aminoterminus of the protein.[212] While dictated by the underlying polypeptide, these antigenic variations also require the sialylated glycans, generating the antibodies that interact between humans and affect blood transfusion occasionally. Similar considerations apply to some other blood group antibodies.[213]

Neu5Gc-based blood groups in cats

As with humans, there is one major antibody system that appears to restrict blood transfusion within cats. However, the two major blood groups in cats (A and B) were shown to be due to antibodies in B cats against a sialylated glycolipid on the red blood cells of A cats (Fig. 2J). The difference appears to be the presence or absence of Neu5Gc in the ganglioside GD3.[214] We do not as yet know if the differential expression of Neu5Gc in the red blood cells of these cats extends to other tissues of the animal. There is evidence that changes in the promoter region of the *CMAH* gene might explain the differential expression in red blood cells.[215] There are similar erythrocyte Neu5Gc polymorphisms in dogs,[216] but evidence of anti-Neu5Gc antibodies has not been reported. In both instances, strain differences would be worth studying further. Thus, it is possible that Sias can exist as alloantigens within populations of

the same species as well as xenoantigens within different species.

Modulation of immunity by sialic acid *O*-acetylation

Given variable expression of *O*-acetyl groups on Sias and their diverse effects in immunity, a separate section on this modification seems justified. We have already mentioned the impact on factor H recognition, the relative resistance to bacterial sialidases to Sias with this modification, the blockade of binding of some virus hemagglutinins, and the facilitation of binding of others. Sias on certain bacterial polysaccharides can be *O*-acetylated. Surprisingly, this modification is actually detrimental to the bacterium in the host–pathogen interaction, either reducing recognition by CD33rSiglecs and/or enhancing immunogenicity.[217] The logical explanation is that these modifications assist the bacteria in surviving in other situations, such as protection from other microbial sialidases, and/or bacteriophage-binding proteins. The exception is *O*-acetyl blockade of recognition by sialoadhesin,[218] which could be beneficial to the bacterium, avoiding phagocytosis.[32,217]

Unlike the case with the selectins, sialic acid–binding by Siglecs almost invariably requires recognition of the C7–C9 side-chain of the molecule. This is exemplified by the loss of recognition upon mild periodate oxidation of this side-chain.[62,63,99] In view of this, it is not surprising that addition of an *O*-acetyl group to the side-chain blocks the binding of all Siglecs studied to date.[218,219] A Siglec selectively recognizing *O*-acetylated Sias has yet to be found. Thus sialic acid *O*-acetylation seemed a logical candidate for regulation of Siglec function (Fig. 2K). This was indeed shown to be the case in mice with a defect in a sialic acid–specific esterase (SIAE), which normally downregulates sialic acid *O*-acetylation on B cells.[220] The mutant mice thus have overreactive B cells, apparently due to lack of proper SAMP ligands for CD22 and Siglec-G.[221] In keeping with this, humans with autoimmune diseases have a higher frequency of harboring mutations in the *SIAE* gene.[79,222]

In another setting, *O*-acetylation of the outer sialic acid of the ganglioside GD3 was first reported as a melanoma-specific antigen not found in other normal tissues.[201] However, it later turned out that normal T cells can also express this structure.[223–225]

Table 1. Cluster of differentiation (CD) numbers related to sialic acid biology

CD number	Common name(s)	Roles in immunity
CD15s	Sialyl Lewis (sLeX)	Key component of sialylated selectin ligands and of preferred ligands for some Siglecs.
CD22	Siglec-2 (sialic acid–binding Ig-like lectin 2)	Dampens B cell reactivity via selective recognition of $\alpha2$–6–linked Sias.
CD24	Heat stable antigen	Heavily glycosylated, sialylated GPI-anchored molecule; some glycoforms may be ligands for Siglec-10 or P-selectin.
CD33	Siglec-3 (sialic acid–binding Ig-like lectin 3)	Myeloid lineage marker; can downregulate reactivity of innate immune cells.
CD34	Hematopoietic progenitor cell antigen	Heavily glycosylated and sialylated cell surface mucin-like protein; stem cell marker; some glycoforms can be L-selectin ligands.
CD43	Leukosialin (leucocyte sialoglycoprotein) (sialophorin)	Heavily sialylated major cell surface mucin-like protein; modulates immune cell responses.
CD45	CD45 leukocyte common antigen	Differing glycoforms in different immune cell types due to alternate splicing; tyrosine phosphatase; possible CD22 ligand.
CD52	CAMPATH-1	Heavily glycosylated and sialylated glycosylphosphoinositol (GPI)-anchored cell surface molecule.
CD56	Neural cell adhesion molecule 1	NK cell marker. Can carry polysialic acid, which can be recognized by Siglec-7 and Siglec-11.
CD60a	GD3 ganglioside	Glycolipid with two sialic acids; human T cell marker; promotes apoptosis?; Siglec-7 ligand?
CD60b	9-O-acetyl-GD3 ganglioside	Presence of 9-O-acetyl group on outer sialic acid of GD3; protects from apoptosis?
CD60c	7-O-acetyl-GD3 ganglioside	Converted to 9-O-acetyl-GD3 over time due to nonenzymatic migration of the O-acetyl group.
CD62E	E-selectin (endothelial leukocyte adhesion molecule 1)	Mediates leukocyte adhesion and rolling on endothelium; expression upregulated by inflammatory cytokines.
CD62L	L-selectin (lymph node homing receptor)	Mediates leukocyte adhesion and rolling on endothelium, including lymphocyte trafficking.
CD62P	P-selectin (granule membrane protein 140, GMP-140)	Mediates platelet and endothelial interactions with leukocytes, via correctly modified PSGL-1.
CD68	Macrosialin (Gp110)	Marker of macrophages and few other cells; receptor for oxidized low-density lipoprotein?
CD75s	$\alpha2$–6–sialylated lactosamines	Cluster of $Sia\alpha2$–6Galβ1–4GlcNAcβ1-R units produced by ST6Gal-I, mainly on N-glycans.
CD169	Sialoadhesin (sialic acid–binding Ig-like lectin 1) (Siglec-1)	Macrophage subset marker; recognizes sialic acids on endogenous ligands and on microbes.
CD170	Sialic acid–binding Ig-like lectin 5 (Siglec-5)	Inhibitory Siglecs on human innate immune cells, and also on lymphocytes in "great apes."
CD175s	Sialyl-Tn	$Sia\alpha2$–6GalNAcα1-Ser/Thr; truncated O-glycan mostly found on malignant cells.

Continued

Table 1. *Continued*

CD number	Common name(s)	Roles in immunity
CD176s	Sialylated form of Thomsen–Friedenreich (T) antigen	Siaα2–3Gal1–3GalNAcα1-Ser/Thr; common *O*-glycan on many cell types.
CD227	MUC-1, polymorphic epithelial mucin, episialin	Major cell surface mucin on epithelial cells; on activated T cells, can send inhibitory signals.
CD235a	Glycophorin-A (PAS-2) (MN sialoglycoprotein)	Major sialic acid carrier on RBCs; target for binding by malarial merozoite via EBA-175.
CD235b	Glycophorin-B (PAS-3) (sialoglycoprotein delta)	Major sialic acid carrier on RBCs; target for binding by malarial merozoite via EBL-1.
CD236	Glycophorin-C (PAS-2′) (glycoprotein beta)	Major sialic acid carrier on RBCs. Target for binding by malarial merozoite via EBA-140.
CD327	Sialic acid–binding Ig-like lectin 6 (Siglec-6)	Inhibitory Siglecs on B cells in primates, and also on placental trophoblast (in humans only).
CD328	Sialic acid–binding Ig-like lectin 7 (Siglec-7); AIRM1	Inhibitory Siglecs on human NK cells; lower levels on monocytes and macrophages.
CD329	Sialic acid–binding Ig-like lectin 9 (Siglec-9);	Inhibitory Siglecs found on human neutrophils; monocytes and macrophages.

For more information, see http://www.hcdm.org/MoleculeInformation/tabid/54/Default.aspx.

Indeed, this is the basis of the CD60 group of antigens (see Table 1). Interestingly, while GD3 is pro-apoptotic, *O*-acetyl-GD3 has opposite effects.[226,227] The claimed mechanisms for these effects are fascinating, involving mitochondrial and other apoptotic pathways. However, there are topological issues that remain unresolved.[228]

Alteration of IgG Fc function by α2–6–sialylation

Recent studies have shown that the minor subset of circulating IgG that has α2–6–linked Sias terminating its N-glycan has an inhibitory potential, working through the human DC-SIGN receptor on a regulatory macrophage population to upregulate FcRγ-IIB on other macrophages, and thereby dampen immune responses (Refs. 229–231; Fig. 2L). This is also suggested to be the mechanism of action of intravenous pooled human IgG (IVIG) that is used for immune suppression in the clinic. While the data are consistent and compelling, the work mostly involves a single model system for autoimmune disease. It also assumes that the other suggested mechanisms of IVIG action (e.g., scavenging of activated complement,[232] and anti-Siglec antibodies[233]) do not contribute significantly. This is a very interesting avenue for future research, especially given than the

relevant sialyltransferase (ST6Gal-I) is highly regulated in response to inflammation.[185] and alters cellular activation and proliferation.[196,234]

Sialylated molecules or sialic acid–binding proteins as cluster of differentiation (CD) markers

It would be incomplete to discuss the immune system without mentioning CD markers. There is a long list of CD molecules that are either sialylated or are involved in sialic acid recognition. Space does not allow a full discussion of each of these antigens, but a brief summary can be found in Table 1.

Conclusions and perspectives

As the outermost "onion layer" on all vertebrate cells types, Sias were predestined to play roles as the "molecular frontier" in ongoing evolutionary arms races, both as targets for attack and as SAMPs. Sialome patterns evolve rapidly, likely because they are prone to being exploited by rapidly evolving pathogens and parasites. Meanwhile, changes in the intrinsic "landscape" of self-Sias have to be closely tracked by the intrinsic lectins such as Siglecs in order to maintain homeostasis, even while they themselves are being exploited by pathogens expressing Sias. On the background of such ongoing "molecular dialectics" (revolution and counterrevolution), major sialome changes at the level of a species, such as the wholesale loss of Neu5Gc in humans likely required major system-wide accommodations. Meanwhile, vertebrate hosts are "locked in" to maintaining the numerous intrinsic roles of Sias in reproduction, development, and normal physiology. Vertebrate host species have evolved a precarious compromise of using the presence of self Sias as "self-associated patterns," even while discriminating against close mimics, as well as against molecular surfaces lacking self Sias. Several other cell surface glycans such as galactose, fucose, and glycosaminoglycans also mediate immunity-related functions, including some of the ones discussed here. However, given their shear abundance, structural diversity, and vertebrate lineage-defining nature, Sias have been recruited for multifarious roles in immunity, only some of which we have addressed here.

Acknowledgments

We thank Shiv Pillai and Takashi Angata for critical comments and suggestions. This work was supported by grants from the NIH and by the Mathers Foundation of New York.

Conflicts of interest

The authors declare no conflicts of interest.

References

1. Schauer, R., G.V. Srinivasan, D. Wipfler, *et al.* 2011. O-Acetylated sialic acids and their role in immune defense. *Adv. Exp. Med. Biol.* **705:** 525–548.
2. Schauer, R. 2009. Sialic acids as regulators of molecular and cellular interactions. *Curr. Opin. Struct. Biol.* **19:** 507–514.
3. Lewis, A.L., N. Desa, E.E. Hansen, *et al.* 2009. Innovations in host and microbial sialic acid biosynthesis revealed by phylogenomic prediction of nonulosonic acid structure. *Proc. Natl. Acad. Sci. USA* **106:** 13552–13557.
4. Varki, A. & R. Schauer. 2009. Sialic Acids. In *Essentials of Glycobiology*. 2nd ed. A. Varki, R.D. Cummings, J.D. Esko, H.H. Freeze, P. Stanley, C.R. Bertozzi, G.W. Hart & M.E. Etzler, Eds.: 199–218. Cold Spring Harbor Laboratory Press. Cold Spring Harbor, NY.
5. Severi, E., D.W. Hood & G.H. Thomas. 2007. Sialic acid utilization by bacterial pathogens. *Microbiology* **153:** 2817–2822.
6. Troy, F.A. 1992. Polysialylation: from bacteria to brains. *Glycobiology* **2:** 5–23.
7. Vimr, E.R., K.A. Kalivoda, E.L. Deszo & S.M. Steenbergen. 2004. Diversity of microbial sialic acid metabolism. *Microbiol. Mol. Biol. Rev.* **68:** 132–153.
8. Byrne, B., G.G. Donohoe & R. O'Kennedy. 2007. Sialic acids: carbohydrate moieties that influence the biological and physical properties of biopharmaceutical proteins and living cells. *Drug Discov. Today* **12:** 319–326.
9. Rutishauser, U. 1998. Polysialic acid at the cell surface: biophysics in service of cell interactions and tissue plasticity. *J. Cell. Biochem.* **70:** 304–312.
10. Collins, B.E., O. Blixt, A.R. DeSieno, *et al.* 2004. Masking of CD22 by cis ligands does not prevent redistribution of CD22 to sites of cell contact. *Proc. Natl. Acad. Sci. USA* **101:** 6104–6109.
11. Pilatte, Y., J. Bignon & C.R. Lambré. 1993. Sialic acids as important molecules in the regulation of the immune system: pathophysiological implications of sialidases in immunity. *Glycobiology* **3:** 201–218.
12. Gahmberg, C.G. & L.C. Andersson. 1977. Selective radioactive labeling of cell surface sialoglycoproteins by periodate-tritiated borohydride. *J. Biol. Chem.* **252:** 5888–5894.
13. Dehoux-zenou, S.M., M. Guenounou, H. Zinbi, *et al.* 1987. Behavior of aldehyde moieties involved in the activation of suppressor cells by sodium periodate. *J. Immunol.* **138:** 1157–1163.
14. Jacobs, C.L., K.J. Yarema, L.K. Mahal, *et al.* 2000. Metabolic labeling of glycoproteins with chemical tags through unnatural sialic acid biosynthesis. *Methods Enzymol.* **327:** 260–275.
15. Du, J., M.A. Meledeo, Z. Wang, *et al.* 2009. Metabolic glycoengineering: sialic acid and beyond. *Glycobiology* **19:** 1382–1401.
16. Prescher, J.A. & C.R. Bertozzi. 2006. Chemical technologies for probing glycans. *Cell* **126:** 851–854.
17. Varki, A. 2010. Colloquium paper: uniquely human evolution of sialic acid genetics and biology. *Proc. Natl. Acad. Sci. USA* **107**(Suppl. 2): 8939–8946.
18. Fearon, D.T. 1978. Regulation by membrane sialic acid of beta1H-dependent decay-dissociation of amplification C3 convertase of the alternative complement pathway. *Proc. Natl. Acad. Sci. USA* **75:** 1971–1975.
19. Pangburn, M.K. & H.J. Muller-Eberhard. 1978. Complement C3 convertase: cell surface restriction of beta1H control and generation of restriction on neuraminidase-treated cells. *Proc. Natl. Acad. Sci. USA* **75:** 2416–2420.
20. Meri, S. & M.K. Pangburn. 1990. Discrimination between activators and nonactivators of the alternative pathway of complement: regulation via a sialic acid /polyanion binding site on factor H. *Proc. Natl. Acad. Sci. USA* **87:** 3982–3986.
21. Ram, S., A.K. Sharma, S.D. Simpson, *et al.* 1998. A novel sialic acid-binding site on factor H mediates serum resistance of sialylated Neisseria gonorrhoeae. *J. Exp. Med.* **187:** 743–752.
22. Varki, A. 2011. Since there are PAMPs and DAMPs, there must be SAMPs? Glycan "self-associated molecular patterns" dampen innate immunity, but pathogens can mimic them. *Glycobiology* **21:** 1121–1124.
23. Johnston, J.W., N.P. Coussens, S. Allen, *et al.* 2008. Characterization of the N-acetyl-5-neuraminic acid-binding site of the extracytoplasmic solute receptor (SiaP) of nontypeable haemophilus influenzae strain 2019. *J. Biol. Chem.* **283:** 855–865.

24. Ram, S., L.A. Lewis & S. Agarwal. 2011. Meningococcal group w-135 and y capsular polysaccharides paradoxically enhance activation of the alternative pathway of complement. *J. Biol. Chem.* **286:** 8297–8307.

25. Zipfel, P.F., C. Skerka, J. Hellwage, *et al.* 2002. Factor H family proteins: on complement, microbes and human diseases. *Biochem. Soc. Trans.* **30:** 971–978.

26. Kajander, T., M.J. Lehtinen, S. Hyvarinen, *et al.* 2011. Dual interaction of factor H with C3d and glycosaminoglycans in host-nonhost discrimination by complement. *Proc. Natl. Acad. Sci. USA* **108:** 2897–2902.

27. Morgan, H.P., J. Jiang, A.P. Herbert, *et al.* 2011. Crystallographic determination of the disease-associated T1184R variant of complement regulator factor H. *Acta. Crystallogr. D Biol. Crystallogr.* **67:** 593–600.

28. Shaughnessy, J., S. Ram, A. Bhattacharjee, *et al.* 2011. Molecular characterization of the interaction between sialylated Neisseria gonorrhoeae and factor H. *J. Biol. Chem.* **286:** 22235–22242.

29. Atkinson, J.P. & T.H. Goodship. 2007. Complement factor H and the hemolytic uremic syndrome. *J. Exp. Med.* **204:** 1245–1248.

30. Donoso, L.A., T. Vrabec & H. Kuivaniemi. 2010. The role of complement Factor H in age-related macular degeneration: a review. *Surv. Ophthalmol.* **55:** 227–246.

31. Nydegger, U.E., D.T. Fearon & K.F. Austen. 1978. Autosomal locus regulates inverse relationship between sialic acid content and capacity of mouse erythrocytes to activate human alternative complement pathway. *Proc. Natl. Acad. Sci. USA* **75:** 6078–6082.

32. Shi, W.X., R. Chammas, N.M. Varki, *et al.* 1996. Sialic acid 9-O-acetylation on murine erythroleukemia cells affects complement activation, binding to I-type lectins, and tissue homing. *J. Biol. Chem.* **271:** 31526–31532.

33. Arming, S., D. Wipfler, J. Mayr, *et al.* 2011. The human Cas1 protein: a sialic acid-specific O-acetyltransferase? *Glycobiology* **21:** 553–564.

34. Kjaer, T.R., A.G. Hansen, U.B. Sorensen, *et al.* 2011. Investigations on the pattern recognition molecule M-ficolin: quantitative aspects of bacterial binding and leukocyte association. *J. Leukoc. Biol.* **90:** 425–437.

35. Honore, C., S. Rorvig, T. Hummelshoj, *et al.* 2010. Tethering of Ficolin-1 to cell surfaces through recognition of sialic acid by the fibrinogen-like domain. *J. Leukoc. Biol.* **88:** 145–158.

36. Gout, E., V. Garlatti, D.F. Smith, *et al.* 2010. Carbohydrate recognition properties of human ficolins: glycan array screening reveals the sialic acid-binding specificity of M-ficolin. *J. Biol. Chem.* **285:** 6612–6622.

37. Rosen, S.D., M.S. Singer, T.A. Yednock & L.M. Stoolman. 1985. Involvement of sialic acid on endothelial cells in organ-specific lymphocyte recirculation. *Science* **228:** 1005–1007.

38. McEver, R.P. 1991. Selectins: novel receptors that mediate leukocyte adhesion during inflammation. *Thromb. Haemost.* **65:** 223–228.

39. Cummings, R.D. & D.F. Smith. 1992. The selectin family of carbohydrate-binding proteins: structure and importance of carbohydrate ligands for cell adhesion. *BioEssays* **14:** 849–856.

40. Rosen, S.D. 1993. L-selectin and its biological ligands. *Histochemistry* **100:** 185–191.

41. Varki, A. 1994. Selectin ligands. *Proc. Natl. Acad. Sci. USA* **91:** 7390–7397.

42. Lasky, L.A. 1995. Selectin-carbohydrate interactions and the initiation of the inflammatory response. *Annu. Rev. Biochem.* **64:** 113–139.

43. Picker, L.J., R.A. Warnock, A.R. Burns, *et al.* 1991. The neutrophil selectin LECAM-1 presents carbohydrate ligands to the vascular selectins ELAM-1 and GMP-140. *Cell* **66:** 921–933.

44. Zhou, Q., K.L. Moore, D.F. Smith, *et al.* 1991. The selectin GMP-140 binds to sialylated, fucosylated lactosaminoglycans on both myeloid and nonmyeloid cells. *J. Cell Biol.* **115:** 557–564.

45. Tyrrell, D., P. James, N. Rao, *et al.* 1991. Structural requirements for the carbohydrate ligand of E-selectin. *Proc. Natl. Acad. Sci. USA* **88:** 10372–10376.

46. Handa, K., E.D. Nudelman, M.R. Stroud, *et al.* 1991. Selectin GMP-140 (CD62; PADGEM) binds to sialosyl-Lea and sialosyl-Lex, and sulfated glycans modulate this binding. *Biochem. Biophys. Res. Commun.* **181:** 1223–1230.

47. Berg, E.L., J. Magnani, R.A. Warnock, *et al.* 1992. Comparison of L-selectin and E-selectin ligand specificities: the L-selectin can bind the E-selectin ligands sialyl Lex and sialyl Lea. *Biochem. Biophys. Res. Commun.* **184:** 1048–1055.

48. Foxall, C., S.R. Watson, D. Dowbenko, *et al.* 1992. The three members of the selectin receptor family recognize a common carbohydrate epitope, the sialyl Lewisx oligosaccharide. *J. Cell Biol.* **117:** 895–902.

49. Larsen, G.R., D. Sako, T.J. Ahern, *et al.* 1992. P-selectin and E-selectin. Distinct but overlapping leukocyte ligand specificities. *J. Biol. Chem.* **267:** 11104–11110.

50. Moore, K.L., N.L. Stults, S. Diaz, *et al.* 1992. Identification of a specific glycoprotein ligand for P-selectin (CD62) on myeloid cells. *J. Cell Biol.* **118:** 445–456.

51. Yeh, J.C., N. Hiraoka, B. Petryniak, *et al.* 2001. Novel sulfated lymphocyte homing receptors and their control by a core1 extension beta1,3-N-acetylglucosaminyltransferase. *Cell* **105:** 957–969.

52. Leppänen, A., P. Mehta, Y.B. Ouyang, *et al.* 1999. A novel glycosulfopeptide binds to P-selectin and inhibits leukocyte adhesion to P-selectin. *J. Biol. Chem.* **274:** 24838–24848.

53. Larkin, M., T.J. Ahern, M.S. Stoll, *et al.* 1992. Spectrum of sialylated and nonsialylated fuco-oligosaccharides bound by the endothelial-leukocyte adhesion molecule E-selectin. Dependence of the carbohydrate binding activity on E-selectin density. *J. Biol. Chem.* **267:** 13661–13668.

54. Norgard-Sumnicht, K.E., N.M. Varki & A. Varki. 1993. Calcium-dependent heparin-like ligands for L-selectin in nonlymphoid endothelial cells. *Science* **261:** 480–483.

55. Nelson, R.M., O. Cecconi, W.G. Roberts, *et al.* 1993. Heparin oligosaccharides bind L- and P-selectin and inhibit acute inflammation. *Blood* **82:** 3253–3258.

56. Norgard, K.E., H. Han, L. Powell, *et al.* 1993. Enhanced interaction of L-selectin with the high endothelial venule ligand via selectively oxidized sialic acids. *Proc. Natl. Acad. Sci. USA* **90:** 1068–1072.

57. Rosen, S.D. 2004. Ligands for L-selectin: homing, inflammation, and beyond. *Annu. Rev. Immunol.* **22:** 129–156.

58. Crocker, P.R. & S. Gordon. 1986. Properties and distribution of a lectin-like hemagglutinin differentially expressed by murine stromal tissue macrophages. *J. Exp. Med.* **164:** 1862–1875.

59. Crocker, P.R., S. Kelm, C. Dubois, *et al.* 1991. Purification and properties of sialoadhesin, a sialic acid-binding receptor of murine tissue macrophages. *EMBO J* **10:** 1661–1669.

60. Stamenkovic, I., D. Sgroi, A. Aruffo, *et al.* 1991. The B lymphocyte adhesion molecule CD22 interacts with leukocyte common antigen CD45RO on T cells and alpha 2–6 sialyltransferase, CD75, on B cells. *Cell* **66:** 1133–1144.

61. Sgroi, D., A. Varki, S. Braesch-Andersen & I. Stamenkovic. 1993. CD22, a B cell-specific immunoglobulin superfamily member, is a sialic acid-binding lectin. *J. Biol. Chem.* **268:** 7011–7018.

62. Powell, L.D., D. Sgroi, E.R. Sjoberg, *et al.* 1993. Natural ligands of the B cell adhesion molecule CD22beta carry N-linked oligosaccharides with alpha-2,6-linked sialic acids that are required for recognition. *J. Biol. Chem.* **268:** 7019–7027.

63. Powell, L.D. & A. Varki. 1994. The oligosaccharide binding specificities of CD22beta, a sialic acid-specific lectin of B cells. *J. Biol. Chem.* **269:** 10628–10636.

64. Crocker, P.R., S. Mucklow, V. Bouckson, *et al.* 1994. Sialoadhesin, a macrophage sialic acid-binding receptor for haemopoietic cells with 17 immunoglobulin-like domains. *EMBO J.* **13:** 4490–4503.

65. Kelm, S., A. Pelz, R. Schauer, *et al.* 1994. Sialoadhesin, myelin-associated glycoprotein and CD22 define a new family of sialic acid-dependent adhesion molecules of the immunoglobulin superfamily. *Curr. Biol.* **4:** 965–972.

66. Freeman, S.D., S. Kelm, E.K. Barber & P.R. Crocker. 1995. Characterization of CD33 as a new member of the sialoadhesin family of cellular interaction molecules. *Blood* **85:** 2005–2012.

67. Kelm, S., R. Schauer & P.R. Crocker. 1996. The sialoadhesins: a family of sialic acid-dependent cellular recognition molecules within the immunoglobulin superfamily. *Glycoconjugate J.* **13:** 913–926.

68. Collins, B.E., M. Kiso, A. Hasegawa, *et al.* 1997. Binding specificities of the sialoadhesin family of I-type lectins: sialic acid linkage and substructure requirements for binding of myelin-associated glycoprotein, Schwann cell myelin protein, and sialoadhesin. *J. Biol. Chem.* **272:** 16889–16895.

69. Powell, L.D. & A. Varki. 1995. I-type lectins. *J. Biol. Chem.* **270:** 14243–14246.

70. Crocker, P.R., E.A. Clark, M. Filbin, *et al.* 1998. Siglecs: a family of sialic-acid binding lectins [letter]. *Glycobiology* **8:** v.

71. Angata, T., E.H. Margulies, E.D. Green & A. Varki. 2004. Large-scale sequencing of the CD33-related Siglec gene cluster in five mammalian species reveals rapid evolution by multiple mechanisms. *Proc. Natl. Acad. Sci. USA* **101:** 13251–13256.

72. Angata, T. 2006. Molecular diversity and evolution of the Siglec family of cell-surface lectins. *Mol. Divers.* **10:** 555–566.

73. Crocker, P.R., J.C. Paulson & A. Varki. 2007. Siglecs and their roles in the immune system. *Nat. Rev. Immunol.* **7:** 255–266.

74. von Gunten, S. & B.S. Bochner. 2008. Basic and clinical immunology of Siglecs. *Ann. N. Y. Acad. Sci.* **1143:** 61–82.

75. Cao, H., B. de Bono, K. Belov, *et al.* 2009. Comparative genomics indicates the mammalian CD33rSiglec locus evolved by an ancient large-scale inverse duplication and suggests all Siglecs share a common ancestral region. *Immunogenetics* **61:** 401–417.

76. O'Reilly, M.K. & J.C. Paulson. 2010. Multivalent ligands for Siglecs. *Methods Enzymol.* **478:** 343–363.

77. Cao, H. & P.R. Crocker. 2011. Evolution of CD33-related Siglecs: regulating host immune functions and escaping pathogen exploitation? *Immunology* **132:** 18–26.

78. Park, C.S. & B.S. Bochner. 2011. Potential targeting of Siglecs, mast cell inhibitory receptors, in interstitial cystitis. *Int. Neurourol. J.* **15:** 61–63.

79. Pillai, S., I.A. Netravali, A. Cariappa & H. Mattoo. 2011. Siglecs and immune regulation. *Annu. Rev. Immunol.* **30.** [Epub ahead of print.] PMID: 22224769.

80. Yu, Z.B., M. Maoui, L.T. Wu, *et al.* 2001. mSiglec-E, a novel mouse CD33-related Siglec (sialic acid-binding immunoglobulin-like lectin) that recruits Src homology 2 (SH2)-domain-containing protein tyrosine phosphatases SHP-1 and SHP-2. *Biochem. J.* **353:** 483–492.

81. Whitney, G., S.L. Wang, H. Chang, *et al.* 2001. A new Siglec family member, Siglec-10, is expressed in cells of the immune system and has signaling properties similar to CD33. *Eur. J. Biochem.* **268:** 6083–6096.

82. Angata, T., S.C. Kerr, D.R. Greaves, *et al.* 2002. Cloning and characterization of human Siglec-11. A recently evolved signaling molecule that can interact with SHP-1 and SHP-2 and is expressed by tissue macrophages, including brain microglia. *J. Biol. Chem.* **277:** 24466–24474.

83. Avril, T., H. Floyd, F. Lopez, *et al.* 2004. The membrane-proximal immunoreceptor tyrosine-based inhibitory motif is critical for the inhibitory signaling mediated by Siglecs-7 and -9, CD33-related Siglecs expressed on human monocytes and NK cells. *J. Immunol.* **173:** 6841–6849.

84. Nitschke, L. 2009. CD22 and Siglec-G: B-cell inhibitory receptors with distinct functions. *Immunol. Rev.* **230:** 128–143.

85. Liu, Y., G.Y. Chen & P. Zheng. 2009. CD24-Siglec G/10 discriminates danger- from pathogen-associated molecular patterns. *Trends Immunol.* **30:** 557–561.

86. Avril, T., S.D. Freeman, H. Attrill, *et al.* 2005. Siglec-5 (CD170) can mediate inhibitory signalling in the absence of immunoreceptor tyrosine-based inhibitory motif phosphorylation. *J. Biol. Chem.* **280:** 19843–19851.

87. Mitsuki, M., K. Nara, T. Yamaji, *et al.* 2010. Siglec-7 mediates nonapoptotic cell death independently of its immunoreceptor tyrosine-based inhibitory motifs in monocytic cell line U937. *Glycobiology* **20:** 395–402.

88. Angata, T., T. Hayakawa, M. Yamanaka, *et al.* 2006. Discovery of Siglec-14, a novel sialic acid receptor undergoing concerted evolution with Siglec-5 in primates. *FASEB J.* **20:** 1964–1973.

89. Angata, T., Y. Tabuchi, K. Nakamura & M. Nakamura. 2007. Siglec-15: an immune system Siglec conserved throughout vertebrate evolution. *Glycobiology* **17**: 838–846.

90. Yasui, K., T. Angata, N. Matsuyama, *et al.* 2011. Detection of anti-Siglec-14 alloantibodies in blood components implicated in nonhaemolytic transfusion reactions. *Br. J. Haematol.* **153**: 794–796.

91. Hiruma, Y., T. Hirai & E. Tsuda. 2011. Siglec-15, a member of the sialic acid-binding lectin, is a novel regulator for osteoclast differentiation. *Biochem. Biophys. Res. Commun.* **409**: 424–429.

92. Cao, H., U. Lakner, B. de Bono, *et al.* 2008. SIGLEC16 encodes a DAP12-associated receptor expressed in macrophages that evolved from its inhibitory counterpart SIGLEC11 and has functional and non-functional alleles in humans. *Eur. J. Immunol.* **38**: 2303–2315.

93. Varki, A. & T. Angata. 2006. Siglecs–the major subfamily of I-type lectins. *Glycobiology* **16**: 1R-27R.

94. Crocker, P.R. & P. Redelinghuys. 2008. Siglecs as positive and negative regulators of the immune system. *Biochem. Soc. Trans.* **36**: 1467–1471.

95. O'Reilly, M.K. & J.C. Paulson. 2009. Siglecs as targets for therapy in immune-cell-mediated disease. *Trends Pharmacol. Sci.* **30**: 240–248.

96. Varki, A. 2009. Natural ligands for CD33-related Siglecs? *Glycobiology* **19**: 810–812.

97. Jandus, C., H.U. Simon & S. von Gunten. 2011. Targeting Siglecs: a novel pharmacological strategy for immuno- and glycotherapy. *Biochem. Pharmacol.* **82**: 323–332.

98. Magesh, S., H. Ando, T. Tsubata, *et al.* 2011. High-affinity ligands of Siglec receptors and their therapeutic potentials. *Curr. Med. Chem.* **18**: 3537–3550.

99. Razi, N. & A. Varki. 1998. Masking and unmasking of the sialic acid-binding lectin activity of CD22 (Siglec-2) on B lymphocytes. *Proc. Natl. Acad. Sci. USA* **95**: 7469–7474.

100. Kawasaki, Y., A. Ito, D.A. Withers, *et al.* 2010. Ganglioside DSGb5, preferred ligand for Siglec-7, inhibits NK cell cytotoxicity against renal cell carcinoma cells. *Glycobiology* **20**: 1373–1379.

101. Blixt, O., S. Han, L. Liao, *et al.* 2008. Sialoside analogue arrays for rapid identification of high affinity Siglec ligands. *J. Am. Chem. Soc.* **130**: 6680–6681.

102. Cui, L., P.I. Kitov, G.C. Completo, *et al.* 2011. Supramolecular complexing of membane Siglec CD22 mediated by a polyvalent heterobifunctional ligand that templates on IgM. *Bioconjug. Chem.* **22**: 546–550.

103. Vinson, M., P.A. Van der Merwe, S. Kelm, *et al.* 1996. Characterization of the sialic acid-binding site in sialoadhesin by site-directed mutagenesis. *J. Biol. Chem.* **271**: 9267–9272.

104. Van der Merwe, P.A., P.R. Crocker, M. Vinson, *et al.* 1996. Localization of the putative sialic acid-binding site on the immunoglobulin superfamily cell-surface molecule CD22. *J. Biol. Chem.* **271**: 9273–9280.

105. Crocker, P.R., M. Vinson, S. Kelm & K. Drickamer. 1999. Molecular analysis of sialoside binding to sialoadhesin by NMR and site-directed mutagenesis. *Biochem. J.* **341**: 355–361.

106. Angata, T. & A. Varki. 2000. Cloning, characterization, and phylogenetic analysis of Siglec-9, a new member of the CD33-related group of Siglecs: evidence for co-evolution with sialic acid synthesis pathways. *J. Biol. Chem.* **275**: 22127–22135.

107. Angata, T., N.M. Varki & A. Varki. 2001. A second uniquely human mutation affecting sialic acid biology. *J. Biol. Chem.* **276**: 40282–40287.

108. Mandal, C. 1990. Sialic acid-binding lectins. *Experientia* **46**: 433–441.

109. Traving, C. & R. Schauer. 1998. Structure, function and metabolism of sialic acids. *Cell Mol Life Sci* **54**: 1330–1349.

110. Isa, P., C.F. Arias & S. Lopez. 2006. Role of sialic acids in rotavirus infection. *Glycoconj. J.* **23**: 27–37.

111. Gee, G.V., A.S. Dugan, N. Tsomaia, *et al.* 2006. The role of sialic acid in human polyomavirus infections. *Glycoconj. J.* **23**: 19–26.

112. Lehmann, F., E. Tiralongo & J. Tiralongo. 2006. Sialic acid-specific lectins: occurrence, specificity and function. *Cell Mol. Life Sci.* **63**: 1331–1354.

113. Taylor, G. 1996. Sialidases: structures, biological significance and therapeutic potential. *Curr. Opin. Struct. Biol.* **6**: 830–837.

114. von Itzstein, M. 2007. The war against influenza: discovery and development of sialidase inhibitors. *Nat. Rev. Drug Discov.* **6**: 967–974.

115. Herrler, G., R. Rott, H.D. Klenk, *et al.* 1985. The receptor-destroying enzyme of influenza C virus is neuraminate-O-acetylesterase. *EMBO J* **4**: 1503–1506.

116. Rogers, G.N., G. Herrler, J.C. Paulson & H.D. Klenk. 1986. Influenza C virus uses 9-O-acetyl-N-acetylneuraminic acid as a high affinity receptor determinant for attachment to cells. *J. Biol. Chem.* **261**: 5947–5951.

117. Vlasak, R., M. Krystal, M. Nacht & P. Palese. 1987. The influenza C virus glycoprotein (HE) exhibits receptor-binding (hemagglutinin) and receptor-destroying (esterase) activities. *Virology* **160**: 419–425.

118. Vlasak, R., W. Luytjes, J. Leider, *et al.* 1988. The E3 protein of bovine coronavirus is a receptor-destroying enzyme with acetylesterase activity. *J. Virol.* **62**: 4686–4690.

119. Herrler, G. & H.-D. Klenk. 1991. Structure and function of the HEF glycoprotein of influenza C virus. *Adv. Virus Res.* **40**: 213–234.

120. Pleschka, S., H.D. Klenk & G. Herrler. 1995. The catalytic triad of the influenza C virus glycoprotein HEF esterase: characterization by site-directed mutagenesis and functional analysis. *J. Gen. Virol.* **76**: 2529–2537.

121. Cornelissen, L.A.H.M., C.M.H. Wierda, d. M.F.J. van, *et al.* 1997. Hemagglutinin-esterase: a novel structural protein of torovirus. *J. Virol.* **71**: 5277–5286.

122. Wurzer, W.J., K. Obojes & R. Vlasak. 2002. The sialate-4-O-acetylesterases of coronaviruses related to mouse hepatitis virus: a proposal to reorganize group 2 Coronaviridae. *J. Gen. Virol.* **83**: 395–402.

123. de Groot, R.J. 2006. Structure, function and evolution of the hemagglutinin-esterase proteins of corona- and toroviruses. *Glycoconj. J.* **23**: 59–72.

124. Schwegmann-Wessels, C. & G. Herrler. 2006. Sialic acids as receptor determinants for coronaviruses. *Glycoconj. J.* **23**: 51–58.

125. Langereis, M.A., A.L. van Vliet, W. Boot & R.J. de Groot. 2010. Attachment of mouse hepatitis virus to O-acetylated sialic acid is mediated by hemagglutinin-esterase and not by the spike protein. *J. Virol.* **84:** 8970–8974.

126. Orlandi, P.A., F.W. Klotz & J.D. Haynes. 1992. A malaria invasion receptor, the 175-kilodalton erythrocyte binding antigen of Plasmodium falciparum recognizes the terminal Neu5Ac(alpha 2–3)Gal- sequences of glycophorin A. *J. Cell Biol.* **116:** 901–909.

127. Dolan, S.A., J.L. Proctor, D.W. Alling, *et al.* 1994. Glycophorin B as an EBA-175 independent Plasmodium falciparum receptor of human erythrocytes. *Mol. Biochem. Parasitol.* **64:** 55–63.

128. DeLuca, G.M., M.E. Donnell, D.J. Carrigan & D.P. Blackall. 1996. Plasmodium falciparum merozoite adhesion is mediated by sialic acid. *Biochem. Biophys. Res. Commun.* **225:** 726–732.

129. Baum, J., R.H. Ward & D.J. Conway. 2002. Natural selection on the erythrocyte surface. *Mol. Biol. Evol.* **19:** 223–229.

130. Byres, E., A.W. Paton, J.C. Paton, *et al.* 2008. Incorporation of a non-human glycan mediates human susceptibility to a bacterial toxin. *Nature* **456:** 648–652.

131. Roggentin, P., R. Schauer, L.L. Hoyer & E.R. Vimr. 1993. The sialidase superfamily and its spread by horizontal gene transfer. *Mol. Microbiol.* **9:** 915–921.

132. Vimr, E., C. Lichtensteiger & S. Steenbergen. 2000. Sialic acid metabolism's dual function in Haemophilus influenzae. *Mol. Microbiol.* **36:** 1113–1123.

133. Johnston, J.W., A. Zaleski, S. Allen, *et al.* 2007. Regulation of sialic acid transport and catabolism in Haemophilus influenzae. *Mol. Microbiol.* **66:** 26–39.

134. Trappetti, C., A. Kadioglu, M. Carter, *et al.* 2009. Sialic acid: a preventable signal for pneumococcal biofilm formation, colonization, and invasion of the host. *J. Infect. Dis.* **199:** 1497–1505.

135. Corfield, A.P., W.M. Sander, R.W. Veh, *et al.* 1986. The action of sialidases on substrates containing O-acetylsialic acids. *Biol. Chem. Hoppe Seyler* **367:** 433–439.

136. Matzinger, P. 2002. The danger model: a renewed sense of self. *Science* **296:** 301–305.

137. Chen, G.Y. & G. Nunez. 2010. Sterile inflammation: sensing and reacting to damage. *Nat. Rev. Immunol.* **10:** 826–837.

138. Miyagi, T., T. Wada, K. Yamaguchi & K. Hata. 2004. Sialidase and malignancy: a minireview. *Glycoconj. J.* **20:** 189–198.

139. Magesh, S., V. Savita, S. Moriya, *et al.* 2009. Human sialidase inhibitors: design, synthesis, and biological evaluation of 4-acetamido-5-acylamido-2-fluoro benzoic acids. *Bioorg. Med. Chem.* **17:** 4595–4603.

140. Monti, E., E. Bonten, A. D'Azzo, *et al.* 2010. Sialidases in vertebrates a family of enzymes tailored for several cell functions. *Adv. Carbohydr. Chem. Biochem.* **64:** 403–479.

141. Carrillo, M.B., C.M. Milner, S.T. Ball, *et al.* 1997. Cloning and characterization of a sialidase from the murine histocompatibility-2 complex: low levels of mRNA and a single amino acid mutation are responsible for reduced sialidase activity in mice carrying the Neu1a allele. *Glycobiology* **7:** 975–986.

142. Rottier, R.J., E. Bonten & A. D'Azzo. 1998. A point mutation in the neu-1 locus causes the neuraminidase defect in the SM/J mouse. *Hum. Mol. Genet.* **7:** 313–321.

143. Van, d.S.A., E. Bonten & A. D'Azzo. 1998. Transport of human lysosomal neuraminidase to mature lysosomes requires protective protein cathepsin A. *EMBO J.* **17:** 1588–1597.

144. Amith, S.R., P. Jayanth, S. Franchuk, *et al.* 2009. Dependence of pathogen molecule-induced toll-like receptor activation and cell function on Neu1 sialidase. *Glycoconj. J.* **26:** 1197–1212.

145. Hinek, A., T.D. Bodnaruk, S. Bunda, *et al.* 2008. Neuraminidase-1, a subunit of the cell surface elastin receptor, desialylates and functionally inactivates adjacent receptors interacting with the mitogenic growth factors PDGF-BB and IGF-2. *Am. J. Pathol.* **173:** 1042–1056.

146. Feng, C., L. Zhang, L. Almulki, *et al.* 2011. Endogenous PMN sialidase activity exposes activation epitope on CD11b/CD18 which enhances its binding interaction with ICAM-1. *J. Leukoc. Biol.* **90:** 313–321.

147. Amith, S.R., P. Jayanth, S. Franchuk, *et al.* 2010. Neu1 desialylation of sialyl alpha-2,3-linked beta-galactosyl residues of TOLL-like receptor 4 is essential for receptor activation and cellular signaling. *Cell. Signal.* **22:** 314–324.

148. Nan, X., I. Carubelli & N.M. Stamatos. 2007. Sialidase expression in activated human T lymphocytes influences production of IFN-{gamma}. *J. Leukoc. Biol.* **81:** 284–296.

149. Seyrantepe, V., A. Iannello, F. Liang, *et al.* 2010. Regulation of phagocytosis in macrophages by neuraminidase 1. *J. Biol. Chem.* **285:** 206–215.

150. Toscano, M.A., G.A. Bianco, J.M. Ilarregui, *et al.* 2007. Differential glycosylation of TH1, TH2 and TH-17 effector cells selectively regulates susceptibility to cell death. *Nat. Immunol.* **8:** 825–834.

151. Monti, E., M.T. Bassi, N. Papini, *et al.* 2000. Identification and expression of NEU3, a novel human sialidase associated to the plasma membrane. *Biochem. J.* **349:** 343–351.

152. Wang, Y., K. Yamaguchi, T. Wada, *et al.* 2002. A close association of the ganglioside-specific sialidase Neu3 with caveolin in membrane microdomains. *J. Biol. Chem.* **277:** 26252–26259.

153. Vimr, E. & C. Lichtensteiger. 2002. To sialylate, or not to sialylate: that is the question. *Trends Microbiol.* **10:** 254–257.

154. Schoenhofen, I.C., D.J. McNally, J.R. Brisson & S.M. Logan. 2006. Elucidation of the CMP-pseudaminic acid pathway in Helicobacter pylori: synthesis from UDP-N-acetylglucosamine by a single enzymatic reaction. *Glycobiology* **16:** 8C-14C.

155. Schoenhofen, I.C., E. Vinogradov, D.M. Whitfield, *et al.* 2009. The CMP-legionaminic acid pathway in Campylobacter: biosynthesis involving novel GDP-linked precursors. *Glycobiology* **19:** 715–725.

156. Marques, M.B., D.L. Kasper, M.K. Pangburn & M.R. Wessels. 1992. Prevention of C3 deposition by capsular polysaccharide is a virulence mechanism of type III group B streptococci. *Infect. Immun.* **60:** 3986–3993.

157. Campagnari, A.A., M.R. Gupta, K.C. Dudas, *et al.* 1987. Antigenic diversity of lipooligosaccharides of nontypable Haemophilus influenzae. *Infect. Immun.* **55:** 882–887.

158. Jones, C., M. Virji & P.R. Crocker. 2003. Recognition of sialylated meningococcal lipopolysaccharide by Siglecs expressed on myeloid cells leads to enhanced bacterial uptake. *Mol. Microbiol.* **49:** 1213–1225.

159. Avril, T., E.R. Wagner, H.J. Willison & P.R. Crocker. 2006. Sialic acid-binding immunoglobulin-like lectin 7 mediates selective recognition of sialylated glycans expressed on Campylobacter jejuni lipooligosaccharides. *Infect. Immun.* **74:** 4133–4141.

160. Khatua, B., A. Ghoshal, K. Bhattacharya, *et al.* 2010. Sialic acids acquired by Pseudomonas aeruginosa are involved in reduced complement deposition and Siglec mediated host-cell recognition. *FEBS Lett.* **584:** 555–561.

161. Carlin, A.F., S. Uchiyama, Y.C. Chang, *et al.* 2009. Molecular mimicry of host sialylated glycans allows a bacterial pathogen to engage neutrophil Siglec-9 and dampen the innate immune response. *Blood* **113:** 3333–3336.

162. Vanderheijden, N., P.L. Delputte, H.W. Favoreel, *et al.* 2003. Involvement of sialoadhesin in entry of porcine reproductive and respiratory syndrome virus into porcine alveolar macrophages. *J. Virol.* **77:** 8207–8215.

163. Rempel, H., C. Calosing, B. Sun & L. Pulliam. 2008. Sialoadhesin expressed on IFN-induced monocytes binds HIV-1 and enhances infectivity. *PLoS ONE* **3:** e1967.

164. Jiang, H.R., L. Hwenda, K. Makinen, *et al.* 2006. Sialoadhesin promotes the inflammatory response in experimental autoimmune uveoretinitis. *J. Immunol.* **177:** 2258–2264.

165. Oetke, C., M.C. Vinson, C. Jones & P.R. Crocker. 2006. Sialoadhesin-deficient mice exhibit subtle changes in B- and T-cell populations and reduced immunoglobulin m levels. *Mol. Cell. Biol.* **26:** 1549–1557.

166. Ip, C.W., A. Kroner, P.R. Crocker, *et al.* 2007. Sialoadhesin deficiency ameliorates myelin degeneration and axonopathic changes in the CNS of PLP overexpressing mice. *Neurobiol. Dis.* **25:** 105–111.

167. Wu, C., U. Rauch, E. Korpos, *et al.* 2009. Sialoadhesin-positive macrophages bind regulatory T cells, negatively controlling their expansion and autoimmune disease progression. *J. Immunol.* **182:** 6508–6516.

168. Hallenbeck, P.C., E.R. Vimr, F. Yu, *et al.* 1987. Purification and properties of a bacteriophage-induced endo-N-acetylneuraminidase specific for poly-alpha-2,8-sialosyl carbohydrate units. *J. Biol. Chem.* **262:** 3553–3561.

169. Gerardy-Schahn, R., A. Bethe, T. Brennecke, *et al.* 1995. Molecular cloning and functional expression of bacteriophage PK1E-encoded endoneuraminidase Endo NE. *Mol. Microbiol.* **16:** 441–450.

170. Cieslewicz, M.J., D. Chaffin, G. Glusman, *et al.* 2005. Structural and genetic diversity of group B streptococcus capsular polysaccharides. *Infect. Immun.* **73:** 3096–3103.

171. Mushtaq, N., M.B. Redpath, J.P. Luzio & P.W. Taylor. 2004. Prevention and cure of systemic Escherichia coli K1 infection by modification of the bacterial phenotype. *Antimicrob. Agents Chemother.* **48:** 1503–1508.

172. Waldor, M.K. & D.I. Friedman. 2005. Phage regulatory circuits and virulence gene expression. *Curr Opin Microbiol* **8:** 459–465.

173. Rutishauser, U. 2008. Polysialic acid in the plasticity of the developing and adult vertebrate nervous system. *Nat. Rev. Neurosci.* **9:** 26–35.

174. Weinhold, B., R. Seidenfaden, I. Rockle, *et al.* 2005. Genetic ablation of polysialic acid causes severe neurodevelopmental defects rescued by deletion of the neural cell adhesion molecule. *J. Biol. Chem.* **280:** 42971–42977.

175. Bax, M., S.J. van Vliet, M. Litjens, *et al.* 2009. Interaction of polysialic acid with CCL21 regulates the migratory capacity of human dendritic cells. *PLoS ONE* **4:** e6987.

176. Rey-Gallardo, A., C. Delgado-Martin, R. Gerardy-Schahn, *et al.* 2011. Polysialic acid is required for neuropilin-2a/b-mediated control of CCL21-driven chemotaxis of mature dendritic cells and for their migration in vivo. *Glycobiology* **21:** 655–662.

177. Drake, P.M., J.K. Nathan, C.M. Stock, *et al.* 2008. Polysialic acid, a glycan with highly restricted expression, is found on human and murine leukocytes and modulates immune responses. *J. Immunol.* **181:** 6850–6858.

178. Drake, P.M., C.M. Stock, J.K. Nathan, *et al.* 2009. Polysialic acid governs T-cell development by regulating progenitor access to the thymus. *Proc. Natl. Acad. Sci. USA* **106:** 11995–12000.

179. Gagneux, P. & A. Varki. 1999. Evolutionary considerations in relating oligosaccharide diversity to biological function. *Glycobiology* **9:** 747–755.

180. Bishop, J.R. & P. Gagneux. 2007. Evolution of carbohydrate antigens–microbial forces shaping host glycomes? *Glycobiology* **17:** 23R–34R.

181. An, G., B. Wei, B. Xia, *et al.* 2007. Increased susceptibility to colitis and colorectal tumors in mice lacking core 3-derived O-glycans. *J. Exp. Med.* **204:** 1417–1429.

182. Ghate, A.A. & G.M. Air. 1999. Influenza type B neuraminidase can replace the function of type A neuraminidase. *Virology* **264:** 265–277.

183. Adams, J.H., B.K. Sim, S.A. Dolan, *et al.* 1992. A family of erythrocyte binding proteins of malaria parasites. *Proc. Natl. Acad. Sci. USA* **89:** 7085–7089.

184. Persson, K.E., F.J. McCallum, L. Reiling, *et al.* 2008. Variation in use of erythrocyte invasion pathways by Plasmodium falciparum mediates evasion of human inhibitory antibodies. *J. Clin. Invest.* **118:** 342–351.

185. Dalziel, M., S. Lemaire, J. Ewing, *et al.* 1999. Hepatic acute phase induction of murine beta-galactoside:alpha2,6 sialyltransferase (ST6Gal I) is IL-6 dependent and mediated by elevation of Exon H-containing class of transcripts. *Glycobiology* **9:** 1003–1008.

186. Schauer, R. 1985. Sialic acids and their role as biological masks. *Trends Biochem. Sci.* **10:** 357–360.

187. Weiss, P. & G. Ashwell. 1989. The asialoglycoprotein receptor: properties and modulation by ligand. *Prog. Clin. Biol. Res.* **300:** 169–184.

188. Braun, J.R., T.E. Willnow, S. Ishibashi, *et al.* 1996. The major subunit of the asialoglycoprotein receptor is expressed on the hepatocellular surface in mice lacking the minor receptor subunit. *J. Biol. Chem.* **271:** 21160–21166.

189. Grewal, P.K., S. Uchiyama, D. Ditto, *et al.* 2008. The Ashwell receptor mitigates the lethal coagulopathy of sepsis. *Nat. Med.* **14:** 648–655.

190. Elward, K. & P. Gasque. 2003. "Eat me" and "don't eat me" signals govern the innate immune response and tissue repair in the CNS: emphasis on the critical role of the complement system. *Mol. Immunol.* **40:** 85–94.

191. Liu, F.T. 2000. Galectins: a new family of regulators of inflammation. *Clin. Immunol.* **97:** 79–88.

192. Rabinovich, G.A., N. Rubinstein & M.A. Toscano. 2002. Role of galectins in inflammatory and immunomodulatory processes. *Biochim. Biophys. Acta. Gen. Subj.* **1572:** 274–284.

193. Lau, K.S., E.A. Partridge, A. Grigorian, *et al.* 2007. Complex N-glycan number and degree of branching cooperate to regulate cell proliferation and differentiation. *Cell* **129:** 123–134.

194. Mendelsohn, R., P. Cheung, L. Berger, *et al.* 2007. Complex N-glycan and metabolic control in tumor cells. *Cancer Res.* **67:** 9771–9780.

195. Bi, S. & L.G. Baum. 2009. Sialic acids in T cell development and function. *Biochim. Biophys. Acta* **1790:** 1599–1610.

196. Zhuo, Y. & S.L. Bellis. 2011. Emerging role of alpha2,6-sialic acid as a negative regulator of galectin binding and function. *J. Biol. Chem.* **286:** 5935–5941.

197. Ideo, H., T. Matsuzaka, T. Nonaka, *et al.* 2011. Galectin-8-N-domain recognition mechanism for sialylated and sulfated glycans. *J. Biol. Chem.* **286:** 11346–11355.

198. Yoshida, H., M. Teraoka, N. Nishi, *et al.* 2010. X-ray structures of human galectin-9 C-terminal domain in complexes with a biantennary oligosaccharide and sialyllactose. *J. Biol. Chem.* **285:** 36969–36976.

199. Fukushi, Y., E. Nudelman, S.B. Levery, *et al.* 1986. A novel disialoganglioside (IV3NeuAcIII6NeuAcLc4) of human adenocarcinoma and the monoclonal antibody (FH9) defining this disialosyl structure. *Biochemistry* **25:** 2859–2866.

200. Song, Y., K. Kitajima & Y. Inoue. 1993. Monoclonal antibody specific to alpha-2->3-linked deaminated neuraminyl beta-galactosyl sequence. *Glycobiology* **3:** 31–36.

201. Cheresh, D.A., R.A. Reisfeld & A. Varki. 1984. O-acetylation of disialoganglioside GD3 by human melanoma cells creates a unique antigenic determinant. *Science* **225:** 844–846.

202. Schauer, R., G.V. Srinivasan, B. Coddeville, *et al.* 2009. Low incidence of N-glycolylneuraminic acid in birds and reptiles and its absence in the platypus. *Carbohydr. Res.* **344:** 1494–1500.

203. Irie, A., S. Koyama, Y. Kozutsumi, *et al.* 1998. The molecular basis for the absence of N-glycolylneuraminic acid in humans. *J. Biol. Chem.* **273:** 15866–15871.

204. Chou, H.H., H. Takematsu, S. Diaz, *et al.* 1998. A mutation in human CMP-sialic acid hydroxylase occurred after the Homo-Pan divergence. *Proc. Natl. Acad. Sci. USA* **95:** 11751–11756.

205. Fujii, Y., H. Higashi, K. Ikuta, *et al.* 1982. Specificities of human heterophilic Hanganutziu and Deicher (H-D) antibodies and avian antisera against H-D antigen-active glycosphingolipids. *Mol. Immunol.* **19:** 87–94.

206. Naiki, M., Y. Fujii, K. Ikuta, *et al.* 1982. Expression of Hanganutziu and Deicher type heterophile antigen on the cell surface of Marek's disease lymphoma. *Adv. Exp. Med. Biol.* **152:** 445–456.

207. Tangvoranuntakul, P., P. Gagneux, S. Diaz, *et al.* 2003. Human uptake and incorporation of an immunogenic nonhuman dietary sialic acid. *Proc. Natl. Acad. Sci. USA* **100:** 12045–12050.

208. Taylor, R.E., C.J. Gregg, V. Padler-Karavani, *et al.* 2010. Novel mechanism for the generation of human xeno-autoantibodies against the nonhuman sialic acid N-glycolylneuraminic acid. *J. Exp. Med.* **207:** 1637–1646.

209. Merrick, J.M., K. Zadarlik & F. Milgrom. 1978. Characterization of the Hanganutziu-Deicher (serum-sickness) antigen as gangliosides containing N-glycolylneuraminic acid. *Int. Arch. Allergy Appl. Immunol.* **57:** 477–480.

210. Higashi, H., M. Naiki, S. Matuo & K. Okouchi. 1977. Antigen of "serum sickness" type of heterophile antibodies in human sera: indentification as gangliosides with N-glycolylneuraminic acid. *Biochem. Biophys. Res. Commun.* **79:** 388–395.

211. Ghaderi, D., S.A. Springer, F. Ma, *et al.* 2011. Sexual selection by female immunity against paternal antigens can fix loss of function alleles. *Proc. Natl. Acad. Sci. USA* **108:** 17743–17748.

212. Sadler, J.E., J.C. Paulson & R.L. Hill. 1979. The role of sialic acid in the expression of human MN blood group antigens. *J. Biol. Chem.* **254:** 2112–2119.

213. Uemura, K., D. Roelcke, Y. Nagai & T. Feizi. 1984. The reactivities of human erythrocyte autoantibodies anti-Pr2, anti-Gd, Fl and Sa with gangliosides in a chromatogram binding assay. *Biochem. J.* **219:** 865–874.

214. Andrews, G.A., P.S. Chavey, J.E. Smith & L. Rich. 1992. N-glycolylneuraminic acid and N-acetylneuraminic acid define feline blood group A and B antigens. *Blood* **79:** 2485–2491.

215. Bighignoli, B., T. Niini, R.A. Grahn, *et al.* 2007. Cytidine monophospho-N-acetylneuraminic acid hydroxylase (CMAH) mutations associated with the domestic cat AB blood group. *BMC Genet.* **8:** 27.

216. Yamakawa, T., A. Suzuki & Y. Hashimoto. 1986. Genetic control of glycolipid expression. *Chem. Phys. Lipids* **42:** 75–90.

217. Weiman, S., S. Uchiyama, F.Y. Lin, *et al.* 2010. O-acetylation of sialic acid on group B streptococcus inhibits neutrophil suppression and virulence. *Biochem. J.* **428:** 163–168.

218. Kelm, S., R. Schauer, J.C. Manuguerra, *et al.* 1994. Modifications of cell surface sialic acids modulate cell adhesion mediated by sialoadhesin and CD22. *Glycoconjugate J.* **11:** 576–585.

219. Sjoberg, E.R., L.D. Powell, A. Klein & A. Varki. 1994. Natural ligands of the B cell adhesion molecule CD22beta can be masked by 9-O-acetylation of sialic acids. *J. Cell Biol.* **126:** 549–562.

220. Cariappa, A., H. Takematsu, H. Liu, *et al.* 2009. B cell antigen receptor signal strength and peripheral B cell development are regulated by a 9-O-acetyl sialic acid esterase. *J. Exp. Med.* **206:** 125–138.

221. Pillai, S., A. Cariappa & S.P. Pirnie. 2009. Esterases and autoimmunity: the sialic acid acetylesterase pathway and the regulation of peripheral B cell tolerance. *Trends Immunol.* **30:** 488–493.

222. Surolia, I., S.P. Pirnie, V. Chellappa, *et al.* 2010. Functionally defective germline variants of sialic acid acetylesterase in autoimmunity. *Nature* **466:** 243–247.

223. Kniep, B., W.A. Flegel, H. Northoff & E.P. Rieber. 1993. CDw60 glycolipid antigens of human leukocytes: structural characterization and cellular distribution. *Blood* **82:** 1776–1786.

224. Rieber, E.P. & G. Rank. 1994. CDw60: a marker for human CD8+ T helper cells. *J. Exp. Med.* **179:** 1385–1390.

225. Fox, D.A., X. He, A. Abe, T. Hollander, *et al.* 2001. The T lymphocyte structure CD60 contains a sialylated carbohydrate epitope that is expressed on both gangliosides and glycoproteins. *Immunol. Invest.* **30:** 67–85.

226. Malisan, F., L. Franchi, B. Tomassini, *et al.* 2002. Acetylation suppresses the proapoptotic activity of GD3 ganglioside. *J. Exp. Med.* **196:** 1535–1541.

227. Birks, S.M., J.O. Danquah, L. King, *et al.* 2011. Targeting the GD3 acetylation pathway selectively induces apoptosis in glioblastoma. *Neuro. Oncol.* **13:** 950–960.

228. Chen, H.Y. & A. Varki. 2002. O-acetylation of GD3: an enigmatic modification regulating apoptosis? *J. Exp. Med.* **196:** 1529–1533.

229. Kaneko, Y., F. Nimmerjahn & J.V. Ravetch. 2006. Anti-inflammatory activity of immunoglobulin G resulting from Fc sialylation. *Science* **313:** 670–673.

230. Anthony, R.M. & J.V. Ravetch. 2010. A novel role for the IgG Fc glycan: the anti-inflammatory activity of sialylated IgG Fcs. *J. Clin. Immunol.* **30**(Suppl. 1): S9–S14.

231. Anthony, R.M., T. Kobayashi, F. Wermeling & J.V. Ravetch. 2011. Intravenous gammaglobulin suppresses inflammation through a novel T(H)2 pathway. *Nature* **475:** 110–113.

232. Durandy, A., S.V. Kaveri, T.W. Kuijpers, M. Basta, *et al.* 2009. Intravenous immunoglobulins: understanding properties and mechanisms. *Clin. Exp. Immunol.* **158**(Suppl. 1): 2–13.

233. von Gunten, S. & H.U. Simon. 2010. Cell death modulation by intravenous immunoglobulin. *J. Clin. Immunol.* **30**(Suppl. 1): S24–S30.

234. Jones, M.B., M. Nasirikenari, L. Feng, *et al.* 2010. Role for hepatic and circulatory ST6Gal-1 sialyltransferase in regulating myelopoiesis. *J. Biol. Chem.* **285:** 25009–25017.

Ann. N.Y. Acad. Sci. ISSN 0077-8923

ANNALS OF THE NEW YORK ACADEMY OF SCIENCES

Issue: *Glycobiology of the Immune Response*

Siglecs as sensors of self in innate and adaptive immune responses

James C. Paulson, Matthew S. Macauley, and Norihito Kawasaki

Departments of Chemical Physiology and Molecular Biology, The Scripps Research Institute, La Jolla, California

Address for correspondence: James C. Paulson, The Scripps Research Institute, 10550 North Torrey Pines Road, MEM-L71, La Jolla, CA 92037. jpaulson@scripps.edu

Siglecs are expressed on most white blood cells of the immune system and are known to modulate the activity of cell signaling receptors via regulatory motifs in their cytoplasmic domains. This immunoglobulin subfamily of coreceptors recognize sialic acid containing glycans as ligands, which are found on glycoproteins and glycolipids of all mammalian cells. By virtue of their ability to recognize this common structural element, siglecs are increasingly recognized for their ability to help immune cells distinguish between self and nonself, and dampen autoimmune responses.

Keywords: sialic acid; siglec; B cell; macrophage; toll like receptors

Introduction

The sialic acid binding immunoglobulin lectins (siglecs) comprise a family of 15 human and 9 murine cell surface receptors that are expressed on various white blood cells of the immune system with the notable exception of most T cells in mouse and man.[1,2] They have in common an N-terminal "V-set" Ig domain that binds sialic acid-containing ligands, and a variable number of "C2-set" Ig domains that extend the ligand-binding site away from surface of the membrane. Many siglecs have cytoplasmic tyrosine motifs, including ITIM (immunoreceptor tyrosine-based inhibitory motif) and ITIM-like motifs, commonly found in coreceptors involved in regulation of cell signaling. Several other siglecs (human Siglecs-14–16 and murine Siglec-H) have no tyrosine motifs, but contain a positively charged trans-membrane spanning region that permits association with adapter proteins such as DNAX-activating protein 12 kDa (DAP-12), which imparts both positive and negative signals.[2,3] While the functions of the siglecs are still being elucidated, there is growing evidence that the majority are endocytic coreceptors that contribute to the regulation of cell signaling in immune cells that mediate innate and adaptive immunity.[1–7]

As inferred from their name, siglecs bind to sialic acid containing glycans of cell surface glycoproteins and glycolipids that are found on all mammalian cells. Thus, in contrast to some immune cell receptors, such as Toll like receptors (TLRs) that recognize danger associated molecular patterns (DAMPS) in pathogens and damaged cells, siglecs recognize ligands that are determinants of "self." Siglecs are documented to interact with sialylated ligands on the same cell in *cis* and on adjacent cells in *trans*, which can modulate their activities in cell signaling and cell–cell interactions (Fig. 1).[1,2,4,5,8–10] In addition, some microbial pathogens cloak themselves with sialic acid containing glycans that mimic self, and as a result can exploit the normal functions of siglecs to downregulate an immune response against them.[1,2,11–15]

The idea that siglecs sense self through recognition of sialoside ligands was first proposed for CD22,[16–19] and was linked to the observation that CD22 plays a major role in the tolerance of peripheral B cells to self antigens.[6,8,19–24] This concept has now been expanded to several other siglecs. Siglec-10/G interactions with self ligands can also contribute to tolerizing B cells to self antigens,[5,23,25,26] and aids in suppression of excessive inflammatory

doi: 10.1111/j.1749-6632.2011.06362.x

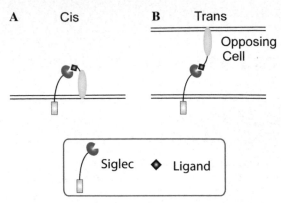

Figure 1. Interactions of siglecs with sialoside ligands in *cis* (A) and *trans* (B).

response to tissue damage by dendritic cells.[4,27–29] Similarly, several siglecs have been demonstrated to bind pathogens that carry sialylated glycans that mimic self, and to modulate immune responses against them.[13,15,30–32]

Here we review evidence that a major role of the siglecs is to assist immune cells in sensing self in both innate and adaptive immune responses, and provide insights into the roles of siglec ligands in mediating these functions. For a more comprehensive perspective of siglecs and their functions we refer the reader to other excellent reviews on the siglec family.[1–7,33,34]

Siglecs recognize sialylated glycans as self-ligands

Soon after the identification of siglecs as sialic acid binding proteins, it was discovered that many of them bind to glycan ligands on the same cell— "in *cis*"—effectively masking their binding to synthetic sialoside ligand probes, unless the sialic acids were first "destroyed" enzymatically with sialidase or chemically with periodate.[35–43] These seminal observations established that siglecs constitutively recognize endogenous self-ligands, and set the stage for investigations into roles of ligands in their function. If siglecs are masked by *cis* ligands, how can they interact with ligands in *trans* (Fig. 1)? This is due in part to the fact that siglecs have relatively low affinity for their sialoside ligands but are masked, or partially masked, due to the high concentration of sialosides on the cell surface.[44–47] Thus, while *cis* ligands may set a threshold to block binding

of soluble glycoproteins, they do not prevent binding to high density ligands in *trans* on adjacent cells,[2,48–53] or to high avidity multivalent ligand-based probes.[54–57]

A more detailed look at the specificity of siglecs toward sequences that terminate glycans of glycoproteins and glycolipids provides additional insight into the potential for these receptors to sense self. For selected human and murine siglecs, Table 1 summarizes their reported expression in the major white blood cell types that mediate innate and adaptive immunity,[1,2,27] and their reported specificity against a panel of common sialosides.[46,58–65] To facilitate comparisons across species, the human and murine orthologs/paralogs are listed together and given the same hybrid name (e.g., hSiglec-8 and mSiglec-F; hSiglec-9 and mSiglec-E).

It is evident that siglecs exhibit varied specificities for sialoside ligands, with some exhibiting high specificity and others exhibiting quite broad specificity. For example, CD22 on B cells exhibits strong specificity for sialosides terminating in the Siaα2-6Gal linkage (Table 1,[1–3]), while Siglec-8/F on eosinophils exhibits high preference for a sialylated, sulfated structure, 6′-sulfo-NeuAcα2-3Galβ1-4GlcNAc (Table 1,[7,8]). It is notable, however, that when two or more siglecs are expressed on the same cell their combined specificities cover the most common classes of sialoside sequences in the mammalian glycome (Table 1). This may provide immune cells with the capacity to sense self regardless of the repertoire of sialosides expressed by the various cell types they encounter.

Differences in the specificity of human and murine siglec orthologs/paralogs also reflect adaptations to recognize self-ligands.[2,66] In particular, murine CD22 preferentially recognizes NeuGc containing α2-6 sialosides (Table 1,[2]) with over 10-fold higher affinity than NeuAc (Table 1,[1]), but human CD22 exhibits equal affinity for both, consistent with the fact that mouse B cells preferentially express NeuGc, while human B cells express only NeuAc.[46,59,67,68] Another difference is that human CD22 exhibits highest affinity for the 6-sulfo-NeuAcα2-6Galβ1-4GlcNAc (Table 1,[3]).[61,69,70] Despite these differences, activation of B cells in both species results in down regulation of the highest affinity ligand. In murine B cells, activation causes de novo synthesis of sialosides with NeuAc instead of NeuGc through down regulation of CMP-sialic

Table 1. Cell type expression and sialoside preference of selected human and murine siglecs

Siglec[1]	Cell type Expression[2]							Glycan Preference[3]										
	Mac	B	Eo	Mo	DC	N	NK	1	2	3	4	5	6	7	8	9	10	11
Sn	X			lo				●	○	*	●	●	●	*	*	*	*	*
hCD22		X						●	●	●	-	-	-	-	-	-	-	-
mCD22		X						○	●	○	-	-	-	-	-	-	-	-
hSig-10	X	lo		lo				○	●	*	●	●	*	*	*	*	*	*
mSig-G	X	lo		lo				-	●	*	-	●	*	*	*	*	*	*
hSig-8		X						-	-	-	-	-	-	●	●	-	-	-
mSig-F		X						○	-	-	○	-	○	●	●	○	○	-
Sig-5		lo		X	X			●	●	*	●	●	●	*	*	*	*	*
Sig-7				lo	X		X	○	○	-	○	○	●	*	●	*	●	●
hSig-9				X	X	X	lo	○	○	*	○	○	○	*	-	*	●	-
mSig-E				X	X	X	X	○	-	●	●	-	●	*	●	*	●	●

[1]Selected murine (shaded) and human (clear) siglecs (Sig) are represented by abbreviations of their common (Sn, sialoadhesin/Siglec-1; CD22, CD22/Siglec-2) or systematic names (Sig-1-10 for human; Sig-E, F, G for murine). Where orthologs/paralogs of each siglec are known, they are annotated as pairs (e.g., human and murine orthologs/paralogs: Sig-10 and Sig-G; Sig-8 and Sig-F; and Sig-9 and Sig-E; and Sn for both murine and human Sn).
[2]Siglec expression reported for macrophages (Mac): B cells (B), eosinophils (Eo), monocytes (Mo), dendritic cells (DC), neutrophils (N) and natural killer cells (NK).[2,33] Indicated is strong expression on the majority of cells (X) or low expression and/or expression only on subsets of cells *(lo)*.
[3]Glycan structures 1–11 represent terminal sequences of glycans of mammalian glycoproteins and glycolipids that interact with siglecs. Structures are grouped and color coded by common structural class including α2-6 sialosides *(green)*, α2-3 sialosides *(orange)*, sulfated α2-3 sialosides *(pink)* and an α2-8 sialoside. Relative preferences of each siglec are compiled from glycan array data from the Consortium for Functional Glycomics (http://www.functionalglycomics.org),[2,60,69,70,131] from competitive inhibition studies,[45,46,54] and from studies using synthetic multivalent ligands.[23,42,55,62–64,67,81,132–134] Relative strengths of interactions are expressed as strong *(large filled circle)*, moderate (small filled circle), weak (small open circle) or not detectable (-). The *symbol indicates not tested.

acid hydroxylase,[68] while in human B cells, differentiation of B cells in germinal centers coincides with loss of the sulfate from the high affinity sulfated ligand (Table 1,[3]).[70] Recent reports also document that 9-*O*-acetyl substitutions of sialic acids also play an important role in regulating the association of CD22 with cis ligands, which is an element of specificity conserved across the two species.[6,71–73]

Although this review focuses on sialosides as self-ligands, for a glycoprotein to be a physiologically relevant *cis* or *trans* ligand of a siglec, carrying glycans with the optimal sialoside sequence is just one criterion that must be met. For example, while most B cell glycoproteins that contain sialic acids coprecipitate with CD22-Fc chimera,[74,75] only a few glycoproteins were detected *in situ* as *cis* and *trans* ligands of CD22 in experiments involving intact B cells engineered to contain sialic acids with a photo-crosslinking substituent installed at C-9 or C-5.[49,74,76] Similarly, several other reports have documented the binding of siglecs to protein ligands in a nonsialic acid-dependent manner.[13,39,77,78] Identification of the physiologically relevant ligands of the siglecs will undoubtedly provide major insights into their detailed functions.

Roles of siglecs in adaptive immunity

B and T lymphocytes are the central players of the adaptive immune system, responsible for the humoral and cell mediated arms of the immune response, respectively. In mice and man, siglecs are well documented to play a major role in the regulation of B cell signaling.[2,5] In contrast, only minor subsets of T cells have been observed to express siglecs,[79–81] and because their functional significance is not understood, they will not be considered further here.

B cells of humans and mice express two major siglecs, CD22 and Siglec-10/G (murine ortholog is Siglec-G), both of which are documented to play important roles in regulation of B cell receptor signaling.[2,5,8,82,83] B cells from CD22 and Siglec-G knock-out mice exhibit hallmarks of hypersensitivity to BCR ligation, consistent with negative regulation of BCR signaling by recruitment of the phosphatase SHP-1 via their cytoplasmic ITIM motifs.[24,84–87] While antibodies to auto-antigens have been variably detected in aging CD22 and Siglec-G null mice,[26,88] double KO mice missing both siglecs have a consistent autoimmune phenotype.[5,26] Thus, CD22 and Siglec-G appear to synergistically contribute to peripheral B cell tolerance. Based on the incomplete penetrance of an autoimmune phenotype in the single KO mice, it has been suggested that mutations in these siglecs or changes in their ligands may act in concert with other susceptibility markers in the induction of autoimmune diseases.[6,8,20–22,73,89–92]

Because of the emerging roles of CD22 and Siglec-G in peripheral tolerance, understanding the role of ligand binding in the regulation of BCR signaling is of great interest. Studies to date have been primarily on CD22. Although the regulation of BCR signaling by CD22 is multi-factorial, involving six cytoplasmic tyrosine motifs, and multiple kinases and phosphases, the predominant negative regulation of BCR signaling is mediated by recruitment of the phosphatase SHP-1.[5,8,83,90,93–97] Current thinking suggests that regulation involves physical proximity of CD22 and the BCR because local recruitment of SHP-1 is required to dephosphorylate BCR signaling components and thereby dampen BCR signaling (Fig. 2A). Indeed, studies involving direct antibody-mediated cross-linking of the BCR and CD22, or conversely, sequestration of CD22 from the BCR,

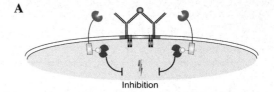

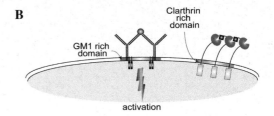

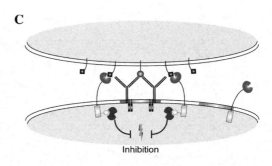

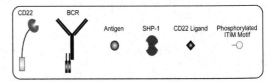

Figure 2. CD22 is negative regulator of B cell activation. (A) Negative regulation of B cell signaling requires that the CD22-SHP-1 complex be in close proximity to the BCR. (B) On resting B cells, CD22 and the B cell receptor (BCR) are in different sub-membrane compartments: while CD22 is located in clathrin rich domains, the BCR resides in GM1 rich domains. CD22 interacts with *cis* ligands on other CD22 molecules, which help maintain it in clathrin rich domains and away from the BCR, allowing for full B cell activation. (C) When the BCR encounters a cell surface autoantigen, CD22 becomes juxtaposed to the BCR through interactions with *trans* ligands. Recruitment of CD22 to the site of cell contact will result in phosphorylation of the ITIM motifs on the cytoplasmic tail of CD22, recruitment of the phosphate SHP-1, and, in turn, dampening of B cell activation. In such a manner, CD22 may inhibit B cell activation toward self and play a key role in preventing autoimmunity.

have demonstrated the importance of proximity in regulation of BCR signaling by CD22.[93,98] An important question, therefore, is how and when do CD22 and the BCR become juxtaposed under normal physiological conditions and, of primary importance to this review, what is the role for CD22 ligand binding in this process?

Defining the physiologically relevant glycoprotein ligands that interact with CD22 in both *cis* and *trans* has been the objective of several studies. These studies have been influential in the understanding of how CD22 and IgM become juxtaposed. Early *in vitro* studies provided evidence that CD22 and IgM interact in a sialic acid-dependent manner,[99,100] which has been the basis for an attractive model where interaction of CD22 with *cis* ligands on IgM itself draws the two molecules together to set a threshold for BCR signaling.[101] This model has provided a satisfying explanation for the effects of small molecule inhibitors of CD22 on *in vitro* alterations in BCR activation and CD22 micro-domain localization.[102,103]

However, recent studies that directly interrogate *cis* ligand interactions *in situ* have brought into question the physiological relevance of this model. In particular, *in situ* glycan–protein crosslinking experiments have detected no sialic acid-dependent *cis* interactions between CD22 and IgM.[49,74] Instead, CD22 appears to prefer sialoglycans on other CD22 molecules as its preferred *cis* ligand and is present as homomultimeric complexes in clathrin-rich domains, presumably due to the high concentration of CD22 in these microdomains. In contrast, the BCR predominately resides in non-clathrin membrane microdomains in resting B cells (Fig. 2B).[91,104] Furthermore, in the initial events following ligation of the BCR, CD22 is largely excluded from GM1-rich activation rafts.[105] Within a few minutes of B cell activation, however, the BCR and CD22 colocalize in clathrin-coated pits prior to endocytosis in a process that appears to be ligand-independent,[91,104] and may instead be mediated by the protein–protein interactions between CD22 and IgM.[39] The fact the BCR and CD22 largely reside in separate microdomains until after the initial activation events might account for the modest increased hyper-responsiveness of CD22-deficient B cells that are consistently observed in response to anti-IgM ligation of the BCR.[24,84–86,91,104,106]

Although interactions of CD22 with *cis* ligands cannot explain how CD22 and IgM become juxtaposed, an intriguing model is beginning to emerge as to how this may be accomplished by *trans* ligands. An elegant study by Lanoue *et al.* described how *trans* CD22 ligands on a cell displaying a membrane-bound antigen (hen egg lysozyme; HEL) significantly inhibited activation of HEL-specific B cells, compared to cells expressing antigen in the absence of CD22 ligands.[18] It was subsequently shown that *trans* ligands could effectively cause CD22 to redistribute to the site of cell–cell contact, providing a mechanistic basis for colocalization of CD22 with the BCR engaged with antigen on the opposing cell.[51] More recently, two key studies using chemical biology approaches have elaborated this finding.[23,107] In both studies, a polymer was used that displayed both an antigen (nitrophenyl; NP) and CD22 ligand in a multivalent manner. The outcome was similar in both cases; inhibition of B cell activation was observed *in vitro* relative to a polymer that displayed antigen alone. The study by Duong *et al.* went one step further by immunizing mice with these polymers.[23] Remarkably, antibody production to NP was nearly completely inhibited by the polymer with ligand and antigen and mice appeared to acquire tolerance to NP since there was a failure to respond to the antigen upon a secondary challenge several weeks later. Several experiments in this study suggested that the mode of tolerance was through apoptosis. Together, these three studies support a model in which *trans* ligands draw CD22 to the site of cell contact, along with the BCR, to recognize its antigen and dampen BCR signaling (Fig. 2C).

Maintaining peripheral B cell tolerance is of major importance since it is estimated that between 20% and 50% of B cells that emerge from the bone marrow can recognize self-antigens.[108–110] Failure to maintain B cells tolerance in the periphery results in autoimmune diseases.[92] Since sialic acid residues are present on the surface of all cells in higher eukaryotes at high density, recognition of sialic acids by inhibitory receptors on B cells may provide a convenient way to downregulate an autoimmune response when a self-antigen is encountered. On murine B cells, CD22 and Siglec-G may work together in this regard since double KO mice exhibit a more profound autoimmune phenotype that found in mice

Table 2. Modulation of TLR signaling by siglecs

Molecule	Cell type	Observed phenotype	TLR ligands used
CD22	B	Enhanced proliferation of CD22 KO B cells[26,106]	TLR3, 4, 7, and 9
Siglec-G	B DC	Enhanced proliferation of Siglec-G KO B cells[26] Enhanced TNF-α production in Siglec-G KO DCs[28]	TLR3, 4, 7, and 9 HMGB1
Siglec-E	Mac	Reduced IL-12 production by cross-linking with Abs[14]	TLR4
Siglec-H	pDC	Reduced IFN-α production by cross-linking with Abs[120]	TLR9
Siglec-5	Mac	Reduced TNF-α and enhanced IL-10 production by overexpression[124]	TLR2, 3, 4, and 9
Siglec-9	Mac	Reduced TNF-α and enhanced IL-10 production by overexpression[124]	TLR2, 3, 4, and 9
Siglec-11	Mac	Reduced IL-1β transcript by cross-linking with Abs[123]	TLR4
Siglec-14	Mac	Augmented TNF-α production by overexpression[125]	TLR4

missing only one of the two siglecs.[87] Indeed, the fact that CD22 and Siglec-G prefer different sialic acid linkages (Table 1), would ensure that a cell would be seen as self regardless of which type of sialic acid linkage predominated. In fact, the results by Duong *et al.* suggest that Siglec-G does participate in tolerizing to T-independent antigens.[23] Further studies are needed to help clarify the extent to which siglecs play in peripheral B cell tolerance.

Siglecs in regulation of the innate immune response

The innate immune system comprises white blood cells that sense microenvironments and distinguish between self and nonself via germ line encoded pattern recognition receptors, such as TLRs and C-type lectins.[111,112] Because siglecs are also widely expressed on these cells, they are believed to play critical roles in innate immune functions, and their precise roles are beginning to emerge.[1,2,29] Here we review accumulating evidence that suggests siglecs can dampen excessive innate immune responses of TLRs through recognition of the sialic acids, a signature of self. We also briefly review recent findings that pathogenic molecules decorated with sialic acids to mimic self to dampen the innate immune response via siglecs.[113,114]

Siglecs as inhibitory coreceptors for TLRs

In the innate immune system, a number of siglecs have been identified as inhibitory receptors based on the ability to recruit phosphatases SHP-1 and 2 that dampen signaling of ITAM-bearing activation receptors.[2,3] Several reports have demonstrated that crosslinking Siglec-7 or Siglec-9 to activation receptors results in inhibition of the cytolytic activity of NK cells against tumor cells and release of chemical mediators from mast cells, respectively.[115,116] Recent studies have found that siglecs also function as inhibitory receptors for TLRs, although in several cases, the mechanism by which siglecs regulate TLRs has yet to be established (Table 2).

Additional evidence for the involvement of siglecs in regulating TLRs comes from siglec KO animals. In particular, two siglecs expressed in B cells, CD22 and Siglec-G, are considered to be negative regulators for TLR signaling. B cells from either CD22 or Siglec-G deficient mouse show hyper-activation in response to poly (I:C), LPS, R848, and CpG, TLR 3, 4, 7, and 9 ligands, respectively.[26,106] Induction of suppressors of cytokine signaling (SOCS) 1 and 3, known to be regulators of TLR signaling,[117,118] is impaired in B cells of CD22 KO mice, and may account for the hyperactivation.[106] Although Siglec-G KO B cells showed augmented NFATc1 expression, the

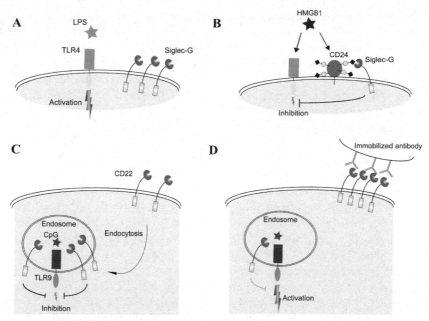

Figure 3. Regulation of TLR signaling by siglecs. (A) Siglec-G does not affect LPS-induced TLR signaling on DCs. (B) Endogenous TLR ligands such as HSPs and HMGB1 activate TLR4, resulting in an inflammatory response to the tissue damage. CD24–Siglec-G/10 complex suppresses the HMGB1-induced TLR activation upon binding of HMGB1 via CD24. (C) CD22 is an endocytic receptor that is localized in endosomes where endosomal TLRs reside. (D) Sequestration of CD22 may alter endosomal TLR signaling by reducing the local concentration of the siglecs in endosomes.

inhibitory mechanism of TLR signaling by Siglec-G in B cells remains to be elucidated.[25]

In myeloid cells, Siglec-G–deficient dendritic cells (DC)s exhibit augmented TNF-α and IL-6 production in response to endogenous TLR4 ligands such as HMGB1 released from necrotic cells.[28] Liu *et al.* demonstrated that Siglec-G forms an inhibitory complex with CD24 on DCs to suppress the TLR signaling by endogenous TLR4 ligands. CD24 is a sialylated cell surface glycoprotein on DCs and is bound by Siglec-G on the same cell (*cis*). This CD24–Siglec-G complex is required to inhibit HMGB1-induced, but not LPS-induced, cytokine production from DCs, which suggests that Siglec-G is expressed in DCs to dampen innate immune response toward self-molecules (Figs. 3A and B).[27,28]

Several reports have shown that cross-linking of siglecs by immobilized antibody dramatically affects TLR signaling. These results have been interpreted as a sequestering effect, where antibody sequestration can reduce the local concentration of siglecs in close proximity with TLR receptor, resulting in altered responsiveness to TLR ligands (Figs. 3C and

D). Specifically, immobilized anti-CD22 antibody resulted in the augmented proliferation of B cells in response to CpG, suggesting that the endocytic activity of CD22 is tied to its regulatory function for TLR signaling.[106] Consistent with this observation, it has been shown that CD22 is an endocytic receptor that recycles between the cell surface and endosomes where TLR 3, 7, and 9 are localized.[55,119] In contrast, in the case of Siglec-H, an endocytic receptor in plasmacytoid DCs, sequestration results in the inhibition of IFN-α production in response to CpG.[120] Although Siglec-H is thought to be an activation receptor due to its accessory molecule DAP-12, an ITAM-bearing adaptor protein,[120–122] it has been postulated that sequestration results in segregation of signaling components on cell surface, which leads to a shortage of the same components for endosomal TLRs.[121] Similarly, cross-linking of Siglec-E and Siglec-11 by immobilized antibody resulted in inhibition of cytokine production in response to LPS in macrophages.[14,123] Such experiments suggest that sequestering or altering the cellular localization of siglecs may disrupt or modulate TLR signaling networks. These results may have strong biological consequences since other cells express siglec ligands and may similarly sequester siglecs under the appropriate conditions.[51] Further studies are needed to test if the mode of inhibition of TLRs by siglecs is cell type specific.

Studies that have overexpressed siglecs also support a model in which siglecs play a key role in regulating the TLR signaling, and suggest the potential for *cis* ligand interactions to be a mode of inhibition for TLR signaling. Ectopic expression of Siglec-5 and Siglec-9 in a macrophage cell line has been shown to inhibit TNF-α production and enhances IL-10 production in response to peptideglycan (a TLR 2 ligand), LPS, and CpG.[124] The inhibition of TNF-α by Siglec-9 is ITIM-dependent because an ITIM-deficient mutant of Siglec-9 does not inhibit TLR signaling.[124] In another macrophage cell line, overexpression of Siglec-14, a DAP-12 associated human Siglec, was shown to augment the LPS-induced TNF-α production in macrophages, suggesting that DAP-12 associated siglecs can synergize with TLR signaling in macrophages.[125] Regulated expression of siglecs may indeed be a mechanism for affecting TLR signaling, as evidenced by LPS-induced Siglec-E expression in macrophages.[126] In this case, Siglec-E expression resulted in recruitment of SHP-1 and 2, presumably through its cytoplasmic ITIM motifs, suggesting that Siglec-E is installed to prevent excessive activation by TLRs in macrophages.[126] It will be interesting to see how the siglec and sialoside expression is regulated upon TLR activation of immune cells to address how the *cis*-interaction between siglecs and sialosides regulate innate immune response.[127,128]

Although numerous reports now link the functions of siglecs and TLRs, an important caveat is that many of the studies to date employ nonphysiological methods to perturb siglec function, such as antibody cross-linking or overexpression, and may therefore be subject to interpretation. Additional studies are required to confirm and further define the interactions between siglecs and TLRs in immune cell function.

Sialylated pathogens dampen an immune response via siglecs

Many pathogens, including membrane-enveloped viruses, bacteria, and parasites, are coated with sialylated glycans that mimic self, and have the potential to be recognized as ligands of siglecs, which can in turn dampen an immune response.[113,114] As an example, Group B *Streptococcus* contain NeuAcα2-3Galβ1-4GlcNAc residues on the capsular polysaccharides and recruit Siglec-9 on neutrophils, resulting in suppression of microbicidal function of neutrophils,[13,129] although at present it is not known which activation receptor(s) is inhibited by Siglec-9. *Trypanosoma cruzi* contains α2,3-linked sialic acids and inhibits IL-12p70 production and enhances IL-10 secretion from dendritic cells, although the involvement of siglecs in this suppression is still unclear.[14] A recent mechanistic study by van Kooyk *et al.* using structurally defined sialylated lipooligosaccharides (LOS)s from *Campylobacter jejuni* suggested that Siglec-1 and Siglec-7, or perhaps even another siglecs on dendritic cells, may modulate DC function through juxtaposition with TLR4.[32] The LOS terminated with α2-3 linked NeuAc generates Th2-prone DCs with augmented expression of OX-40L and suppressed IL-12p70, sensitizing T cells toward a Th2 phenotype. On the other hand, α2-8 linked NeuAc terminated LOS results in induction of Th1 response by IL-12p70–producing DCs. Although it had previously been shown that Siglec-1 and Siglec-7 bind to the LOS decorated with α2-3– and α2-8–linked NeuAc, respectively,[11,30] this recent study suggests a mechanistic consequence for this difference in ligand specificity. The observations relating to Siglec-1 are surprising in view of the fact that the cytoplasmic domain is devoid of known signaling motifs.[2] Further studies will be required to test if the binding of the LOS to DCs is siglec-mediated and if so, what is the signaling mechanism mediated by these siglecs. These observations may have important consequences for understanding the demyelinating neuropathy Guillain-Barre syndrome.[130] In this disease, sialylated LOS from *C. jejuni* is thought to cause autoreactive antiglycolipid antibodies that attack the gangliosides in the nervous system. It, therefore, remains a possibility that DC activation via siglec recognition of self-mimicked LOS modulates antiglycolipid antibody production and subsequent disease onset and severity.

Summary

Here we have summarized evidence suggesting that siglecs are sentinals of self. By recognition of sialoside ligands expressed on glycoproteins of mammalian cells they serve to dampen immune innate and adaptive immune responses against self. Because a detailed understanding of the functions of many siglecs is just beginning to emerge, it will be of interest to determine the extent to which this is a major function of the siglec family. Such information is also likely to suggest ways to exploit the activity of siglecs to modulate immune responses.

Acknowledgments

We would like to thank Anna Tran-Crie for her help in manuscript preparation. M.S.M. was supported by the HFSP fellowship. This work was supported by NIH grants AI050143 and CA138891 to J.C.P.

Conflicts of interest

The authors declare no conflicts of interest.

References

1. Cao, H. & P.R. Crocker. 2011. Evolution of CD33-related siglecs: regulating host immune functions and escaping pathogen exploitation? *Immunology* **132:** 18–26.

2. Crocker, P.R., J.C. Paulson & A. Varki. 2007. Siglecs and their roles in the immune system. *Nat. Rev. Immunol.* **7:** 255–266.

3. Crocker, P.R. & P. Redelinghuys. 2008. Siglecs as positive and negative regulators of the immune system. *Biochem. Soc. Trans.* **36:** 1467–1471.

4. Liu, Y., G.Y. Chen & P. Zheng. 2011. Sialoside-based pattern recognitions discriminating infections from tissue injuries. *Curr. Opin. Immunol.* **23:** 41–45.

5. Nitschke, L. 2009. CD22 and Siglec-G: B-cell inhibitory receptors with distinct functions. *Immunol. Rev.* **230:** 128–143.

6. Pillai, S., A. Cariappa & S.P. Pirnie. 2009. Esterases and autoimmunity: the sialic acid acetylesterase pathway and the regulation of peripheral B cell tolerance. *Trends Immunol.* **30:** 488–493.

7. von Gunten, S. & B.S. Bochner. 2008. Basic and clinical immunology of Siglecs. *Ann. N.Y. Acad. Sci.* **1143:** 61–82.

8. Walker, J.A. & K.G. Smith. 2008. CD22: an inhibitory enigma. *Immunology* **123:** 314–325.

9. Varki, A. 2009. Natural ligands for CD33-related Siglecs? *Glycobiology* **19:** 810–812.

10. Lajaunias, F., J.M. Dayer & C. Chizzolini. 2005. Constitutive repressor activity of CD33 on human monocytes requires sialic acid recognition and phosphoinositide 3-kinase-mediated intracellular signaling. *Eur. J. Immunol.* **35:** 243–251.

11. Avril, T. *et al.* 2006. Sialic acid-binding immunoglobulin-like lectin 7 mediates selective recognition of sialylated glycans expressed on Campylobacter jejuni lipooligosaccharides. *Infect. Immun.* **74:** 4133–4141.

12. Brinkman-Van der Linden, E.C., *et al.* 2003. CD33/Siglec-3 binding specificity, expression pattern, and consequences of gene deletion in mice. *Mol. Cell Biol.* **23:** 4199–4206.

13. Carlin, A.F. *et al.* 2009. Molecular mimicry of host sialylated glycans allows a bacterial pathogen to engage neutrophil Siglec-9 and dampen the innate immune response. *Blood* **113:** 3333–3336.

14. Erdmann, H. *et al.* 2009. Sialylated ligands on pathogenic Trypanosoma cruzi interact with Siglec-E (sialic acid-binding Ig-like lectin-E). *Cell Microbiol.* **11:** 1600–1611.

15. Khatua, B. *et al.* 2010. Sialic acids acquired by Pseudomonas aeruginosa are involved in reduced complement deposition and siglec mediated host-cell recognition. *FEBS Lett.* **584:** 555–561.

16. Sinclair, N.R. 1999. Why so many coinhibitory receptors? *Scand. J. Immunol.* **50:** 10–13.

17. Huck, S. *et al.* 2001. Expression of B cell receptor-associated signaling molecules in human lupus. *Autoimmunity* **33:** 213–224.

18. Lanoue, A. *et al.* 2002. Interaction of CD22 with alpha2,6-linked sialoglycoconjugates: innate recognition of self to dampen B cell autoreactivity? *Eur. J. Immunol.* **32:** 348–355.

19. Pritchard, N.R. & K.G. Smith. 2003. B cell inhibitory receptors and autoimmunity. *Immunology* **108:** 263–273.

20. Youinou, P. & Y. Renaudineau. 2004. The antiphospholipid syndrome as a model for B cell-induced autoimmune diseases. *Thromb. Res.* **114:** 363–369.

21. Ferry, H. *et al.* 2005. Analysis of Lyn/CD22 double-deficient B cells in vivo demonstrates Lyn- and CD22-independent pathways affecting BCR regulation and B cell survival. *Eur. J. Immunol.* **35:** 3655–3663.

22. Youinou, P. *et al.* 2006. B lymphocytes on the front line of autoimmunity. *Autoimmun. Rev.* **5:** 215–221.

23. Duong, B.H. *et al.* 2010. Decoration of T-independent antigen with ligands for CD22 and Siglec-G can suppress immunity and induce B cell tolerance in vivo. *J. Exp. Med.* **207:** 173–187.

24. O'Keefe, T.L. *et al.* 1996. Hyperresponsive B cells in CD22-deficient mice. *Science* **274:** 798–801.

25. Jellusova, J. *et al.* 2010. Siglec-G regulates B1 cell survival and selection. *J. Immunol.* **185:** 3277–3284.

26. Jellusova, J. *et al.* 2010. CD22 x Siglec-G double-deficient mice have massively increased B1 cell numbers and develop systemic autoimmunity. *J. Immunol.* **184:** 3618–3627.

27. Chen, G.Y. *et al.* 2011. Amelioration of sepsis by inhibiting sialidase-mediated disruption of the CD24-SiglecG interaction. *Nat. Biotechnol.* **29:** 428–435.

28. Chen, G.Y. *et al.* 2009. CD24 and Siglec-10 selectively repress tissue damage-induced immune responses. *Science* **323:** 1722–1725.

29. Liu, Y., G.Y. Chen & P. Zheng. 2009. CD24-Siglec G/10 discriminates danger- from pathogen-associated molecular patterns. *Trends. Immunol.* **30:** 557–561.

30. Jones, C., M. Virji & P.R. Crocker. 2003. Recognition of sialylated meningococcal lipopolysaccharide by siglecs expressed on myeloid cells leads to enhanced bacterial uptake. *Mol. Microbiol.* **49:** 1213–1225.

31. Vanderheijden, N. *et al.* 2003. Involvement of sialoadhesin in entry of porcine reproductive and respiratory syndrome virus into porcine alveolar macrophages. *J. Virol.* **77:** 8207–8215.

32. Bax, M. *et al.* 2011. Campylobacter jejuni lipooligosaccharides modulate dendritic cell-mediated T cell polarization in a sialic acid linkage-dependent manner. *Infect. Immun.* **79:** 2681–2689.

33. O'Reilly, M.K. & J.C. Paulson. 2009. Siglecs as targets for therapy in immune-cell-mediated disease. *Trends Pharmacol. Sci.* **30:** 240–248.

34. von Gunten, S. & H.U. Simon. 2008. Natural anti-Siglec autoantibodies mediate potential immunoregulatory mechanisms: implications for the clinical use of intravenous immunoglobulins (IVIg). *Autoimmun. Rev.* **7:** 453–456.

35. Razi, N. & A. Varki. 1998. Masking and unmasking of the sialic acid-binding lectin activity of CD22 (Siglec-2) on B lymphocytes. *Proc. Natl. Acad. Sci. USA* **95:** 7469–7474.

36. Razi, N. & A. Varki. 1999. Cryptic sialic acid binding lectins on human blood leukocytes can be unmasked by sialidase treatment or cellular activation. *Glycobiology* **9:** 1225–1234.

37. Nakamura, K. *et al.* 2002. Lymph node macrophages, but not spleen macrophages, express high levels of unmasked sialoadhesin: implication for the adhesive properties of macrophages in vivo. *Glycobiology* **12:** 209–216.

38. Brinkman-Van der Linden, E.C. & A. Varki. 2003. Probing for masked and unmasked siglecs on cell surfaces. *Methods Enzymol.* **363:** 113–120.

39. Zhang, M. & A. Varki. 2004. Cell surface sialic acids do not affect primary CD22 interactions with CD45 and surface IgM nor the rate of constitutive CD22 endocytosis. *Glycobiology* **14:** 939–949.

40. Attrill, H. *et al.* 2006. The structure of siglec-7 in complex with sialosides: leads for rational structure-based inhibitor design. *Biochem. J.* **397:** 271–278.

41. Avril, T. *et al.* 2006. Probing the cis interactions of the inhibitory receptor Siglec-7 with alpha2,8-disialylated ligands on natural killer cells and other leukocytes using glycan-specific antibodies and by analysis of alpha2,8-sialyltransferase gene expression. *J. Leukoc. Biol.* **80:** 787–796.

42. Hashimoto, Y. *et al.* 1998. A streptavidin-based neoglycoprotein carrying more than 140 GT1b oligosaccharides: quantitative estimation of the binding specificity of murine sialoadhesin expressed on CHO cells. *J. Biochem.* **123:** 468–478.

43. Freeman, S.D. *et al.* 1995. Characterization of CD33 as a new member of the sialoadhesin family of cellular interaction molecules. *Blood* **85:** 2005–2012.

44. Collins, B.E. & J.C. Paulson. 2004. Cell surface biology mediated by low affinity multivalent protein–glycan interactions. *Curr. Opin. Chem. Biol.* **8:** 617–625.

45. Bakker, T.R. *et al.* 2002. Comparison of CD22 binding to native CD45 and synthetic oligosaccharide. *Eur. J. Immunol.* **32:** 1924–1932.

46. Blixt, O. *et al.* 2003. Sialoside specificity of the siglec family assessed using novel multivalent probes: identification of potent inhibitors of myelin-associated glycoprotein. *J. Biol. Chem.* **278:** 31007–31019.

47. Powell, L.D. *et al.* 1995. Characterization of sialyloligosaccharide binding by recombinant soluble and native cell-associated CD22. Evidence for a minimal structural recognition motif and the potential importance of multisite binding. *J. Biol. Chem.* **270:** 7523–7532.

48. Floyd, H., L. Nitschke & P.R. Crocker. 2000. A novel subset of murine B cells that expresses unmasked forms of CD22 is enriched in the bone marrow: implications for B-cell homing to the bone marrow. *Immunology* **101:** 342–347.

49. Ramya, T.N. *et al.* 2010. In situ trans ligands of CD22 identified by glycan-protein photocross-linking-enabled proteomics. *Mol. Cell Proteomics.* **9:** 1339–1351.

50. Danzer, C.P. *et al.* 2003. Transitional and marginal zone B cells have a high proportion of unmasked CD22: implications for BCR signaling. *Int. Immunol.* **15:** 1137–1147.

51. Collins, B.E. *et al.* 2004. Masking of CD22 by cis ligands does not prevent redistribution of CD22 to sites of cell contact. *Proc. Natl. Acad. Sci. U S A.* **101:** 6104–6109.

52. Nicoll, G. *et al.* 2003. Ganglioside GD3 expression on target cells can modulate NK cell cytotoxicity via siglec-7-dependent and -independent mechanisms. *Eur. J. Immunol.* **33:** 1642–1648.

53. Crocker, P.R. *et al.* 1990. Ultrastructural localization of a macrophage-restricted sialic acid binding hemagglutinin, SER, in macrophage-hematopoietic cell clusters. *Blood* **76:** 1131–1138.

54. Collins, B.E. *et al.* 2006. High-affinity ligand probes of CD22 overcome the threshold set by cis ligands to allow for binding, endocytosis, and killing of B cells. *J. Immunol.* **177:** 2994–3003.

55. Tateno, H. *et al.* 2007. Distinct endocytic mechanisms of CD22 (Siglec-2) and Siglec-F reflect roles in cell signaling and innate immunity. *Mol. Cell. Biol.* **27:** 5699–5710.

56. Chen, W.C. *et al.* 2010. In vivo targeting of B-cell lymphoma with glycan ligands of CD22. *Blood* **115:** 4778–4786.

57. O'Reilly, M.K. *et al.* 2008. Bifunctional CD22 ligands use multimeric immunoglobulins as protein scaffolds in assembly of immune complexes on B cells. *J. Am. Chem. Soc.* **130:** 7736–7745.

58. Powell, L.D. & A. Varki. 1994. The oligosaccharide binding specificities of CD22 beta, a sialic acid-specific lectin of B cells. *J. Biol. Chem.* **269:** 10628–10636.

59. Kelm, S. *et al.* 1994. Modifications of cell surface sialic acids modulate cell adhesion mediated by sialoadhesin and CD22. *Glycoconj. J.* **11:** 576–585.

60. Bochner, B.S. *et al.* 2005. Glycan array screening reveals a candidate ligand for Siglec-8. *J. Biol. Chem.* **280:** 4307–4312.

61. Consortium for Functional Glycomics. http://www.functionalglycomics.org.

62. Angata, T. *et al.* 2001. Cloning and characterization of a novel mouse Siglec, mSiglec-F: differential evolution of the mouse and human (CD33) Siglec-3-related gene clusters. *J. Biol. Chem.* **276:** 45128–45136.

63. Angata, T. & A. Varki. 2000. Cloning, characterization, and phylogenetic analysis of siglec-9, a new member of the CD33-related group of siglecs. Evidence for co-evolution with sialic acid synthesis pathways. *J. Biol. Chem.* **275:** 22127–22135.

64. Li, N. *et al.* 2001. Cloning and characterization of Siglec-10, a novel sialic acid binding member of the Ig

superfamily, from human dendritic cells. *J. Biol. Chem.* **276:** 28106–28112.

65. Yu, Z. *et al.* 2001. mSiglec-E, a novel mouse CD33-related siglec (sialic acid-binding immunoglobulin-like lectin) that recruits Src homology 2 (SH2)-domain-containing protein tyrosine phosphatases SHP-1 and SHP-2. *Biochem. J.* **353:** 483–492.

66. Varki, A. 2010. Colloquium paper: uniquely human evolution of sialic acid genetics and biology. *Proc. Natl. Acad. Sci. U S A.* **107**(Suppl 2): 8939–8946.

67. Brinkman-Van der Linden, E.C., *et al.* 2000. Loss of *N*-glycolylneuraminic acid in human evolution. Implications for sialic acid recognition by siglecs. *J. Biol. Chem.* **275:** 8633–8640.

68. Naito, Y. *et al.* 2007. Germinal center marker GL7 probes activation-dependent repression of *N*-glycolylneuraminic acid, a sialic acid species involved in the negative modulation of B-cell activation. *Mol. Cell Biol.* **27:** 3008–3022.

69. Blixt, O. *et al.* 2004. Printed covalent glycan array for ligand profiling of diverse glycan binding proteins. *Proc. Natl. Acad. Sci. USA* **101:** 17033–17038.

70. Kimura, N. *et al.* 2007. Human B-lymphocytes express alpha2-6-sialylated 6-sulfo-*N*-acetyllactosamine serving as a preferred ligand for CD22/Siglec-2. *J. Biol. Chem.* **282:** 32200–32207.

71. Sjoberg, E.R. *et al.* 1994. Natural ligands of the B cell adhesion molecule CD22 beta can be masked by 9-O-acetylation of sialic acids. *J. Cell Biol.* **126:** 549–562.

72. Cariappa, A. *et al.* 2009. B cell antigen receptor signal strength and peripheral B cell development are regulated by a 9-O-acetyl sialic acid esterase. *J. Exp. Med.* **206:** 125–138.

73. Surolia, I. *et al.* 2010. Functionally defective germline variants of sialic acid acetylesterase in autoimmunity. *Nature* **466:** 243–247.

74. Han, S. *et al.* 2005. Homomultimeric complexes of CD22 in B cells revealed by protein–glycan cross-linking. *Nat. Chem. Biol.* **1:** 93–97.

75. Sgroi, D. *et al.* 1993. CD22, a B cell-specific immunoglobulin superfamily member, is a sialic acid-binding lectin. *J. Biol. Chem.* **268:** 7011–7018.

76. Tanaka, Y. & J.J. Kohler. 2008. Photoactivatable crosslinking sugars for capturing glycoprotein interactions. *J. Am. Chem. Soc.* **130:** 3278–3279.

77. Santos, L. *et al.* 2008. Dendritic cell-dependent inhibition of B cell proliferation requires CD22. *J. Immunol.* **180:** 4561–4569.

78. Martinez-Pomares, L. *et al.* 1999. Cell-specific glycoforms of sialoadhesin and CD45 are counter-receptors for the cysteine-rich domain of the mannose receptor. *J. Biol. Chem.* **274:** 35211–35218.

79. Ikehara, Y., S.K. Ikehara & J.C. Paulson. 2004. Negative regulation of T cell receptor signaling by Siglec-7 (p70/AIRM) and Siglec-9. *J. Biol. Chem.* **279:** 43117–43125.

80. Zhang, J.Q. *et al.* 2000. Siglec-9, a novel sialic acid binding member of the immunoglobulin superfamily expressed broadly on human blood leukocytes. *J. Biol. Chem.* **275:** 22121–22126.

81. Nicoll, G. *et al.* 1999. Identification and characterization of a novel siglec, siglec-7, expressed by human natural killer cells and monocytes. *J. Biol. Chem.* **274:** 34089–34095.

82. Tedder, T.F., J.C. Poe & K.M. Haas. 2005. CD22: a multifunctional receptor that regulates B lymphocyte survival and signal transduction. *Adv. Immunol.* **88:** 1–50.

83. Poe, J.C. *et al.* 2004. CD22 regulates B lymphocyte function in vivo through both ligand-dependent and ligand-independent mechanisms. *Nat. Immunol.* **5:** 1078–1087.

84. Nitschke, L. *et al.* 1997. CD22 is a negative regulator of B-cell receptor signalling. *Curr. Biol.* **7:** 133–143.

85. Otipoby, K.L. *et al.* 1996. CD22 regulates thymus-independent responses and the lifespan of B cells. *Nature* **384:** 634–637.

86. Sato, S. *et al.* 1996. CD22 is both a positive and negative regulator of B lymphocyte antigen receptor signal transduction: altered signaling in CD22-deficient mice. *Immunity* **5:** 551–562.

87. Hoffmann, A. *et al.* 2007. Siglec-G is a B1 cell-inhibitory receptor that controls expansion and calcium signaling of the B1 cell population. *Nat. Immunol.* **8:** 695–704.

88. O'Keefe, T.L. *et al.* 1999. Deficiency in CD22, a B cell-specific inhibitory receptor, is sufficient to predispose to development of high affinity autoantibodies. *J. Exp. Med.* **189:** 1307–1313.

89. Mary, C. *et al.* 2000. Dysregulated expression of the Cd22 gene as a result of a short interspersed nucleotide element insertion in Cd22a lupus-prone mice. *J. Immunol.* **165:** 2987–2996.

90. Cornall, R.J. *et al.* 1998. Polygenic autoimmune traits: Lyn, CD22, and SHP-1 are limiting elements of a biochemical pathway regulating BCR signaling and selection. *Immunity* **8:** 497–508.

91. Grewal, P.K. *et al.* 2006. ST6Gal-I restrains CD22-dependent antigen receptor endocytosis and Shp-1 recruitment in normal and pathogenic immune signaling. *Mol. Cell Biol.* **26:** 4970–4981.

92. Nussenzweig, M.C. *et al.* 2005. Defective B cell tolerance checkpoints in systemic lupus erythematosus. *J. Exp.l Med.* **201:** 703–711.

93. Doody, G.M. *et al.* 1995. A role in B cell activation for CD22 and the protein tyrosine phosphatase SHP. *Science* **269:** 242–244.

94. Otipoby, K.L., K.E. Draves & E.A. Clark. 2001. CD22 regulates B cell receptor-mediated signals via two domains that independently recruit Grb2 and SHP-1. *J. Biol. Chem.* **276:** 44315–44322.

95. Tuscano, J.M. *et al.* 1996. Involvement of p72syk kinase, p53/56lyn kinase and phosphatidyl inositol-3 kinase in signal transduction via the human B lymphocyte antigen CD22. *Eur. J. Immunol.* **26:** 1246–1252.

96. Tedder, T.F. *et al.* 2000. CD22 forms a quaternary complex with SHIP, Grb2, and Shc—a pathway for regulation of B lymphocyte antigen receptor-induced calcium flux. *J.Biol. Chem.* **275:** 17420–17427.

97. Smith, K.G. *et al.* 1998. Inhibition of the B cell by CD22: a requirement for Lyn. *J. Exp. Med.* **187:** 807–811.

98. Tooze, R.M., G.M. Doody & D.T. Fearon. 1997. Counterregulation by the coreceptors CD19 and CD22 of MAP

kinase activation by membrane immunoglobulin. *Immunity* **7:** 59–67.

99. Leprince, C. *et al.* 1993. CD22 associates with the human surface IgM-B-cell antigen receptor complex. *Proc. Natl. Acad. Sci. USA.* **90:** 3236–3240.

100. Peaker, C.J. & M.S. Neuberger. 1993. Association of CD22 with the B cell antigen receptor. *Eur. J. Immunol.* **23:** 1358–1363.

101. Cyster, J.G. & C.C. Goodnow. 1997. Tuning antigen receptor signaling by CD22: integrating cues from antigens and the microenvironment. *Immunity* **6:** 509–517.

102. Kelm, S. *et al.* 2002. The ligand-binding domain of CD22 is needed for inhibition of the B cell receptor signal, as demonstrated by a novel human CD22-specific inhibitor compound. *J. Exp. Med.* **195:** 1207–1213.

103. Yu, J. *et al.* 2007. Synthetic glycan ligand excludes CD22 from antigen receptor-containing lipid rafts. *Biochem. Biophys. Res. Commun.* **360:** 759–764.

104. Collins, B.E. *et al.* 2006. Ablation of CD22 in ligand-deficient mice restores B cell receptor signaling. *Nat. Immunol.* **7:** 199–206.

105. Pierce, S.K. 2002. Lipid rafts and B-cell activation. *Nat. Rev. Immunol.* **2:** 96–105.

106. Kawasaki, N., C. Rademacher & J.C. Paulson. 2011. CD22 regulates adaptive and innate immune responses of B cells. *J. Innate. Immun.* **3:** 411–419.

107. Courtney, A.H. *et al.* 2009. Sialylated multivalent antigens engage CD22 in trans and inhibit B cell activation. *Proc. Natl. Acad. Sci. USA.* **106:** 2500–2505.

108. von Boehmer, H. & F. Melchers. 2010. Checkpoints in lymphocyte development and autoimmune disease. *Nat. Immunol.* **11:** 14–20.

109. Wardemann, H. & M.C. Nussenzweig. 2007. B-cell self-tolerance in humans. *Adv. Immunol.* **95:** 83–110.

110. Basten, A. & P.A. Silveira. 2010. B-cell tolerance: mechanisms and implications. *Curr. Opin. Immunol.* **22:** 566–574.

111. Osorio, F. & C. Reis e Sousa. 2011. Myeloid C-type lectin receptors in pathogen recognition and host defense. *Immunity* **34:** 651–664.

112. Kawai, T. & S. Akira. 2011. Toll-like receptors and their crosstalk with other innate receptors in infection and immunity. *Immunity* **34:** 637–650.

113. Severi, E., D.W. Hood & G.H. Thomas. 2007. Sialic acid utilization by bacterial pathogens. *Microbiology* **153:** 2817–2822.

114. Vimr, E. & C. Lichtensteiger. 2002. To sialylate, or not to sialylate: that is the question. *Trends Microbiol.* **10:** 254–257.

115. Avril, T. *et al.* 2004. The membrane-proximal immunoreceptor tyrosine-based inhibitory motif is critical for the inhibitory signaling mediated by Siglecs-7 and -9, CD33-related Siglecs expressed on human monocytes and NK cells. *J. Immunol.* **173:** 6841–6849.

116. Falco, M. *et al.* 1999. Identification and molecular cloning of p75/AIRM1, a novel member of the sialoadhesin family that functions as an inhibitory receptor in human natural killer cells. *J. Exp. Med.* **190:** 793–802.

117. Naka, T. *et al.* 2005. Negative regulation of cytokine and TLR signalings by SOCS and others. *Adv. Immunol.* **87:** 61–122.

118. Dalpke, A.H. *et al.* 2001. Suppressors of cytokine signaling (SOCS)-1 and SOCS-3 are induced by CpG-DNA and modulate cytokine responses in APCs. *J. Immunol.* **166:** 7082–7089.

119. O'Reilly, M.K., H. Tian & J.C. Paulson. 2011. CD22 is a recycling receptor that can shuttle cargo between the cell surface and endosomal compartments of B cells. *J. Immunol.* **186:** 1554–1563.

120. Blasius, A. *et al.* 2004. A cell-surface molecule selectively expressed on murine natural interferon-producing cells that blocks secretion of interferon-alpha. *Blood* **103:** 4201–4206.

121. Blasius, A.L. & M. Colonna. 2006. Sampling and signaling in plasmacytoid dendritic cells: the potential roles of Siglec-H. *Trends Immunol.* **27:** 255–260.

122. Zhang, J. *et al.* 2006. Characterization of Siglec-H as a novel endocytic receptor expressed on murine plasmacytoid dendritic cell precursors. *Blood* **107:** 3600–3608.

123. Wang, Y. & H. Neumann. 2010. Alleviation of neurotoxicity by microglial human Siglec-11. *J. Neurosci.* **30:** 3482–3488.

124. Ando, M. *et al.* 2008. Siglec-9 enhances IL-10 production in macrophages via tyrosine-based motifs. *Biochem. Biophys. Res. Commun.* **369:** 878–883.

125. Yamanaka, M. *et al.* 2009. Deletion polymorphism of SIGLEC14 and its functional implications. *Glycobiology* **19:** 841–846.

126. Boyd, C.R. *et al.* 2009. Siglec-E is up-regulated and phosphorylated following lipopolysaccharide stimulation in order to limit TLR-driven cytokine production. *J. Immunol.* **183:** 7703–7709.

127. Bax, M. *et al.* 2007. Dendritic cell maturation results in pronounced changes in glycan expression affecting recognition by siglecs and galectins. *J. Immunol.* **179:** 8216–8224.

128. Lock, K. *et al.* 2004. Expression of CD33-related siglecs on human mononuclear phagocytes, monocyte-derived dendritic cells and plasmacytoid dendritic cells. *Immunobiology* **209:** 199–207.

129. Cieslewicz, M.J. *et al.* 2005. Structural and genetic diversity of group B streptococcus capsular polysaccharides. *Infect. Immun.* **73:** 3096–3103.

130. Aspinall, G.O. *et al.* 1994. Lipopolysaccharides from Campylobacter jejuni associated with Guillain–Barre syndrome patients mimic human gangliosides in structure. *Infect. Immun.* **62:** 2122–2125.

131. Tateno, H., P.R. Crocker & J.C. Paulson. 2005. Mouse Siglec-F and human Siglec-8 are functionally convergent paralogs that are selectively expressed on eosinophils and recognize 6′-sulfo-sialyl Lewis X as a preferred glycan ligand. *Glycobiology* **15:** 1125–1135.

132. Hartnell, A. *et al.* 2001. Characterization of human sialoadhesin, a sialic acid binding receptor expressed by resident and inflammatory macrophage populations. *Blood* **97:** 288–296.

133. Brinkman-Van der Linden, E.C. & A. Varki. 2000. New aspects of siglec binding specificities, including the significance of fucosylation and of the sialyl-Tn epitope. Sialic acid-binding immunoglobulin superfamily lectins. *J. Biol. Chem.* **275:** 8625–8632.

134. Yamaji, T. *et al.* 2002. A small region of the natural killer cell receptor, Siglec-7, is responsible for its preferred binding to alpha 2,8-disialyl and branched alpha 2,6-sialyl residues. A comparison with Siglec-9. *J. Biol. Chem.* **277:** 6324–6332.

Ann. N.Y. Acad. Sci. ISSN 0077-8923

ANNALS OF THE NEW YORK ACADEMY OF SCIENCES
Issue: *Glycobiology of the Immune Response*

Interleukin-2, Interleukin-7, T cell-mediated autoimmunity, and N-glycosylation

Ani Grigorian,[1,3,*] Haik Mkhikian,[2,3,*] and Michael Demetriou[1,2,3]

[1]Department of Neurology, University of California, Irvine, California [2]Department of Microbiology and Molecular Genetics, University of California, Irvine, California. [3]Institute for Immunology, University of California, Irvine, California.

Address for correspondence: Michael Demetriou, Departments of Neurology and Microbiology and Molecular Genetics, Institute for Immunology, 250 Sprague Hall, University of California, Irvine, CA 92697. mdemetri@uci.edu

T cell activation and self-tolerance are tightly regulated to provide effective host defense against foreign pathogens while deflecting inappropriate autoimmune responses. Golgi Asn (N)-linked protein glycosylation coregulates homeostatic set points for T cell growth, differentiation, and self-tolerance to influence risk of autoimmune disorders such as multiple sclerosis (MS). Human autoimmunity is a complex trait that develops from intricate and poorly understood interactions between an individual's genetics and their environmental exposures. Recent evidence from our group suggests that in MS, additive and/or epistatic interactions between multiple genetic and environmental risk factors combine to dysregulate a common biochemical pathway, namely Golgi N-glycosylation. Here, we review the multiple regulatory mechanisms controlling *N*-glycan branching in T cells and autoimmunity, focusing on recent data implicating a critical role for interleukin-2 (IL-2) and IL-7 signaling.

Keywords: autoimmunity; T cells; N-glycosylation; IL-2; IL-7

Introduction

T cells are the master regulators of the antigen-specific adaptive immune response and play a central role in the delicate balance between protective immunity and tolerance of self-antigen. T cell activation and self-tolerance are tightly regulated and have evolved to provide effective host defense against foreign antigens while deflecting inappropriate inflammatory and autoimmune responses directed against the host.[1,2] The T cell growth cycle is defined by distinct temporal phases comprising (1) basal signaling in the absence of antigen, (2) antigen—major histocompatibility complex (MHC)—induced activation signaling by the T cell receptor (TCR), (3) multiple rounds of cell division, (4) growth arrest by cytotoxic T lymphocyte antigen-4 (CTLA-4) and transforming growth factor-β receptor (TGFβR), (5) differentiation into effector T cells, and (6) apoptosis. Dysregulation at any of these phases can lead to aberrant T cell function and promote autoimmune pathogenesis. The common cytokine receptor γ-chain (γ_c) family cytokines have important functions in T cell proliferation, differentiation, and survival. Interleukin-2 (IL-2) and IL-7 are both potent T cell growth factors and have been widely considered to be key cytokines controlling the effector and regulatory arms of the immune system, yet their underlying mechanisms are not entirely clear.[3,4]

Although T cell fate and function are largely considered to be dictated by changes in gene expression and protein production, alterations in protein glycosylation is increasingly recognized as a major regulatory mechanism. Here, we review evidence suggesting that cytokine signaling by IL-2 and IL-7 regulates T cell function and self-tolerance through changes in protein N-glycosylation. Multiple genetic and environmental factors also influence N-glycosylation, providing an example of the complex genetic and environmental interactions converging on a single pathway to influence autoimmune diseases such as multiple sclerosis (MS).

---

*Both the authors contributed equally to this paper.

doi: 10.1111/j.1749-6632.2011.06391.x

Ann. N.Y. Acad. Sci. 1253 (2012) 49–57 © 2012 New York Academy of Sciences.

Regulation of T cells by N-glycosylation

Differential regulation by N-glycan number and branching

Most cell surface receptors and transporters are cotranslationally and posttranslationally modified with asparagine (N)-linked glycans in the endoplasmic reticulum (ER) and Golgi apparatus after sequential but incomplete action of a series of glycohydrolases and glycosyltransferases.[5,6] Eukaryotic cell surfaces are heavily glycosylated and the size, abundance, and complexity of these glycan structures provide information encoding distinct from the genome.[7] In contrast to proteins and nucleic acids, production of complex carbohydrates is not template driven. Rather, N-glycan biosynthesis depends on the nutrient environment of the cell, metabolic supply of substrates, and enzymatic activities of the ER/Golgi enzymes. Cell surface N-glycans serve as ligands for a number of carbohydrate-binding protein families, including galectins, siglecs, and selectins, all of which play important roles in immunity. Here, we focus on the role of N-glycosylation in titrating binding to galectins, interactions that regulate glycoprotein distribution and concentration at the cell surface to affect T cell function and autoimmunity.[8–10]

Galectins are ubiquitously expressed at the cell surface and extracellular matrix, and bind multivalent glycan ligands to form microdomains or molecular "lattices" at the cell surface.[9,11,12] The minimal binding structure for galectins is N-acetyllactosamine (LacNAc; Galactose β1,4-N-acetylglucosamine)[13] and binding avidity for individual glycoproteins increase in proportion to the degree of LacNAc content within the attached N-glycans. LacNAc density is determined by the degree of N-acetylglucosamine (GlcNAc) branching per N-glycan (nutrient/environment dependent) as well as the number of N-glycans per glycoprotein (gene-encoded; occupied N-X-S/T sites).[7] N-glycan branching and structural diversity produced in the Golgi is dependent upon the expression and activity of Golgi α-mannosidases and N-acetylglucosaminyltransferases I, II, IV, and V (encoded by *Mgat*1, 2, 4, and 5) and the metabolic supply of the substrate uridine diphosphate GlcNAc (UDP-GlcNAc) via the hexosamine pathway.[8,14,15]

The multivalent galectin-glycoprotein lattice controls the function of cell surface receptors by regulating membrane localization, lateral mobility, clustering, cell surface retention, and endocytosis rates.[8–10,16–23] Growth-promoting receptors (e.g., TCR and receptor tyrosine kinases) typically have high numbers of N-glycans ($n > 5$) whereas growth inhibitory receptors (e.g., TGF-βR and CTLA-4) have few N-glycans ($n \leq 4$). Higher density of N-glycans within a glycoprotein generates significant avidity for galectins at low levels of branching, resulting in significant surface retention under basal Golgi conditions. In contrast, a low density of N-glycans falls below the threshold for stable association under basal Golgi activity and requires high levels of branching activity to incorporate into the galectin lattice and maintenance at the cell surface. For example, with increased Golgi branching activity, surface levels of glycoproteins with high N-glycan density increase with a steady, modest incline, whereas glycoproteins with low N-glycan density increase with a sharp, steep incline.

Golgi branching activity depends on both the activity of Golgi enzymes and metabolic supply of UDP-GlcNAc. As such, the strength of the galectin–glycoprotein lattice reflects both genetic and metabolic conditions to modulate receptor and transporter surface levels and cellular sensitivity to ligands and cytokines. The distinct N-glycan profile of each glycoprotein allows for differential association with the galectin lattice dependent on Golgi branching activity, thereby determining the intricate balance between growth stimulatory and inhibitory receptors at the cell surface. In this manner, the Golgi can globally set growth characteristics of cells, controlling cellular transitions from growth to arrest/differentiation.[8]

Control of TCR signaling and CTLA-4–mediated growth arrest by N-glycans

T cell expansion and inhibition are tightly regulated to ensure effective immune responses. Full T cell activation requires TCR clustering above a certain threshold determined by engagement of the TCR with a specific MHC-peptide complex and cosignals by CD28. Binding of galectins to the multiple N-glycans of the TCR complex prevents spontaneous TCR oligomerization in the absence of antigen and subsequent filamentous (F)-actin–mediated transfer to lipid microdomains and activation of CD4-lymphocyte protein tyrosine kinase (Lck).[16] Concurrently, galectin binding to the tyrosine phosphatase CD45 blocks F-actin–mediated movement of CD45 out of the lipid microdomain, thus maintaining a negative regulator of TCR–Lck signaling within signaling microdomains. In this manner, *N*-glycan branching directly controls basal TCR signaling. The same molecular mechanisms serve to set T cell activation thresholds in response to peptide-MHC. Galectin binding to the TCR complex inhibits ligand-dependent TCR clustering while concurrently promoting CD45 at the immune synapse.[9,16,22,24] For example, mice deficient in the branching enzyme β1,6-*N*-acetylglucosaminyltransferase V (Mgat5), or in β1,3-*N*-acetylglucosaminyltransferase (β3GnT2), an enzyme that extends branched *N*-glycans with poly-LacNAc, display lower thresholds to TCR clustering, T cell activation, and autoimmune disease.[9,25] Moreover, removal of *N*-glycan sites in the TCR results in increased receptor diffusion, clustering, and activation.[22] Galectin binding to the TCR also inhibits CD8 binding, hindering complex formation.[17]

Inhibitory regulators of the T cell response are critical for T cell growth arrest, differentiation and T cell tolerance. CTLA-4 is an inhibitory receptor that competes with CD28 for CD80/CD86 costimulatory ligand on antigen presenting cells (APCs) and is induced to the cell surface four to five days after T cell activation to initiate growth arrest.[26] The levels of CTLA-4 transcript and intracellular stores of the protein increase with TCR signaling during the early T cell growth phase. However, endosomal trafficking limits CTLA-4 expression at the cell surface, and therefore CTLA-4 is predominantly located in endosomes. Similar to other growth-inhibitory receptors, CTLA-4 has a low number of *N*-glycans (one to two N-X-S/T sites in humans) and is highly sensitive to UDP-GlcNAc and *N*-glycan branching by the Golgi to increase binding avidity for galectins. Integration of CTLA-4 into the lattice enhances surface retention by opposing endocytic loss, resulting in sustained and increased growth arrest signaling. TCR signaling increases metabolic flux to *N*-glycan branching, thus contributing to the incorporation of CTLA-4 into the galectin lattice and promote arrest signaling.[8] T cells deficient in Mgat5 or deficient in the galectin lattice (i.e., chemical inhibitors of branching) have decreased levels of CTLA-4 at the cell surface but high levels inside the cells, indicating a deficiency in surface retention. Cell surface levels can be enhanced by metabolic supplementation of UDP-GlcNAc, as well as maximal CD28 costimulation, which drives metabolism.

Consistent with serving as a critical regulator of both early and late T cell growth, deficiency in *N*-glycan branching promotes spontaneous autoimmunity. On the 129/Sv background, Mgat5 deficiency in mice induces spontaneous kidney autoimmunity and increased susceptibility to experimental autoimmune encephalomyelitis (EAE), a model of MS.[9] Several mouse strains highly susceptible to EAE (PL/J, SJL, and NOD) display intrinsic deficiency in *N*-glycan branching in T cells, when compared with resistant strains (129/Sv, BALB/c, and B10.S).[27] The PL/J strain, with the lowest levels of *N*-glycan branching, contains natural deficiencies in multiple *N*-glycan branching enzymes (i.e., Mgat1, 2, and 5), as observed by mass spectroscopy and enzyme assays. PL/J mice with targeted deficiency in Mgat5 develop a spontaneous, late-onset clinical MS-like disease manifested by inflammatory demyelination and neurodegeneration.[27] A much milder form of disease is observed in wild-type PL/J mice, consistent with the defective *N*-glycan branching inherent to this inbred strain. As discussed below, defective regulation of *N*-glycan branching in human T cells is also directly associated with MS.

Environmental/metabolic regulation of N-*glycan branching*

N-glycan branching in T cells is influenced by metabolism, nutrient status, and environmental factors (Fig. 1), providing a potential explanation and underlying molecular mechanism for the environmental influence of T cell mediated autoimmune disease. Golgi Mgat1, 2, 4, and 5 catalyze *N*-glycan branching using the same sugar-nucleotide donor (i.e., UDP-GlcNAc), but with declining efficiency.[7] Mgat1 and Mgat2 activities are limited by low affinity for *N*-glycan acceptors, whereas Mgat4 and Mgat5 activities are limited by UDP-GlcNAc concentrations. Therefore, initiation of branching by Mgat1 and Mgat2 depends on rates of bulk protein synthesis, whereas branching by Mgat4 and Mgat5 is sensitive to metabolic production of UDP-GlcNAc. Biosynthesis of UDP-GlcNAc by the hexosamine pathway requires highly regulated intermediates of carbohydrate, nitrogen, and fatty acid metabolism.[14] Indeed, increased supply of glucose,

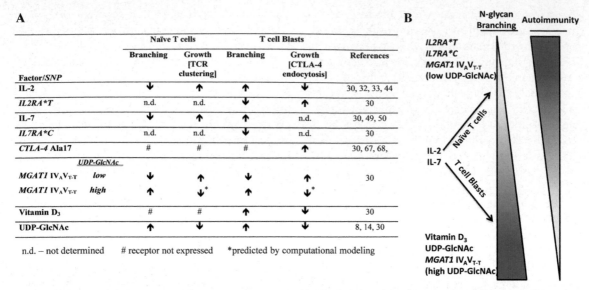

Figure 1. Genetic and environmental factors contribute to varied levels of *N*-glycan branching. Resulting effects on T cell function then determine risk of autoimmunity.

glutamine (a critical nitrogen metabolite), acetyl-CoA (final metabolite of free fatty acids), glucosamine, and GlcNAc increases *N*-glycan branching in T cells in culture, suppressing T cell growth. Oral GlcNAc supplementation in mice also increases *N*-glycan branching and results in inhibition of Th1 and Th17 responses, EAE, and autoimmune diabetes.[14,28] Simon Murch *et al.* observed that oral GlcNAc therapy in children with treatment-resistant inflammatory bowel disease inhibited clinical disease in 8 out of 12 cases.[29] Limited sunlight exposure and associated vitamin D3 deficiency is a well-established environmental risk factor for autoimmune diseases such as MS and type 1 diabetes. We recently reported that vitamin D3 regulates the expression of several Golgi enzymes leading to enhanced *N*-glycan branching and suppressed T cell growth and EAE in mice.[30] Thus, environmental factors such as sunlight/vitamin D3 and metabolism/nutrient supply can directly influence autoimmune pathogenesis by regulating *N*-glycan branching and the effector T cell response (Fig. 1).

IL-2 and IL-7 regulation of T cell growth via *N*-glycan branching

Regulation of T-cell dependent immunity by IL-2 and IL-7

The common cytokine receptor γ_c family cytokines have central roles in the regulation of T cell growth and function, including development, proliferation, survival, and peripheral tolerance. The IL-2 receptor (IL-2R) is the prototypical member of this family and consists of three subunits: the α-chain, the β-chain, and the γ_c-chain that is shared by IL-4R, IL-7R, IL-9R, IL-15R, and IL-21R. IL-2 is a potent T cell growth factor identified nearly 30 years ago[31] that is well-known to induce the proliferation and survival of TCR-activated mouse and human T cells[32,33] and is required to sustain expansion of T cell populations.[34] Its crucial role in promoting T cell-dependent immune responses became questionable when it was unexpectedly found that IL-2 deficiency in mice resulted in lymphoproliferation followed by lethal autoimmunity.[35–37] These findings indicate that a critical function of IL-2 is to regulate peripheral T cell tolerance. Subsequent studies revealed that IL-2/IL-2R signaling is essential for regulatory T cell (T_{reg}) development and peripheral homeostasis.[38–40] The reduction of T_{reg} cells is thought to be a major factor in the autoimmunity associated with IL-2/IL-2R deficiency, as studies show that transfer of T_{reg} cells or reconstitution with IL-2 prevents lymphoproliferation and autoimmunity.[41–43] However, it has also been shown that transgenic overexpression of CTLA-4 can control the lymphoproliferation in IL-2–deficient mice, suggesting an important role for IL-2 in CTLA-4–mediated effector cell growth arrest.[44]

Extensive studies indicate IL-2 is responsible for both generating and, subsequently, limiting the T cell-dependent immune response by regulating clonal expansion, effector, and T_{reg} cell development and contraction of antigen-specific T cells. The use of IL-2 in the past to boost immune responses and the current status of evaluating its use to inhibit immune responses highlights the need to reevaluate the current understanding of the biology of IL-2. It is evident that IL-2 has a wide range of actions with multifaceted roles, such as promoting growth early in T cell activation but limiting expansion later. Yet, the underlying molecular mechanisms of IL-2 effects remain unclear.

The IL-7R, another member of the γ_c family, is a heterodimer of IL-7Rα and the γ_c. IL-7/IL-7R is critical for the development[45] and survival[46,47] of several blood cell types. Mice mutant for IL-7(Ref. 46) or either component of its receptor[45] have major hematopoietic abnormalities and exhibit T cell-negative (T$^-$), B cell-negative (B$^-$), natural killer cell-negative (NK$^-$) severe combined immunodeficiency (SCID). In humans, loss-of-function mutations in the IL-7Rα gene have been found to cause T$^-$ B$^+$ NK$^+$ SCID.[48] In addition, IL-7 has also been found to be critical for the survival and homeostatic cycling of T cells in the periphery,[47,49,50] a process that also depends on signaling through the TCR. However, the effector mechanisms underpinning IL-7 signaling and in particular the crosstalk with TCR signaling remain ill defined.

Our recent data has revealed that IL-2 and IL-7 are major regulators of N-glycan branching in T cells (Fig. 1), suggesting that many of the complex and paradoxical actions of these cytokines may be mediated through changes in branching.[30] IL-2– and IL-7RA–deficient *ex vivo* resting mouse T cells display marked increases in N-glycan branching, up to approximately 2.5-fold. In resting human cells, exogenous IL-2 and IL-7 inhibit N-glycan branching. Alterations in N-glycan branching, significantly smaller than those observed with IL-2/IL-7R deficiency, are sufficient to affect T cell function, growth, and autoimmune risk. Therefore, the much larger changes observed with IL-2 and IL-7 deficiencies are highly significant and imply essential regulatory roles for these cytokines in N-glycosylation. Indeed, our data reveal that IL-2 promotes T cell growth early by lowering N-glycan branching through enhanced TCR signaling, yet later increases growth

arrest of T cell blasts by enhancing branching and CTLA-4 surface retention (Fig. 1).[30]

IL-2 and IL-7 modulate mRNA levels of multiple Golgi branching enzymes, but in opposite direction to the changes induced by TCR signaling, the latter being an enhancer of N-glycan branching.[30,51] However, IL-2 and IL-7 affect N-glycan branching in resting and activated T cells differently, reducing branching in the former while increasing branching in blasting (i.e., activated) T cells. These opposing effects on resting and blasting T cells seem to result from upregulation of Mgat1. Because of large differences in K_m, the Mgat1 enzyme sequesters UDP-GlcNAc from the distal Mgat4 and Mgat5 enzymes, thereby reducing branching when UDP-GlcNAc is limiting.[8] IL-2 signaling increases Mgat1 and decreases Mgat5, and when coupled with low intracellular UDP-GlcNAc levels in resting T cells, results in reduced branching. However, TCR activation signaling increases Mgat5 activity, cell metabolism, and UDP-GlcNAc levels, thereby allowing IL-2–induced increases in Mgat1 in T cell blasts to increase N-glycan branching by supplying more glycoprotein substrate to downstream enzymes. In this manner, IL-2 promotes T cell growth early via reduced N-glycan branching and enhanced TCR clustering/signaling, while later promoting growth arrest by enhancing branching and CTLA-4 surface retention. Thus, through modulation of Golgi enzymes, IL-2 produces opposing effects on T cell growth early and late in activation, mechanisms sensitive to the metabolic state of the cell. Activation-induced cell death (AICD) is also mediated by receptors low in N-glycan number (i.e., Fas, FasL), which are likely to also rely critically on Golgi N-glycan branching for cell surface retention and are therefore also expected to be promoted later by IL-2–induced increases in branching.

IL-2 also plays an important role in the development and frequency of T_{reg} cells. Spontaneous autoimmunity in Mgat5$^{-/-}$ mice develops despite increases in T_{reg} cells, indicating hyperactive effector cell responses and ineffective T_{reg} cell function.[27] Many human autoimmune disorders, including MS, type 1 diabetes, and rheumatoid arthritis, are not associated with reduced T_{reg} cell numbers but rather with alteration in T_{reg} cell function. As IL-2 controls N-glycan branching in T_{reg} cells, suppressor activity and tolerogenic function may also be affected by changes in N-glycan branching and galectin interactions. CTLA-4 is required for T_{reg} cell function and autoimmune suppression in mice,[52] suggesting that dysregulation of N-glycan branching may alter CTLA-4 surface levels and subsequent T_{reg} cell function.

Maintenance of the peripheral T cell pool requires a low level proliferative cycling of naive and memory cells. Under lymphopenic conditions, this proliferative rate is dramatically increased in an effort to reconstitute the T cell pool, and is referred to as homeostatic peripheral expansion (HPE).[53,54] HPE has been shown to be largely deficient in the absence of IL-7.[47,49] Several studies indicate that TCR signal strength is also critical for HPE, with IL-7 neutralization having the greatest effect on the slower proliferating pool that is characteristic of response to low affinity antigens, but little effect on the most rapidly proliferating pool resulting from high affinity TCR.[50,55] This conclusion is consistent with our recent finding that IL-7 decreases N-glycan branching in T cells. In the absence of IL-7, the resulting increase in N-glycan branching increases the threshold of antigen affinity required for TCR clustering, and thus is expected to produce an inhibitory effect that is dominant in low antigen affinity T cells. These findings suggest that IL-7 enhances TCR signaling and activation of peripheral T cells by downregulating N-glycan branching. Thus, IL-7–mediated peripheral T cell survival and homeostatic proliferation may in part be because of promotion of basal TCR signaling through reductions in N-glycan branching.

Although IL-2 and IL-7 have similar effects on N-glycan branching, targeted deficiencies of these pathways in mice paradoxically induce opposing phenotypes: lymphoproliferation/lethal autoimmunity or immunodeficiency, respectively. This is likely because of the expression pattern of their respective receptors, which in turn dictates the dominant effect of deficiency on N-glycan branching. In contrast to IL-2R, whose expression increases with T cell activation and/or cytokine signaling, IL-7Rα is expressed by most resting T cells and is downregulated after T cell activation and/or cytokine signaling.[56,57] Thus, the dominant effect of IL-7 signaling *in vivo* is to reduce branching in resting T cells and in early activation, thus promoting TCR clustering/signaling and T cell growth. The nonredundant role of IL-2 occurs likely later in activation, when it acts to promote growth arrest by upregulating branching and CTLA-4 expression. Continuous IL-7–mediated signaling in naive T cells induces costimulatory and antiapoptotic responses necessary for survival. IL-7Rα expression is low in FOXP3[+] T_{reg} cells, and it is interesting to note that in contrast to non-T_{reg} cell populations activated FOXP3[+] T_{reg} cells upregulate expression of IL-7R.[58] Signaling by IL-2 and IL-7 affects the expression of both their own and other cytokine receptors, allowing for intricate cross-regulation, and for confounding interpretations of studies looking just at cytokine levels. A model that integrates and sums the many inputs affecting N-glycan branching may illuminate seemingly complex and contradictory regulation by these cytokines.

MS risk factors alter N-*glycosylation in T cells*

As with other complex trait diseases, multiple genetic and environmental factors combine to determine disease risk in autoimmune disorders such as MS and type 1 diabetes. However, defining how these factors combine at the molecular level to promote disease has been a great challenge. A role for genetics in MS was defined in studies showing first-degree relatives and identical twins display ~20 to 40-fold and ~300-fold respective increases risk over the general population.[59] Candidate gene studies have validated association of MS with genes in the MHC region, and more recent genome wide association studies (GWAS) have identified a number of potential genes associated with MS, including receptors for IL-7 (IL-7RA) and IL-2 (IL-2RA).[60] IL-7RA and IL-2RA also associate with type 1 diabetes, and the IL-2 and IL-7 pathways have previously been demonstrated to regulate autoimmunity and EAE in animal models.[4,45,61–63] Although, many potential genetic risk factors for MS have been identified, critical issues remain. Most identified variants confer relatively small increases in disease risk and explain only a small proportion of the genetic variance.[64] Furthermore, monozygotic twins codevelop MS only ~30% of the time,[59] which implies that there is direct environmental impact on genetic risk for MS. Indeed, recent studies by Baranzini *et al.* have reported that there is no evidence for genetic, epigenetic, or transcriptome sequence variation to explain disease discordance in monozygotic twins.[65] Despite known genetic and environmental risk factors for MS, there seems to be no obvious shared molecular mechanisms or pathways. These observations may be explained by epistatic and/or additive interactions between multiple, seemingly unrelated, alleles and environmental factors that converge to dysregulate a critical final common pathway. Consistent with this hypothesis, we recently reported that multiple genetic variants (IL-7RA, IL-2RA, MGAT1, and CTLA-4) interact with multiple environmental factors (vitamin D_3 deficiency and metabolism) to dysregulate Golgi N-glycosylation in MS (Fig. 1).[30]

The IL-2RA*T and IL-7RA*C MS risk alleles are associated with enhanced secretion of soluble receptors that block signaling by cognate cytokines. Consistent with the effects of IL-2 and IL-7 on N-glycan branching, the soluble receptors associated with the IL-2RA*T and IL-7RA*C MS risk alleles downregulate MGAT1 mRNA and N-glycan branching in human T cell blasts.[30] On the basis of this, we screened the MGAT1 gene for functional variants and defined an MS-associated gain-of-function haplotype of MGAT1 (IV_A and V_{T-T} polymorphisms) that reduces or enhances N-glycan branching depending on metabolism and on UDP-GlcNAc supply to the Golgi (Fig. 1). The MGAT1 $IV_A V_{T-T}$ variant enhances mRNA levels and enzyme activity twofold to threefold approximately, thereby increasing the N-glycan product of Mgat1 and limiting the UDP-GlcNAc supply to downstream Mgat4 and Mgat5. As Mgat4 and Mgat5 activities are limited by UDP-GlcNAc concentrations, increased Mgat1 protein out-competes Mgat4 and Mgat5 for UDP-GlcNAc under basal UDP-GlcNAc levels, thereby reducing N-glycan branching. However, with increasing UDP-GlcNAc and/or Mgat5 levels, enhanced Mgat1 expression is not as effective in limiting UDP-GlcNAc supply to Mgat4 and Mgat5, allowing these enzymes to act upon the increased supply of N-glycan acceptors to increase N-glycan branching. Thus, the phenotypic effect of the MGAT1 $IV_A V_{T-T}$ variant depends upon the metabolic status of the cell, providing an example of how the same genetic risk factors may promote or inhibit MS based on environmental variation.

Importantly, the IL-2RA*T and IL-7RA*C MS risk variants lower MGAT1 mRNA whereas the MGAT1 $IV_A V_{T-T}$ risk haplotype enhances MGAT1 mRNA. Consistent with these effects, upregulation of Mgat1 by IL-2 and/or IL-7 signaling increases N-glycan branching in MGAT1 $IV_A V_{T-T}$ noncar-

riers, but decreases N-glycan branching in MGAT1 $IV_A V_{T-T}$ carriers.[30] In other words, IL-2 and/or IL-7 signaling upregulates MGAT1 and either enhances or reduces N-glycan branching depending on baseline Mgat1 activity.

The Thr17Ala polymorphism in the human CTLA-4 gene (49A/G) is associated with type 1 diabetes but not MS, encodes a signal peptide variant with inefficient glycosylation (N-glycan occupancy at the two N-X-S/T sites is reduced from two to one), and results in decreased CTLA-4 surface levels to enhance T cell growth.[66–69] The MGAT1 $IV_A V_{T-T}$ variant also limits CTLA-4 surface levels when expressed with the common CTLA-4 allele with two N-glycans, yet combining it with the CTLA-4 Ala17 allele, which has one N-glycan, further reduces CTLA-4 surface levels.[30]

The nonlinear biological interactions observed between IL2RA*T, IL7RA*C, MGAT1 $IV_A V_{TT}$, and CTLA-4 Thr17Ala predict similar genetic interactions in MS. Indeed, the expected genetic interactions observed between these four genetic variants in MS are summarized in Figure 1. It is interesting to note that the active form of vitamin D_3, $1,25(OH)_2 D_3$ enhances MGAT1 mRNA levels similar to the MGAT1 $IV_A V_{T-T}$ variant but opposite of the IL-2RA*T and IL-7RA*C risk alleles, providing a molecular mechanism for the association of vitamin D_3 deficiency in MS.

In summary, complex genetic interactions regulate N-glycan branching, T cell activation thresholds, and CTLA-4 surface levels with sensitivity to metabolic conditions, vitamin D_3, Golgi enzyme activity, cytokine signaling (IL-2, IL-7), and the number of N-glycan sites. Complex trait diseases such as common autoimmune disorders develop from intricate and poorly understood interactions between genetics and the environment. Our and other data suggest that epistatic and additive interactions between multiple, seemingly unrelated, genetic and environmental risk factors combine to dysregulate a common biochemical pathway, namely Golgi N-glycosylation.

Acknowledgments

Research was supported by the National Institutes of Health R01AI053331 and R01AI082266 to M.D. and F32AI081456 to A.G. through the National Institute of Allergy and Infectious Disease and F30HL108451 to H.M. through the National Heart, Lung, and

Blood Institute as well as through a Collaborative Multiple Sclerosis Research Center Award to M.D.

Conflicts of interest

The authors declare no conflicts of interest.

References

1. Abbas, A.K. & C.A. Janeway Jr. 2000. Immunology: improving on nature in the twenty-first century. *Cell.* **100:** 129–138.
2. Medzhitov, R. & C.A. Janeway Jr. 2000. How does the immune system distinguish self from nonself? *Semin. Immunol.* **12:** 185–188; discussion 257–344.
3. Ma, A., R. Koka & P. Burkett. 2006. Diverse functions of IL-2, IL-15, and IL-7 in lymphoid homeostasis. *Annu. Rev. Immunol.* **24:** 657–679.
4. Malek, T.R. 2008. The biology of interleukin-2. *Annu. Rev. Immunol.* **26:** 453–479.
5. Schachter, H. 1991. The 'yellow brick road' to branched complex N-glycans. *Glycobiology.* **1:** 453–461.
6. Kornfeld, R. & S. Kornfeld. 1985. Assembly of asparagine-linked oligosaccharides. *Annu. Rev. Biochem.* **54:** 631–664.
7. Dennis, J.W., I.R. Nabi & M. Demetriou. 2009. Metabolism, cell surface organization, and disease. *Cell.* **139:** 1229–1241.
8. Lau, K.S. *et al.* 2007. Complex N-glycan number and degree of branching cooperate to regulate cell proliferation and differentiation. *Cell.* **129:** 123–134.
9. Demetriou, M., *et al.* 2001. Negative regulation of T-cell activation and autoimmunity by Mgat5 N-glycosylation. *Nature.* **409:** 733–739.
10. Partridge, E.A. *et al.* 2004. Regulation of cytokine receptors by Golgi N-glycan processing and endocytosis. *Science.* **306:** 120–124.
11. Brewer, C.F., M.C. Miceli & L.G. Baum. 2002. Clusters, bundles, arrays and lattices: novel mechanisms for lectin-saccharide-mediated cellular interactions. *Curr. Opin. Struct. Biol.* **12:** 616–623.
12. Ahmad, N. *et al.* 2004. Galectin-3 precipitates as a pentamer with synthetic multivalent carbohydrates and forms heterogeneous cross-linked complexes. *J. Biol. Chem.* **279:** 10841–10847.
13. Hirabayashi, J. *et al.* 2002. Oligosaccharide specificity of galectins: a search by frontal affinity chromatography. *Biochim. Biophys. Acta.* **1572:** 232–254.
14. Grigorian, A. *et al.* 2007. Control of T Cell-mediated autoimmunity by metabolite flux to N-glycan biosynthesis. *J. Biol. Chem.* **282:** 20027–20035.
15. Sasai, K. *et al.* 2002. UDP-GlcNAc concentration is an important factor in the biosynthesis of beta 1,6-branched oligosaccharides: regulation based on the kinetic properties of N-acetylglucosaminyltransferase V. *Glycobiology.* **12:** 119–127.
16. Chen, I.J., H.L. Chen & M. Demetriou. 2007. Lateral compartmentalization of T cell receptor versus CD45 by galectin-N-glycan binding and microfilaments coordinate basal and activation signaling. *J. Biol. Chem.* **282:** 35361–35372.
17. Demotte, N. *et al.* 2008. Restoring the association of the T cell receptor with CD8 reverses anergy in human tumor-infiltrating lymphocytes. *Immunity.* **28:** 414–424.
18. Pace, K.E. *et al.* 1999. Restricted receptor segregation into membrane microdomains occurs on human T cells during apoptosis induced by galectin-1. *J. Immunol.* **163:** 3801–3811.
19. Lajoie, P. *et al.* 2007. Plasma membrane domain organization regulates EGFR signaling in tumor cells. *J. Cell. Biol.* **179:** 341–356.
20. Ohtsubo, K. *et al.* 2005. Dietary and genetic control of glucose transporter 2 glycosylation promotes insulin secretion in suppressing diabetes. *Cell.* **123:** 1307–1321.
21. Cha, S.K. *et al.* 2008. Removal of sialic acid involving Klotho causes cell-surface retention of TRPV5 channel via binding to galectin-1. *Proc. Natl. Acad. Sci. USA* **105:** 9805–9810.
22. Kuball, J. *et al.* 2009. Increasing functional avidity of TCR-redirected T cells by removing defined N-glycosylation sites in the TCR constant domain. *J. Exp. Med.* **206:** 463–475.
23. Grigorian, A., S. Torossian & M. Demetriou. 2009. T-cell growth, cell surface organization, and the galectin-glycoprotein lattice. *Immunol. Rev.* **230:** 232–246.
24. Chung, C.D. *et al.* 2000. Galectin-1 induces partial TCR zeta-chain phosphorylation and antagonizes processive TCR signal transduction. *J. Immunol.* **165:** 3722–3729.
25. Togayachi, A. *et al.* 2007. Polylactosamine on glycoproteins influences basal levels of lymphocyte and macrophage activation. *Proc. Natl. Acad. Sci. U S A.* **104:** 15829–15834.
26. Alegre, M.L., K.A. Frauwirth & C.B. Thompson. 2001. T-cell regulation by CD28 and CTLA-4. *Nat. Rev. Immunol.* **1:** 220–228.
27. Lee, S.U. *et al.* 2007. N-glycan processing deficiency promotes spontaneous inflammatory demyelination and neurodegeneration. *J. Biol. Chem.* **282:** 33725–33734.
28. Grigorian, A. *et al.* 2011. N-Acetylglucosamine Inhibits T-helper 1 (Th1)/T-helper 17 (Th17) Cell Responses and Treats Experimental Autoimmune Encephalomyelitis. *J. Biol. Chem.* **286:** 40133–40141.
29. Salvatore, S. *et al.* 2000. A pilot study of N-acetyl glucosamine, a nutritional substrate for glycosaminoglycan synthesis, in paediatric chronic inflammatory bowel disease. *Aliment. Pharmacol. Ther.* **14:** 1567–1579.
30. Mkhikian, H. *et al.* 2011. Genetics and the environment converge to dysregulate N-glycosylation in multiple sclerosis. *Nat. Commun.* **2:** 334.
31. Taniguchi, T. *et al.* 1983. Structure and expression of a cloned cDNA for human interleukin-2. *Nature.* **302:** 305–310.
32. Kim, H.P., J. Imbert & W.J. Leonard. 2006. Both integrated and differential regulation of components of the IL-2/IL-2 receptor system. *Cytokine Growth Factor Rev.* **17:** 349–366.
33. Morgan, D.A., F.W. Ruscetti & R. Gallo. 1976. Selective in vitro growth of T lymphocytes from normal human bone marrows. *Science.* **193:** 1007–1008.
34. D'Souza, W.N. & L. Lefrancois. 2003. IL-2 is not required for the initiation of CD8 T cell cycling but sustains expansion. *J. Immunol.* **171:** 5727–5735.
35. Sadlack, B. *et al.* 1993. Ulcerative colitis-like disease in mice with a disrupted interleukin-2 gene. *Cell.* **75:** 253–261.

36. Sadlack, B. *et al.* 1995. Generalized autoimmune disease in interleukin-2-deficient mice is triggered by an uncontrolled activation and proliferation of CD4+ T cells. *Eur. J. Immunol.* **25:** 3053–3059.

37. Sadlack, B. *et al.* 1994. Development and proliferation of lymphocytes in mice deficient for both interleukins-2 and -4. *Eur. J. Immunol.* **24:** 281–284.

38. Setoguchi, R. *et al.* 2005. Homeostatic maintenance of natural Foxp3(+) CD25(+) CD4(+) regulatory T cells by interleukin (IL)-2 and induction of autoimmune disease by IL-2 neutralization. *J. Exp. Med.* **201:** 723–735.

39. Fontenot, J.D. *et al.* 2005. A function for interleukin 2 in Foxp3-expressing regulatory T cells. *Nat. Immunol.* **6:** 1142–1151.

40. D'Cruz, L.M. & L. Klein. 2005. Development and function of agonist-induced CD25+Foxp3+ regulatory T cells in the absence of interleukin 2 signaling. *Nat. Immunol.* **6:** 1152–1159.

41. Malek, T.R. *et al.* 2002. CD4 regulatory T cells prevent lethal autoimmunity in IL-2Rbeta-deficient mice. Implications for the nonredundant function of IL-2. *Immunity.* **17:** 167–178.

42. Almeida, A.R. *et al.* 2002. Homeostasis of peripheral CD4+ T cells: IL-2R alpha and IL-2 shape a population of regulatory cells that controls CD4+ T cell numbers. *J. Immunol.* **169:** 4850–4860.

43. Furtado, G.C. *et al.* 2002. Interleukin 2 signaling is required for CD4(+) regulatory T cell function. *J. Exp. Med.* **196:** 851–857.

44. Hwang, K.W. *et al.* 2004. Transgenic expression of CTLA-4 controls lymphoproliferation in IL-2-deficient mice. *J. Immunol.* **173:** 5415–5424.

45. Peschon, J.J. *et al.* 1994. Early lymphocyte expansion is severely impaired in interleukin 7 receptor-deficient mice. *J. Exp. Med.* **180:** 1955–1960.

46. von Freeden-Jeffry, U. *et al.* 1995. Lymphopenia in interleukin (IL)-7 gene-deleted mice identifies IL-7 as a nonredundant cytokine. *J. Exp. Med.* **181:** 1519–1526.

47. Maraskovsky, E. *et al.* 1996. Impaired survival and proliferation in IL-7 receptor-deficient peripheral T cells. *J. Immunol.* **157:** 5315–5323.

48. Puel, A. *et al.* 1998. Defective IL7R expression in T(-)B(+)NK(+) severe combined immunodeficiency. *Nat. Genet.* **20:** 394–397.

49. Schluns, K.S. *et al.* 2000. Interleukin-7 mediates the homeostasis of naive and memory CD8 T cells in vivo. *Nat. Immunol.* **1:** 426–432.

50. Tan, J.T. *et al.* 2001. IL-7 is critical for homeostatic proliferation and survival of naive T cells. *Proc. Natl. Acad. Sci. U S A.* **98:** 8732–8737.

51. Chen, H.L. *et al.* 2009. T cell receptor signaling co-regulates multiple Golgi genes to enhance N-glycan branching. *J. Biol. Chem.* **284:** 32454–32461.

52. Wing, K. *et al.* 2008. CTLA-4 control over Foxp3 +regulatory T cell function. *Science.* **322:** 271–275.

53. Stutman, O. 1986. Postthymic T-cell development. *Immunol. Rev.* **91:** 159–194.

54. Jameson, S.C. 2002. Maintaining the norm: T-cell homeostasis. *Nat. Rev. Immunol.* **2:** 547–556.

55. Min, B. *et al.* 2003. Neonates support lymphopenia-induced proliferation. *Immunity.* **18:** 131–140.

56. Park, J.H. *et al.* 2004. Suppression of IL7Ralpha transcription by IL-7 and other prosurvival cytokines: a novel mechanism for maximizing IL-7-dependent T cell survival. *Immunity.* **21:** 289–302.

57. Fry, T.J. *et al.* 2003. IL-7 therapy dramatically alters peripheral T-cell homeostasis in normal and SIV-infected nonhuman primates. *Blood.* **101:** 2294–2299.

58. Simonetta, F. *et al.* 2010. Increased CD127 expression on activated FOXP3+CD4+ regulatory T cells. *Eur. J. Immunol.* **40:** 2528–2538.

59. Ebers, G.C. *et al.* 1986. A population-based study of multiple sclerosis in twins. *N. Engl. J. Med.* **315:** 1638–1642.

60. Oksenberg, J.R. & S.E. Baranzini. 2010. Multiple sclerosis genetics–is the glass half full, or half empty? *Nat. Rev. Neurol.* **6:** 429–437.

61. Maier, L.M. *et al.* 2009. IL2RA genetic heterogeneity in multiple sclerosis and type 1 diabetes susceptibility and soluble interleukin-2 receptor production. *PLoS Genet.* **5:** e1000322.

62. Wellcome Trust Case Control Consortium. 2007. Genome-wide association study of 14,000 cases of seven common diseases and 3,000 shared controls. *Nature.* **447:** 661–678.

63. Todd, J.A. *et al.* 2007. Robust associations of four new chromosome regions from genome-wide analyses of type 1 diabetes. *Nat. Genet.* **39:** 857–864.

64. Manolio, T.A. *et al.* 2009. Finding the missing heritability of complex diseases. *Nature.* **461:** 747–753.

65. Baranzini, S.E. *et al.* 2010. Genome, epigenome and RNA sequences of monozygotic twins discordant for multiple sclerosis. *Nature.* **464:** 1351–1356.

66. Anjos, S. *et al.* 2002. A common autoimmunity predisposing signal peptide variant of the cytotoxic T-lymphocyte antigen 4 results in inefficient glycosylation of the susceptibility allele. *J. Biol. Chem.* **277:** 46478–46486.

67. Maurer, M. *et al.* 2002. A polymorphism in the human cytotoxic T-lymphocyte antigen 4 (CTLA4) gene (exon 1 +49) alters T-cell activation. *Immunogenetics.* **54:** 1–8.

68. Kavvoura, F.K. & J.P. Ioannidis. 2005. CTLA-4 gene polymorphisms and susceptibility to type 1 diabetes mellitus: a HuGE Review and meta-analysis. *Am. J. Epidemiol.* **162:** 3–16.

69. Bagos, P.G. *et al.* 2007. No evidence for association of CTLA-4 gene polymorphisms with the risk of developing multiple sclerosis: a meta-analysis. *Mult. Scler.* **13:** 156–168.

Ann. N.Y. Acad. Sci. ISSN 0077-8923

ANNALS OF THE NEW YORK ACADEMY OF SCIENCES
Issue: *Glycobiology of the Immune Response*

T cells modulate glycans on CD43 and CD45 during development and activation, signal regulation, and survival

Mary C. Clark and Linda G. Baum

Department of Pathology and Laboratory Medicine, UCLA School of Medicine, University of California, Los Angeles, California

Address for correspondence: Dr. L. G. Baum, Department of Pathology and Laboratory Medicine, UCLA School of Medicine, University of California, 10833 Le Conte Ave., Los Angeles, CA 90095. lbaum@mednet.ucla.edu

Glycosylation affects many essential T cell processes and is intrinsically controlled throughout the lifetime of a T cell. CD43 and CD45 are the two most abundant glycoproteins on the T cell surface and are decorated with *O*- and *N*-glycans. Global T cell glycosylation and specific glycosylation of CD43 and CD45 are modulated during thymocyte development and T cell activation; T cells control the type and abundance of glycans decorating CD43 and CD45 by regulating expression of glycosyltransferases and glycosidases. Additionally, T cells regulate glycosylation of CD45 by expressing alternatively spliced isoforms of CD45 that have different glycan attachment sites. The glycophenotype of CD43 and CD45 on T cells influences how T cells interact with the extracellular environment, including how T cells interact with endogenous lectins. This review focuses on changes in glycosylation of CD43 and CD45 occurring throughout T cell development and activation and the role that glycosylation plays in regulating T cell processes, such as migration, T cell receptor signaling, and apoptosis.

Keywords: glycosylation; T cell; CD43; CD45

Introduction

Glycosylation of cell surface proteins on lymphocytes is a dynamic process that affects a multitude of cellular processes, including pathogen recognition by the immune system,[1] leukocyte migration,[2] and tumor immunology.[3] The addition of glycans to proteins is modulated at several levels, including cell type–specific, developmental and differentiation state–specific, and functional stimulus–specific glycosylation (reviewed in Refs. 4–9). Genes encoding glycan-modifying proteins—glycosyltransferases and glycosidases—account for 1–2% of the protein-coding genes in eukaryotic genomes,[10,11] and the expression and activity of these proteins depends on the cell type and environmental context. However, simply determining expression and activity levels of glycan-modifying enzymes in a particular cell subset will not definitively reveal whether and how specific glycan structures on specific glycoproteins are modified, because factors, such as glycosyltransferase competition for substrates in the Golgi and expression of alternatively spliced isoforms of proteins, also determine whether and how a protein will be glycosylated.

Thus, to understand the range of glycans that can decorate a specific protein, it is important to consider which T cell subset expresses this glycoprotein, which isoform of this glycoprotein is expressed, and which glycan-modifying enzymes are concurrently expressed.

T cells undergo major changes in glycoprotein glycosylation during development in the thymus and activation in the periphery. Glycans found on CD43 and CD45, two abundant glycoproteins on the T cell surface, are thus dynamically modulated throughout the life of a T cell. These changes in glycosylation are crucial for proper T cell responses. In this review, we will discuss changes in glycosylation on CD43 and CD45 that occur during T cell development, activation, and survival.

Polypeptide structures of CD43 and CD45

CD43 and CD45 are two of the most abundant transmembrane glycoproteins on the T cell surface. These two proteins are expressed throughout all stages of T cell development, from double negative thymocytes to memory T cells,[12] and play vital roles in T cell development, activation, and

doi: 10.1111/j.1749-6632.2011.06304.x

survival.[13–15] CD43 and CD45 are recognized by a variety of counter-receptors, many of which are endogenous lectins produced by other immune cells, endothelial cells, and tumor cells.[16] Thus, glycosylation of CD43 and CD45 is critical in regulating how T cells interact with their environment.

CD43 is a large, mucin-type glycoprotein with a small, globular cytoplasmic domain. The gene that codes for CD43 has only one exon, encoding a 400-amino acid polypeptide.[17] The cytoplasmic domain of CD43 interacts with members of the ezrin–radixin–moesin family of cytoskeleton linker proteins to regulate T cell migration,[18] activation,[19,20] and survival.[21] The extracellular region of CD43 extends ~45 nm from the T cell surface and is decorated by ~80 O-linked glycans dispersed over the entire extracellular domain (Figs. 1 and 2).[22] Human CD43 also has one N-glycan site, the significance of which has not been determined.[23,24] T cells express two glycoforms of CD43, with molecular weights of 115 kDa and 130 kDa due to decoration with either core 1 or core 2 O-glycans[9] (Fig. 2), respectively. Thus, given that the predicted molecular weight of the CD43 polypeptide backbone is ~44 kDa, compared to the observed molecular weight of 115–130 kDa, ~62-66% of the mass of CD43 is glycans.

There are five isoforms of CD45 found on human lymphocytes that extend between ~28–52 nm from the T cell surface (Fig. 3).[25–27] The intracellular region of CD45 is common to all CD45 isoforms and is composed of tandem phosphatase domains that are crucial for T cell receptor signaling.[28,29] Extracellularly, all five CD45 isoforms include three membrane proximal fibronectin type III repeats and a cysteine-rich region; these domains are the primary sites of N-linked glycosylation.[30] The extracellular region of CD45 may contain up to three domains termed A, B, and C, resulting from alternative splicing of exons 4, 5, and 6, respectively, which are the primary sites of O-linked glycosylation.[25] Although there are at least eight possible isoforms of CD45, only RO (containing no alternatively spliced domains), RA, RB, RBC, and RABC polypeptides are found on human and murine lymphocytes.[25,31] Expression of alternatively spliced isoforms of CD45 is modulated throughout T cell development and activation; double negative thymocytes express RA, RB, and RBC, double positive thymocytes express primarily RO, mature single positive thymocytes express RB and

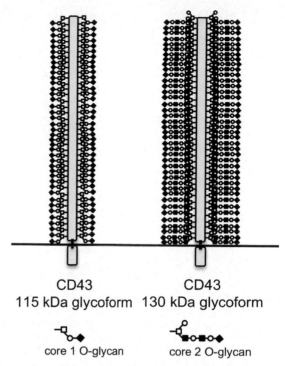

CD43
115 kDa glycoform

CD43
130 kDa glycoform

core 1 O-glycan

core 2 O-glycan

Figure 1. Thymocytes and T cells express two glycoforms of CD43. The CD43 polypeptide backbone extends ~45 nm from the T cell surface and is decorated by ~80 O-linked glycans. Human CD43 also has one N-glycan site, which is not depicted in this schematic, as the significance of this glycan has not been determined. CD43 can be decorated by either sialylated core 1 O-glycans or core 2 O-glycans, generating the 115 kDa and the 130 kDa glycoforms of CD43, respectively. The 115 kDa glycoform of CD43 is the primary glycoform expressed by immature double negative thymocytes, mature CD4 or CD8 single positive thymocytes, and naive peripheral T cells. The 130 kDa glycoform of CD43 is upregulated by immature double positive thymocytes and by activated peripheral T cells.

RBC, naive peripheral T cells express RB, and activated and memory T cells express RO.[31–34]

Glycans also contribute substantially to the overall mass of each CD45 isoform. Given that the predicted molecular weight of the CD45 polypeptide ranges between ~123 kDa and ~141 kDa,[35] depending on the isoform, and the observed molecular weight of CD45 ranges between 180 kDa and 230 kDa, ~32–36% of the mass of CD45 is glycans. Specifically, all isoforms of CD45 are decorated with up to 11 N-glycans in the membrane proximal region, each contributing ~3–4 kDa to the overall molecular weight of CD45. The membrane distal regions of CD45 are decorated with ~8–47 O-glycans, depending on the expression of the

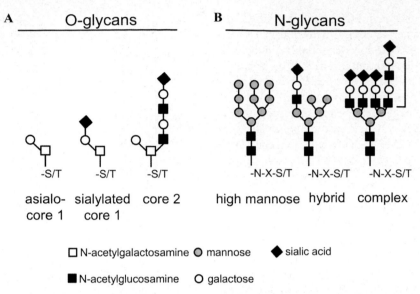

Figure 2. T cells decorate their glycoproteins with asialo-core 1, sialylated core 1, and core 2 O-glycans and high mannose-, hybrid-, and complex-type N-glycans. (A) Core 1 structures can be uncapped (asialo) or α2,3-sialylated by the enzyme ST3Gal1. Core 1 O-glycans can also be extended into core 2 O-glycans by the enzyme C2GnT. Core 2 O-glycans contain lactosamine (Gal-GlcNAc) sequences and may or may not also be capped by ST3Gal1 (this schematic depicts uncapped core 2 O-glycans). (B) High mannose N-glycans are precursors for hybrid-type and complex-type N-glycans. Complex N-glycans on T cells are elaborated by the action of the enzyme GnT-V, which adds a β1,6-linked GlcNAc residue that can be further extended into a polylactosamine sequence (bracket). Complex N-glycans can be capped by α2,3- or α2,6-linked sialic acid.

alternatively spliced domains, each contributing ∼1 kDa to the overall molecular weight of CD45.

Dynamic changes in glycosylation during development, activation, and differentiation into effector cells

O-glycans

Eight core O-glycan structures have been described.[36] Thymocytes and T cells display primarily core 1 and core 2 O-glycan structures (Fig. 2 and Table 1). Core 1 O-glycans can be capped by sialic acid added by β-galactoside α2,3-sialyltransferase 1 (ST3Gal1) and/or extended into core 2 O-glycans by core 2 β-1,6-N-acetylglucosaminyltransferase (C2GnT). C2GnT adds a β1,6-linked N-acetylglucosamine (β1,6-GlcNAc) to the core 1 structure that can be further extended into a polylactosamine sequence. In the thymus, sialylated core 1 structures are found on double negative immature thymocytes and on mature CD4 or CD8 single positive thymocytes, while asialo-core 1 and core 2 structures are found on immature double positive thymocytes.[16,37–39] Peripheral naive T cells in general express sialylated core 1 O-glycans. Activation of CD4+ and CD8+ T cells

results in a general loss of sialylation and an increase in core 2 O-glycans; this appears to be due primarily to *de novo* synthesis of hyposialylated CD43[15] and CD45[40] (see below and Table 1),[33,36] although there may be microheterogeneity in sialylation at specific sites.[37,41,42] In addition, Grabie *et al.* showed that ST3Gal1 is differentially expressed during differentiation of CD4+ T cells into Th1 and Th2 effector populations following an activation stimulus.[43] Specifically, Th2 cells express ST3Gal1 that can sialylate core 1 O-glycans, while Th1 cells do not express ST3Gal1 and have asialo-core 1 O-glycans.[43] Both Th1 and Th2 cells express C2GnT and can bear glycoproteins decorated with core 2 O-glycans.[44]

As previously mentioned, CD43 is decorated with ∼80 O-glycans, and on T cells, CD43 is found primarily in two glycoforms, displaying either core 1 or core 2 O-glycans. Two monoclonal antibodies, S7 and 1B11, that recognize murine CD43 decorated with core 1 or core 2 O-glycans, respectively, have been used to characterize glycoform expression of CD43 throughout thymocyte development and T cell activation. In the thymus, both immature and mature thymocytes are reactive with the S7 antibody and thus express CD43 with core

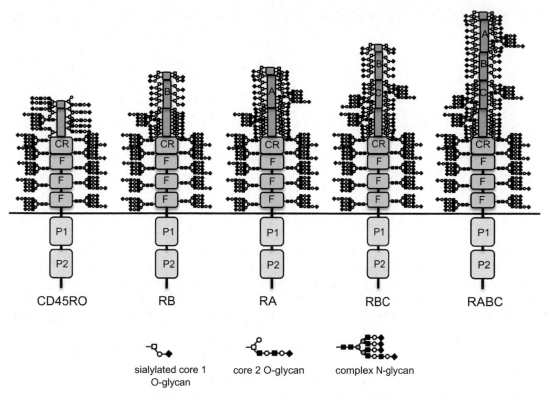

sialylated core 1   core 2 O-glycan   complex N-glycan
O-glycan

Figure 3. There are five isoforms of CD45 on human lymphocytes due to alternative splicing. Intracellularly, all isoforms of CD45 have tandem phosphatase domains (P1 and P2; only P1 is enzymatically active). Extracellularly, all isoforms have three fibronectin type III repeats (F), a cysteine-rich region (CR), and a terminal region. The F, CR, and terminal regions are the primary sites of N-linked glycosylation. Alternative splicing of exons 4, 5, and 6 generates up to three additional extracellular domains—termed A, B, and C, respectively—and a total of five isoforms of CD45: RO (with no alternatively spliced domains), RB (exon 5), RA (exon 4), RBC (exons 5 and 6), and RABC (exons 4–6). The extended region and alternatively spliced regions are the primary sites of O-linked glycosylation, although RA and RC also have canonical sites of N-linked glycosylation. The schematics of the five isoforms of CD45 depict glycans based on the following assumptions: (1) The molecular weights of CD45RO, RB, RA, RBC, and RABC are 180, 190, 210, and 230, respectively; (2) Each N-glycan is complex and contributes ~4 kDa to the overall molecular weight; RO has core 2 O-glycans, each contributing ~1.5 kDa to the overall molecular weight; RB, RA, and RBC have sialylated core 1 O-glycans, each contributing ~0.7 kDa to the overall molecular weight; (3) All N-glycan sites in the F, CR, and extended regions are occupied in all isoforms; (4) Some O-glycan sites may overlap between isoforms, but the total number of O-glycans should be higher for larger isoforms (i.e., RO should have the least number of O-glycans, while RABC should have the most O-glycans); (5) Domains with more serine and threonine residues should have more potential O-glycosylation sites (i.e., RB has the least and RA has the most); and (6) The alternatively spliced domains A, B, and C should have O-glycan occupancy approximately proportional to the number of serine and threonine residues contained in the polypeptide sequences.

1 O-glycans, while immature double positive thymocytes are also reactive with the 1B11 antibody, and therefore also express CD43 decorated with core 2 O-glycans, consistent with the global glycosylation profile of these thymocyte subsets described above.[45,46] In the periphery, both naive and activated T cells express core 1 O-glycosylated CD43, while activated T cells additionally express core 2 O-glycosylated CD43.[37,45,47,48]

The alternatively spliced extracellular domains of CD45 are also decorated with core 1 and core 2 O-

glycans.[24,49] Different isoforms of CD45 differ in the extent of potential O-glycosylation. Unlike S7 and 1B11 monoclonal antibodies for CD43, few CD45 antibodies have been evaluated for specific CD45 glycoform binding,[41,50,51] and most studies use changes in global T cell glycosylation to infer how glycosylation of CD45 changes throughout T cell development and activation. Thus, immature double negative thymocytes and mature CD4 or CD8 single positive thymocytes express CD45RA/RBC/RB and CD45RB/RBC, respectively, with sialylated core 1

Table 1. Global, CD43-, and CD45-specific glycosylation during T cell development

	Thymus		
Cell type:	Double negative	Double positive	CD4 or CD8 single positive
Global *O*-glycans:	Sialylated core 1	Asialo-core 1; core 2	Sialylated core 1
Global *N*-glycans:	High mannose; hybrid	High mannose; hybrid	α2,6-sialylated complex
CD43 *O*-glycans:	Core 1	Core 1; core 2	Core 1
CD45 isoform:	RA, RBC, RB	RO	RB, RBC
CD45 *O*-glycans:	nd[a]	Sialo-core 1; core 2	nd
CD45 *N*-glycans:	High mannose; hybrid	High mannose; hybrid	α2,6-sialylated complex

	Periphery				
Cell type:	Naive CD4$^+$ or CD8$^+$	Activated CD8$^+$	Activated CD4$^+$	Th1	Th2
Global *O*-glycans:	Sialylated core 1	Asialo-core 1; core 2	Asialo-core 1; core 2	Asialo-core 1; core 2	Sialylated core 1
Global *N*-glycans:	α2,6-sialylated complex	Complex	Complex	Complex	α2,6-sialylated complex
CD43 *O*-glycans:	Core 1	Core 1; core 2	Core 1; core 2	Core 1; core 2	nd
CD45 isoform:	RB	RO	RO	RO	RO
CD45 *O*-glycans:	Sialylated core 1	Asialo-core 1	asialo-core 1	nd	nd
CD45 *N*-glycans:	α2,6-sialylated complex	Complex	Complex	Complex	nd

^{a}nd: not determined

O-glycans, while double positive thymocytes express CD45RO decorated with asialo-core 1 and core 2 *O*-glycans. In the periphery, naive T cells express CD45RB and are predicted to be decorated with sialylated core 1 *O*-glycans, while activated T cells, like double positive thymocytes, express CD45RO decorated with asialo-core 1 and core 2 *O*-glycans.[40]

N-*glycans*

N-glycan structures are classified as high-mannose, hybrid-type, and complex-type (Fig. 2). Thymocytes and T cells can express all three types of *N*-glycans (Table 1).[38,52] High-mannose *N*-glycan structures can be trimmed by mannosidases to create substrates for glycosyltransferases that create hybrid and complex *N*-glycan structures. Complex *N*-glycans found on T cells can be tri- or tetra-antennary structures created by the action of *N*-acetylglucosaminyltransferase V (GnT-V). GnT-V adds a β1,6-linked GlcNAc residue to the mannose core, which can be further extended into a polylactosamine sequence and capped by terminal sialic acid residues in either α2,3- or α2,6-linkages. In the thymus, α2,3- and α2,6-linked sialic acid is added to *N*-glycans by ST3Gal4 and ST6Gal1, respectively.[53] Addition of sialic acid is modulated throughout thymocyte development: both cortical and medullary thymocytes bear α2,3-linked sialic acids on *N*-glycans, while mature medullary thymocytes also bear α2,6-linked sialic acids on *N*-glycans.[53] Like mature thymocytes, naive T cells that leave the thymus also express *N*-glycans capped with α2,6-linked sialic acid. Once activated, peripheral T cells upregulate *N*-glycan processing enzymes and thus increase complex *N*-glycans on the cell surface, while simultaneously decreasing levels of α2,6-linked sialic acid.[54–56]

As mentioned above, the membrane proximal domains of CD45 that are common to all isoforms are the primary sites of N-linked glycosylation. Human CD45RO has 11 potential sites of *N*-glycosylation.[25] Early analyses of *N*-glycans on CD45 revealed that roughly half of the *N*-glycans on CD45 on

Table 2. Lectin coreceptors of CD43 and CD45 on thymocytes and T cells

CD43 coreceptor	Glycan	Function	References
E-selectin	Sialyl Lewis x structure on core 2 O-glycans	T cell migration	47,48,66
Galectin-1	Lactosamine[a]	T cell apoptosis	79,84
Galectin-3	Lactosamine	Unknown	80
Macrophage galactose-type lectin	Terminal GalNAc	Unknown	75
Mannose receptor[d]	Mannose[b]	Unknown	60

CD45 coreceptor	Glycan	Function	References
Galectin-1	Lactosamine[a]	T cell apoptosis	78,79
Galectin-3	Lactosamine	T cell apoptosis	80
Placental protein 14	Lactosamine[c]	T cell apoptosis	73,74
CD22/Siglec-2	α2,6-linked sialic acid[d]	T cell signaling	67,68
Macrophage galactose-type lectin	Terminal GalNAc	T cell apoptosis	75
Serum mannan-binding protein	Mannose; GlcNAc	Unknown	57
Mannose receptor	Mannose[b]	Unknown	60

[a]Binding is inhibited by α2,6-linked sialic acid.
[b]Interaction was determined for the cysteine-rich region of the mannose receptor only.[60]
[c]Binding is enhanced by α2,6-linked sialic acid.
[d]Interaction between CD22 and CD45 may also be protein–protein mediated.

thymocytes are high mannose or hybrid, whereas the majority of N-glycans on CD45 on peripheral T cells are complex.[57–60] As mentioned in the previous paragraph, the sialyltransferase ST6Gal1 modifies complex N-glycans.[61,62] CD45 is an established acceptor substrate for this enzyme.[53,63] Thus, both mature medullary thymocytes and naive peripheral T cells express CD45 with N-glycans capped with α2,6-linked sialic acid, whereas CD45 on activated T cells express N-glycans that are not capped with α2,6-linked sialic acid.

Glycan-dependent functions of CD43 and CD45

CD43 and CD45 interact with endogenous lectins in a glycan-dependent manner to regulate several T cell functions (Table 2). Thus, as glycans on CD43 and CD45 are modulated throughout T cell development and activation as described above, so are the interactions with endogenous lectins and potential functional outcomes.

T cell migration
T cells migrate to sites of inflammation to generate effective immune responses. E-selectin is an adhesion protein expressed by activated endothe-lial cells that is important in recruiting T cells to sites of inflammation. CD43 has been identified as an endogenous coreceptor for E-selectin.[64] Specifically, several groups have found that, on T cells, E-selectin binds to the 130 kDa glycoform of CD43-bearing sialyl Lewis x tetrasaccharide (SLe^x) on core 2 O-glycans.[47,48,65,66] As activated T cells upregulate CD43-bearing core 2 O-glycans, and thus the SLe^x glycan ligands for E-selectin, these cells can preferentially bind to E-selectin and migrate to sites of inflammation to generate an immune response.

T cell receptor signaling
T cell receptor signaling is vital for proper positive and negative selection of thymocytes and activation of peripheral T cells. Activity of the CD45 intracellular phosphatase domain is critical for regulating signaling thresholds required for signaling through the T cell receptor.[29,67,68] There is intriguing evidence to suggest a role for glycans in regulating the activity of the CD45 intracellular phosphatase domains. Briefly, bulky N- and O-glycans and negatively charged sialic acid residues on the extracellular domains of CD45 can keep individual CD45 molecules physically separated on the plasma membrane, resulting in increased T cell receptor

signaling, most likely due to activation of the intracellular phosphatase domains of CD45.[69] Conversely, reduction in sialic acid content and/or binding of multivalent lectins to the extracellular domain of CD45 results in clustering or oligomerization of CD45 and decreased T cell receptor signaling;[69–72] oligomerization of CD45 also has been shown to decrease activity of the intracellular phosphatase domains of CD45 in cell extracts.[70] Thus, when CD45 is bound by lectins such as placental protein 14,[73,74] macrophage galactose-type lectin,[75] or galectin-1,[70] clustered CD45 molecules would be predicted to have inactive phosphatase domains and thus negatively regulate T cell receptor signaling (Table 2).

The role of CD43 in T cell receptor signaling is less clear than that of CD45. Historically, it was thought that CD43 negatively regulates T cell activation by sterically hindering interactions of T cells with antigen presenting cells via CD43's large extracellular domain.[76] Recent studies have also explored the participation of the cytoplasmic domain of CD43 in T cell receptor signaling events.[20,77] However, few studies exploring the role of CD43 in T cell receptor signaling have examined the specific role of glycans on CD43 in this process.

T cell apoptosis

Specific thymocyte subsets and activated peripheral T cells undergo apoptosis induced by galectin-1, the prototypic member of the family of β-galactoside-binding lectins.[78] Three glycoprotein counter-receptors, CD43, CD45, and CD7 are involved in T cell death induced by galectin-1.[79–82] Expression of CD7 on human T cells is required for galectin-1–induced apoptosis.[81] Although neither CD43 nor CD45 is required for galectin-1–induced apoptosis, expression of both of these glycoproteins enhances apoptosis induced by galectin-1.[83,84] Galectin-1 binding to CD43 and CD45 is dependent on the relative amount and type of glycans present on these glycoproteins. Although galectin-1 preferentially binds to lactosamine sequences that are not capped by α2,6-linked sialic acid,[63] such as those found on core 2 O-glycans and complex N-glycans, galectin-1 may also bind to low affinity, yet highly abundant glycans, such as core 1 O-glycans, if presented in sufficient density on a polypeptide backbone (i.e., a mucin, such as CD43).[70,84] Core 2 O-glycans on CD45 seem to be required for galectin-1–induced T cell apoptosis,[83,85] while either core

1 or core 2 O-glycans on CD43 are sufficient for galectin-1 binding and apoptotic signaling.[84]

In the thymus, galectin-1 induces apoptosis of immature thymocytes undergoing negative selection.[86] CD43 and CD45 on immature thymocytes are decorated with core 2 O-glycans and can bind galectin-1. In contrast, CD45 on mature thymocytes bears core 1 O-glycans as well as N-glycans capped with α2,6-linked sialic acid. Thus, this "glycophenotype" inhibits galectin-1 binding, due both to loss of core 2 O-glycans and increased α2,6-sialylation. Moreover, this glycosylation renders mature thymocytes resistant to galectin-1–induced apoptosis.[63,70] In the periphery, naive T cells also express CD45 with core 1 O-glycans and α2,6-linked sialic acid, and are likewise resistant to galectin-1–induced apoptosis. Upon activation, CD4[+] and CD8[+] T cells express CD43- and CD45-bearing core 2 O-glycans and complex N-glycans with reduced α2,6-linked sialic acid, thus rendering activated T cells susceptible to galectin-1-induced apoptosis.[70] However, CD4[+] Th2 effector cells express sialylated core 1 O-glycans and α2,6-linked sialic acid on N-glycans, which reduces galectin-1 binding and decreases Th2 cell susceptibility to galectin-1–induced apoptosis.[87] Thus, although thymocytes and T cells are universally exposed to galectin-1 in the thymus and peripheral tissues, specific T cell subsets are resistant to galectin-1–induced apoptosis due to differential glycosylation of CD43 and CD45.

Concluding remarks and remaining questions

T cells intrinsically regulate interactions with the extracellular environment, including those with endogenous lectins, by regulating the type and abundance of glycans on the cell surface glycoproteins CD43 and CD45. Thus, processes such as migration, T cell receptor signaling, and apoptosis, which are all affected by glycans on CD43 and CD45, are "decided" at the level of the T cell. It is, therefore, critical to analyze glycans on CD43 and CD45 at specific points during T cell development and activation to infer how a T cell might function at that point.

Currently, our understanding of thymocyte and T cell glycosylation during development and activation is largely limited to analyses of global cellular glycosylation patterns, with few analyses of glycoprotein-specific glycosylation patterns. Many questions regarding the glycosylation of CD43 and CD45 during the lifetime of a T cell remain unanswered. For instance, how do changes in the glycosylation of the extracellular domain of CD43 affect the intracellular functions of CD43? Unlike CD45, for which a role for glycans in regulating the activity of the intracellular phosphatase domains of CD45 has been suggested, few studies have evaluated the functional role of the glycans on CD43. Secondly, how are each of the individual ~80 O-glycans on CD43 and the ~11 N-glycans and ~8-47 O-glycans on CD45 modulated throughout the lifetime of a T cell? What is the role of microheterogeneity of glycans on these glycoproteins?[41] Finally, how does glycosylation of CD43 and CD45 on thymocytes and T cells compare to that of CD43 and CD45 expressed by other leukocytes, such as B cells or dendritic cells? Answering these questions will bring us closer to understanding how the intricate nature of glycosylation of CD43 and CD45 regulates T cell development and function.

Acknowledgments

The authors would like to thank Drs. Sandra Thiemann, Jenny Davies, and Sam Strom for critical reading of the manuscript and Brian McMorran for assistance with the figures. This work was supported by NIH Grant HL102989 (to LGB) and NIH T32 Grant CA009056:35 (to MCC).

Conflicts of interest

The authors declare no conflicts of interest.

References

1. Doores, K.J. et al. 2010. Envelope glycans of immunodeficiency virions are almost entirely oligomannose antigens. Proc. Natl. Acad. Sci. USA 107: 13800–10655.
2. Bi, S. et al. 2011. Galectin-9 binding to cell surface protein disulfide isomerase regulates the redox environment to enhance T cell migration and HIV entry. Proc. Natl. Acad. Sci. USA 108: 10650–10655.
3. Coelho, V. et al. 2010. Glycosylation of surface Ig creates a functional bridge between human follicular lymphoma and microenvironmental lectins. Proc. Natl. Acad. Sci. USA 107: 18587–18592.
4. Baum, L.G. & P.R. Crocker. 2009. Glycoimmunology: ignore at your peril! Immunol. Rev. 230: 5–8.
5. Hart, G.W. & R.J. Copeland. 2010. Glycomics hits the big time. Cell. 143: 672–676.
6. Stanley, P. 2011. Golgi glycosylation. Cold Spring Harb. Perspect. Biol. 3: 1-13.
7. Larkin, A. & B. Imperiali. 2011. The expanding horizons of asparagine-linked glycosylation. Biochemistry. 50: 4411–4426.
8. Dennis, J.W. et al. 2009. Adaptive regulation at the cell surface by N-glycosylation. Traffic. 10: 1569–1578.
9. Marth, J.D. & P.K. Grewal. 2008. Mammalian glycosylation in immunity. Nat. Rev. Immunol. 8: 874–887.
10. Hashimoto, K. et al. 2009. Comprehensive analysis of glycosyltransferases in eukaryotic genomes for structural and functional characterization of glycans. Carbohydr. Res. 344: 881–887.
11. Hashimoto, K. et al. 2010. Functional states of homooligomers: insights from the evolution of glycosyltransferases. J. Mol. Biol. 399: 196–206.
12. Yang, Q., J. Jeremiah Bell & A. Bhandoola. 2010. T cell lineage determination. Immunol. Rev. 238: 12–22.
13. Perillo, N.L. et al. 1995. Apoptosis of T cells mediated by galectin-1. Nature. 378: 736–739.
14. Lai, J.C. et al. 2010. CD45 regulates migration, proliferation, and progression of double negative 1 thymocytes. J. Immunol. 185: 2059–2070.
15. Onami, T.M. et al. 2002. Dynamic regulation of T cell immunity by CD43. J. Immunol. 168: 6022–6031.
16. Earl, L.A. & L.G. Baum. 2008. CD45 glycosylation controls T cell life and death. Immunol. Cell Biol. 86: 608–615.
17. de Laurentiis, A. et al. 2011. Mass spectrometry-based identification of the tumor antigen UN1 as the transmembrane CD43 sialoglycoprotein. Mol. Cell. Proteomics. 10: M111.007898.
18. Cannon, J.L. et al. 2011. CD43 interaction with ezrin-radixin-moesin (ERM) proteins regulates T cell trafficking and CD43 phosphorylation. Mol. Biol. Cell. 22: 954–963.
19. Allenspach, E.J. et al. 2001. ERM-dependent movement of CD43 defines a novel protein complex distal to the immunological synapse. Immunity. 15: 739–750.
20. Tong, J. et al. 2004. CD43 regulation of T cell activation is not through steric inhibition of T cell-APC interactions but through an intracellular mechanism. J. Exp. Med. 199: 1277–1283.
21. Lee, J.B. & J. Chang. 2010. CD43 expression regulated by IL-12 signaling is associated with survival of CD8 T Cells. Immune Netw. 10: 153–163.
22. Cyster, J.G., D.M. Shotton & A.F. Williams. 1991. The dimensions of the T lymphocyte glycoprotein leukosialin and identification of linear protein epitopes that can be modified by glycosylation. EMBO J. 10: 893–902.
23. Fukuda, M. 1991. Leukosialin, a major O-glycan-containing sialoglycoprotein defining leukocyte differentiation and malignancy. Glycobiology. 1: 347–356.
24. Barran, P. et al. 1997. Modification of CD43 and other lymphocyte O-glycoproteins by core 2 N-acetylglucosaminyltransferase. Glycobiology. 7: 129–136.
25. Streuli, M. et al. 1987. Differential usage of three exons generates at least five different mRNAs encoding human leukocyte common antigens. J. Exp. Med. 166: 1548–1566.
26. Woollett, G.R., A.F. Williams & D.M. Shotton. 1985. Visualisation by low-angle shadowing of the leucocyte-common antigen: a major cell surface glycoprotein of lymphocytes. EMBO J. 4: 2827–2830.
27. McCall, M.N., D.M. Shotton & A.N. Barclay. 1992. Expression of soluble isoforms of rat CD45: analysis by electron

microscopy and use in epitope mapping of anti-CD45R monoclonal antibodies. *Immunology.* **76:** 310–317.

28. Weaver, C.T. *et al.* 1991. CD8+ T cell clones deficient in the expression of the CD45 protein tyrosine phosphatase have impaired responses to T cell receptor stimuli. *Mol. Cell. Biol.* **11:** 4415–4422.

29. Saunders, A. & P. Johnson. 2010. Modulation of immune cell signalling by the leukocyte common tyrosine phosphatase, CD45. *Cell Signal.* **22:** 339–348.

30. Okumura, M. *et al.* 1996. Comparison of CD45 extracellular domain sequences from divergent vertebrate species suggests the conservation of three fibronectin type III domains. *J. Immunol.* **157:** 1569–1575.

31. Hermiston, M.L., Z. Xu & A. Weiss. 2003. CD45: a critical regulator of signaling thresholds in immune cells. *Annu. Rev. Immunol.* **21:** 107–137.

32. Fujii, Y. *et al.* 1992. CD45 isoform expression during T cell development in the thymus. *Eur. J. Immunol.* **22:** 1843–1850.

33. McNeill, L. *et al.* 2004. CD45 isoforms in T cell signalling and development. *Immunol. Lett.* **92:** 125–134.

34. Fukuhara, K. *et al.* 2002. A study on CD45 isoform expression during T cell development and selection events in the human thymus. *Hum. Immunol.* **63:** 394–404.

35. Nam, H.J. *et al.* 2005. Structural basis for the function and regulation of the receptor protein tyrosine phosphatase CD45. *J. Exp. Med.* **201:** 441–452.

36. Gill, D.J., H. Clausen & F. Bard. 2011. Location, location, location: new insights into *O*-GalNAc protein glycosylation. *Trends Cell. Biol.* **21:** 149–158.

37. Piller, F. *et al.* 1988. Human T-lymphocyte activation is associated with changes in *O*-glycan biosynthesis. *J. Biol. Chem.* **263:** 15146–15150.

38. Balcan, E., A. Gümüş & M. Sahin. 2008. The glycosylation status of murine [corrected] postnatal thymus: a study by histochemistry and lectin blotting. *J. Mol. Histol.* **39:** 417–426.

39. Gillespie, W. *et al.* 1993. Regulation of alpha 2,3-sialyltransferase expression correlates with conversion of peanut agglutinin (PNA)+ to PNA– phenotype in developing thymocytes. *J. Biol. Chem.* **268:** 3801–3804.

40. Amado, M. *et al.* 2004. Peanut agglutinin high phenotype of activated CD8+ T cells results from de novo synthesis of CD45 glycans. *J. Biol. Chem.* **279:** 36689–36697.

41. Hernandez, J.D. *et al.* 2007. T cell activation results in microheterogeneous changes in glycosylation of CD45. *Int. Immunol.* **19:** 847–856.

42. Chervenak, R. & J.J. Cohen. 1982. Peanut lectin binding as a marker for activated T-lineage lymphocytes. *Thymus.* **4:** 61–67.

43. Grabie, N. *et al.* 2002. Beta-galactoside alpha2,3-sialyltransferase-I gene expression during Th2 but not Th1 differentiation: implications for core2-glycan formation on cell surface proteins. *Eur. J. Immunol.* **32:** 2766–2772.

44. Lim, Y.C. *et al.* 2001. IL-12, STAT4-dependent upregulation of CD4(+) T cell core 2 beta-1,6-n-acetylglucosaminyltransferase, an enzyme essential for biosynthesis of P-selectin ligands. *J. Immunol.* **167:** 4476–4484.

45. Jones, A.T. *et al.* 1994. Characterization of the activation-associated isoform of CD43 on murine T lymphocytes. *J. Immunol.* **153:** 3426–3439.

46. Ellies, L.G. *et al.* 1996. The CD43 130-kD peripheral T cell activation antigen is downregulated in thymic positive selection. *Blood.* **88:** 1725–1732.

47. Matsumoto, M. *et al.* 2005. CD43 functions as a ligand for E-Selectin on activated T cells. *J. Immunol.* **175:** 8042–8050.

48. Alcaide, P. *et al.* 2007. The 130-kDa glycoform of CD43 functions as an E-selectin ligand for activated Th1 cells in vitro and in delayed-type hypersensitivity reactions in vivo. *J. Invest. Dermatol.* **127:** 1964–1972.

49. Furukawa, K. *et al.* 1998. Structural study of the O-linked sugar chains of human leukocyte tyrosine phosphatase CD45. *Eur. J. Biochem.* **251:** 288–294.

50. Pulido, R. *et al.* 1994. Identification of amino acids at the junction of exons 3 and 7 that are used for the generation of glycosylation-related human CD45RO and CD45RO-like antigen specificities. *J. Exp. Med.* **179:** 1035–1040.

51. Cyster, J.G., D. Fowell & A.N. Barclay. 1994. Antigenic determinants encoded by alternatively spliced exons of CD45 are determined by the polypeptide but influenced by glycosylation. *Int. Immunol.* **6:** 1875–1881.

52. Paessens, L.C. *et al.* 2007. The glycosylation of thymic microenvironments: a microscopic study using plant lectins. *Immunol. Lett.* **110:** 65–73.

53. Baum, L.G. *et al.* 1996. Characterization of terminal sialic acid linkages on human thymocytes. Correlation between lectin-binding phenotype and sialyltransferase expression. *J. Biol. Chem.* **271:** 10793–10799.

54. Chen, H.L. *et al.* 2009. T cell receptor signaling co-regulates multiple Golgi genes to enhance *N*-glycan branching. *J. Biol. Chem.* **284:** 32454–32461.

55. Comelli, E.M. *et al.* 2006. Activation of murine CD4 +and CD8+ T lymphocytes leads to dramatic remodeling of N-linked glycans. *J. Immunol.* **177:** 2431–2440.

56. Kaech, S.M. *et al.* 2002. Molecular and functional profiling of memory CD8 T cell differentiation. *Cell.* **111:** 837–851.

57. Uemura, K. *et al.* 1996. A unique CD45 glycoform recognized by the serum mannan-binding protein in immature thymocytes. *J. Biol. Chem.* **271:** 4581–4584.

58. Sato, T. *et al.* 1993. Structural study of the sugar chains of human leukocyte common antigen CD45. *Biochemistry.* **32:** 12694–12704.

59. Baldwin, T.A. & H.L. Ostergaard. 2001. Developmentally regulated changes in glucosidase II association with, and carbohydrate content of, the protein tyrosine phosphatase CD45. *J. Immunol.* **167:** 3829–3835.

60. Martínez-Pomares, L. *et al.* 1999. Cell-specific glycoforms of sialoadhesin and CD45 are counter-receptors for the cysteine-rich domain of the mannose receptor. *J. Biol. Chem.* **274:** 35211–35218.

61. Joziasse, D.H. *et al.* 1987. Branch specificity of bovine colostrum CMP-sialic acid: gal beta 1—-4GlcNAc-R alpha 2—-6-sialyltransferase. Sialylation of bi-, tri-, and tetraantennary oligosaccharides and glycopeptides of the *N*-acetyllactosamine type. *J. Biol. Chem.* **262:** 2025–2033.

62. Weinstein, J., U. de Souza-e-Silva & J.C. Paulson. 1982. Purification of a Gal beta 1 to 4GlcNAc alpha 2 to 6

sialyltransferase and a Gal beta 1 to 3(4)GlcNAc alpha 2 to 3 sialyltransferase to homogeneity from rat liver. *J. Biol. Chem.* **257:** 13835–13844.

63. Amano, M. *et al.* 2003. The ST6Gal I sialyltransferase selectively modifies *N*-glycans on CD45 to negatively regulate galectin-1-induced CD45 clustering, phosphatase modulation, and T cell death. *J. Biol. Chem.* **278:** 7469–7475.

64. Merzaban, J.S. *et al.* 2011. Analysis of glycoprotein E-selectin ligands on human and mouse marrow cells enriched for hematopoietic stem/progenitor cells. *Blood* **118:** 1774–1783.

65. Fuhlbrigge, R.C. *et al.* 2006. CD43 is a ligand for E-selectin on CLA+ human T cells. *Blood.* **107:** 1421–1426.

66. Matsumoto, M. *et al.* 2007. CD43 collaborates with P-selectin glycoprotein ligand-1 to mediate E-selectin-dependent T cell migration into inflamed skin. *J. Immunol.* **178:** 2499–2506.

67. Stamenkovic, I. *et al.* 1991. The B lymphocyte adhesion molecule CD22 interacts with leukocyte common antigen CD45RO on T cells and alpha 2-6 sialyltransferase, CD75, on B cells. *Cell.* **66:** 1133–1144.

68. Sgroi, D., G.A. Koretzky & I. Stamenkovic. 1995. Regulation of CD45 engagement by the B-cell receptor CD22. *Proc. Natl. Acad. Sci. U.S.A.* **92:** 4026–4030.

69. Xu, Z. & A. Weiss. 2002. Negative regulation of CD45 by differential homodimerization of the alternatively spliced isoforms. *Nat. Immunol.* **3:** 764–771.

70. Earl, L.A., S. Bi & L.G. Baum. 2010. *N*- and *O*-glycans modulate galectin-1 binding, CD45 signaling, and T cell death. *J. Biol. Chem.* **285:** 2232–2244.

71. Desai, D.M. *et al.* 1993. Ligand-mediated negative regulation of a chimeric transmembrane receptor tyrosine phosphatase. *Cell.* **73:** 541–554.

72. Majeti, R. *et al.* 1998. Dimerization-induced inhibition of receptor protein tyrosine phosphatase function through an inhibitory wedge. *Science.* **279:** 88–91.

73. Rachmilewitz, J. *et al.* 2003. Negative regulation of T cell activation by placental protein 14 is mediated by the tyrosine phosphatase receptor CD45. *J. Biol. Chem.* **278:** 14059–14065.

74. Ish-Shalom, E. *et al.* 2006. alpha2,6-Sialylation promotes binding of placental protein 14 via its Ca2+-dependent lectin activity: insights into differential effects on CD45RO and CD45RA T cells. *Glycobiology.* **16:** 173–183.

75. van Vliet, S.J. *et al.* 2006. Regulation of effector T cells by antigen-presenting cells via interaction of the C-type lectin MGL with CD45. *Nat. Immunol.* **7:** 1200–1208.

76. Thurman, E.C. *et al.* 1998. Regulation of in vitro and in vivo T cell activation by CD43. *Int. Immunol.* **10:** 691–701.

77. Mattioli, I. *et al.* 2004. Comparative analysis of T cell costimulation and CD43 activation reveals novel signaling pathways and target genes. *Blood.* **104:** 3302–3304.

78. Perillo, N. *et al.* 1995. Apoptosis of T cells mediated by galectin-1. *Nature.* **378:** 736–739.

79. Pace, K.E. *et al.* 1999. Restricted receptor segregation into membrane microdomains occurs on human T cells during apoptosis induced by galectin-1. *J. Immunol.* **163:** 3801–3811.

80. Stillman, B.N. *et al.* 2006. Galectin-3 and galectin-1 bind distinct cell surface glycoprotein receptors to induce T cell death. *J. Immunol.* **176:** 778–789.

81. Pace, K.E. *et al.* 2000. CD7 delivers a pro-apoptotic signal during galectin-1-induced T cell death. *J. Immunol.* **165:** 2331–2334.

82. Bi, S. *et al.* 2008. Structural features of galectin-9 and galectin-1 that determine distinct T cell death pathways. *J. Biol. Chem.* **283:** 12248–12258.

83. Nguyen, J.T. *et al.* 2001. CD45 modulates galectin-1-induced T cell death: regulation by expression of core 2 *O*-glycans. *J. Immunol.* **167:** 5697–5707.

84. Hernandez, J.D. *et al.* 2006. Galectin-1 binds different CD43 glycoforms to cluster CD43 and regulate T cell death. *J. Immunol.* **177:** 5328–5336.

85. Cabrera, P.V. *et al.* 2006. Haploinsufficiency of C2GnT-I glycosyltransferase renders T lymphoma cells resistant to cell death. *Blood.* **108:** 2399–2406.

86. Perillo, N.L. *et al.* 1997. Galectin-1, an endogenous lectin produced by thymic epithelial cells, induces apoptosis of human thymocytes. *J. Exp. Med.* **185:** 1851–1858.

87. Toscano, M.A. *et al.* 2007. Differential glycosylation of TH1, TH2 and TH-17 effector cells selectively regulates susceptibility to cell death. *Nat. Immunol.* **8:** 825–834.

Ann. N.Y. Acad. Sci. ISSN 0077-8923

ANNALS OF THE NEW YORK ACADEMY OF SCIENCES
Issue: *Glycobiology of the Immune Response*

Interplay between carbohydrate and lipid in recognition of glycolipid antigens by natural killer T cells

Bo Pei,[1] Jose Luis Vela,[1] Dirk Zajonc,[2] and Mitchell Kronenberg[1]

[1]Division of Developmental Immunology, La Jolla Institute for Allergy and Immunology, La Jolla, California. [2]Division of Cell Biology, La Jolla Institute for Allergy and Immunology, La Jolla, California

Address for correspondence: Mitchell Kronenberg, La Jolla Institute for Allergy and Immunology, 9420 Athena Circle, La Jolla, CA 92037. mitch@liai.org

Natural killer T (NKT) cells are a T cell subpopulation that were named originally based on coexpression of receptors found on natural killer (NK) cells, cells of the innate immune system, and by T lymphocytes. The maturation and activation of NKT cells requires presentation of glycolipid antigens by CD1d, a cell surface protein distantly related to the major histocompatibility complex (MHC)-encoded antigen presenting molecules. This specificity distinguishes NKT cells from most CD4[+] and CD8[+] T cells that recognize peptides presented by MHC class I and class II molecules. The rapid secretion of a large amount of both Th1 and Th2 cytokines by activated NKT cells endows them with the ability to play a vital role in the host immune defense against various microbial infections. In this review, we summarize progress on identifying the sources of microbe-derived glycolipid antigens recognized by NKT cells and the biochemical basis for their recognition.

Keywords: microbial infection; immune responses; glycolipid antigens; CD1; NKT cells

Introduction

Natural killer T (NKT) cells have fascinated immunologists because they have a number of unique properties. First, mouse and human NKT cells carry out an immediate, innate-like response that, in some circumstances, leads to the production of diverse cytokines, including IFN-γ, IL-4, and TNF-α.[1,2] Because of this vigorous cytokine production, NKT cells have been reported to influence many types of immune responses and disease processes, particularly in mouse models.[3–5] Second, most NKT cells in mice express an invariant T cell antigen receptor (TCR) α chain, and a homologous invariant α chain is found in many other mammalian species, including humans.[1,2,6] Because the TCR β chain is more variable, the NKT cell TCR is sometimes described as being "semi-invariant." Third, as noted, rather than recognizing peptides, NKT cells recognize glycolipids presented by a CD1d, a nonpolymorphic antigen-presenting molecule.[5,7,8] The trimolecular structures of the NKT T cell TCR bound to glycolipid–CD1d complexes have been solved, and these illustrate

the structure of the CD1d-bound antigen, with the hydrophobic lipid buried in the CD1d antigen-binding groove. Despite this, the buried lipid still contributes to antigen recognition by positioning the sugar above the CD1d surface for the TCR.

Mouse and human NKT cells recognize essentially the same structures: but why has this specificity been conserved over 80 million years of evolution? Glycolipid antigens have been found in several pathogens, and NKT cell responses to these infectious agents are highly protective.[5,9,10] In agreement with others,[11] we propose that the canonical, or invariant, NKT cell TCR acts as a kind of pattern recognition receptor that recognizes widely expressed microbial glycolipids with α-linked hexose sugars. Highly pathogenic bacteria with glycolipid antigens may therefore have evolutionarily conserved the specificity of human and mouse NKT cells.

The CD1 family of antigen presenting molecules

In the battle of the mammalian immune system against invading pathogens, antigen-presenting cells

doi: 10.1111/j.1749-6632.2011.06435.x

(APCs) must capture, process, and present antigens to T lymphocytes. The classical antigen-presenting molecules, the highly polymorphic major histocompatibility complex (MHC) class I and class II polypeptides, bind self or foreign peptides for presentation to CD8[+] or CD4[+] T cells, respectively.[12–14] A third family of antigen-presenting molecules, the CD1 family, was first identified as a thymocyte antigen in 1979, and designated with the name "CD1" in 1984.[15,16] Not highly polymorphic, the CD1 proteins present antigenic glycolipids to diverse T cell subpopulations, including CD4[+], CD8[+], and double-negative αβ T lymphocytes as well as δγ T cells.[6,17] The CD1 family consists of four members, CD1a-CD1d, that present glycolipids on the cell surface. The family is divided into two groups according to the sequence identities of their antigen-binding α1 and α2 domains. CD1a, CD1b, and CD1c constitute group 1; CD1d is the only group 2 member. Humans and most other mammals have all four CD1 proteins, while muridae rodents, including laboratory mice and rats, have only one CD1 molecule, an ortholog of group 2 CD1d.[5,7,8,18] CD1a, CD1b, CD1c, and CD1d are expressed predominantly by professional APCs, especially dendritic cells (DCs), macrophages, and marginal zone B (MZB) cells, as well as by thymocytes,[15,19–24] although expression of CD1d by other cell types, including hepatocytes and intestinal epithelial cells, has been reported. All CD1 molecules are type I integral membrane proteins, with a heavy chain consisting of extracellular α1 and α2 domains that bind hydrophobic antigens, and more membrane-proximal α3 domains that are immunoglobulin (Ig) super family homology unit. The CD1 proteins are highly glycosylated polypeptides, with the heavy chain noncovalently coupled to the light chain β2-microglobulin, also a member of the Ig super family.[25–28] CD1e, the fifth family member localized in lysosomes, facilities the processing and consequent loading of complex glycolipids onto other CD1 proteins.[29]

In acidic endosomal and lysosomal compartments CD1 molecules bind to glycolipid antigens. In some cases, the carbohydrate portions of these antigens are hydrolyzed to their antigenic form by specific glycosidases,[30] and this form of antigen processing and CD1 loading is facilitated by lipid transfer proteins such as CD1e and the saposins.[31–34] After arriving at the cell surface via exocytosis, CD1

molecules present antigens.[30,35] Subpopulations of CD1-reactive T lymphocytes have been intensively investigated during more than a decade of research. T lymphocytes that recognize CD1a, CD1b, and CD1c express TCRs with diverse combinations of α and β chains, and recent evidence suggests that some of these cells, such as those reactive to CD1a, are abundant in the peripheral blood of humans.[36] In addition, CD1-reactive T lymphocytes that express γδ TCRs also have been reported, particularly for recognition of CD1c.[37]

NKT cells are reactive to CD1d

CD1d-reactive T cells were defined as NKT cells because most cells of this population coexpress NK cell makers, such as NK1.1, with their TCR.[38] However, it was found later that expression of NK cell markers is not the exclusive characteristic of this T cell subset, and not all cells in this subset uniformly express NK receptors.[1,2,4,39,40] The discovery of α-galactosylceramide (αGalCer; Fig. 1), a synthetic glycolipid antigen that specifically stimulates CD1d-reactive T cells expressing the semi-invariant TCR, and the development of αGalCer-loaded CD1d tetramers that bind to the semi-invariant TCR greatly contributed to the unambiguous definition and identification of this T cell subset.[40,41] These cells now are most often referred to as invariant NKT cells (*i*NKT cells) or type I NKT cells, to distinguish them from CD1d-reactive T lymphocytes expressing diverse TCRs, or type II NKT cells.[1,2,6,42]

*i*NKT cells exhibit an antigen-experienced, or memory, phenotype. They are further activated when their TCRs are stimulated by glycolipid antigens presented by CD1d, by the synergistic stimulation by glycolipid antigens and cytokines, particularly IL-12, or even by cytokines alone.[43–46] *i*NKT cells are striking because a single cell is able to secrete a large amount of Th1 and Th2 cytokines rapidly after stimulation, consequently activating and influencing various immune cells.[47–51] Therefore, although *i*NKT cells account for a relatively small portion of total T lymphocytes (<1%), they play a critical role in amplifying innate immune responses against infectious pathogens, and they provide a bridge linking innate immunity to acquired immunity.[3,4] In mouse models, *i*NKT cells are involved in surveillance for tumors, prevention of autoimmune diseases, and

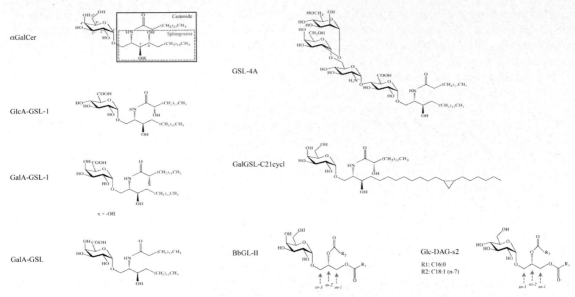

Figure 1. Structures of synthetic αGalCer and natural microbial lipid Ags. αGalCer has a galactosyl sugar head group and a ceramide (enclosed by a larger solid line box) lipid backbone containing a sphingosine base (enclosed in a dot line internal box), connected by a 1–1′ α linkage. The lipid moiety of *B. burgdorferi* Ag BbGL-II is a diacylglycerol, containing two fatty acid chains, R1 and R2, which are described in Table 1. GlcA-GSL-1, GalA-GSL-1, GSL-4A, and BbGL-II are the names given to purified antigens, which in some cases have heterogeneity.

clearance of viral and bacterial pathogens,[10,52–55] and they participate in various inflammatory reactions such as ischemia reperfusion injury, ozone-induced asthma, and formation of atherosclerotic plaques.[56–59] Furthermore, various human diseases have been correlated with reduced *i*NKT cell number or altered function.[60,61]

The remainder of this short review provides an overview on the sources of bacterial glycolipid antigens, their structure and presentation by CD1d, and the biochemical basis for their induction of *i*NKT cell activation.

Synthetic antigens for *i*NKT cells

In 1994, in a screen of marine sponge glycolipids for their antitumor activity, Natori *et al.* identified agelasphins, derived from the marine sponge *Agelas mauritianus*, as novel antitumor and immunostimulatory cerebrosides.[62] Later, some minor structural modifications on the agelasphins were performed that produced α-galactosylceramide (αGalCer), commercially designated as KRN7000 (Fig. 1). αGalCer possesses a ceramide backbone comprised of a C18 phytosphingosine with hydroxyl groups at the 3′ and 4′ carbon positions, and a C26 fatty acid chain. The ceramide lipid was conjugated to

a single galactose carbohydrate head group via an α-anomeric linkage (Fig. 1). In 1997, Kawano *et al.* identified αGalCer as an exceptionally potent ligand for *i*NKT cells. Surface plasmon resonance studies documented the unusually strong affinity of the semi-invariant TCR for αGalCer/CD1d complexes, measured in the 11–350 nM range.[63–68] A striking characteristic of αGalCer is the α-anomeric linkage between the ceramide moiety and the carbohydrate head group, given that mammalian cells only synthesize β-linked glycolipids. The α-linkage is critical for the potent antigenic activity of αGalCer, as the β-anomeric isomer (βGalCer) does not stimulate *i*NKT cells, or stimulates them only weakly.[69]

A number of studies have probed structural variants of αGalCer in order to understand the biochemical basis for the extraordinarily high affinity of the semi-invariant TCR. Not surprisingly, the sugar protruding from the CD1d groove plays a major role in determining TCR affinity. Substitution of glucose for galactose reduced *i*NKT cell TCR affinity by approximately 10-fold,[65,68] and substitution to mannose completely abrogated the stimulation of *i*NKT cells.[41,65] On the galactose sugar, all modifications of the 2′ hydroxyl group of the sugar (Fig. 1) that were tested abrogated the antigenic activity of αGalCer, while modifications of the 3′ or 4′ hydroxyl groups were better tolerated, although they did reduce antigenic potency to some extent[70–72] and had effects on TCR affinity.[68] The 6′-position, by contrast, can accommodate a number of chemical modifications without a requirement for lysosomal antigen processing to remove the modifications needed for recognition by the semi-invariant TCR.[30,73]

In addition to the carbohydrate moiety, the lipid backbone buried in the CD1d antigen-binding groove also contributes to antigenic potency. Removal of both 3′- and 4′-hydroxyl groups on the phytosphingosine base resulted in the complete loss of activity for the compound,[74–78] while removal of the 4′-hydroxyl alone led to some loss of antigenic potency and reduced TCR affinity.[65,68,74] Truncation of the lipid chains, either the phytosphingosine base or the fatty acid, also reduced the ability of the compound to stimulate *i*NKT cells.[75]

Although the α-linkage of the sugar is required for a high degree of antigenic potency, several recent reports have shown that ceramide containing antigens with β-linked sugars can be recognized as well, although they are weaker antigens. For βGalCer and isoglobotrihexosylceramide, structural studies show that TCR binding apparently "squashes" the sugar down so it is in close to an α conformation more parallel to the top of the CD1d molecule.[79,80] It is surprising, however, that recent studies reported that β-glucosylceramide is a potent self-antigen, and that β-mannosylceramide is a novel *i*NKT cell agonist that induces a potent antitumor response *in vivo* dominated by TNF secretion.[81,82] The structural basis for the recognition of these latter two β-linked glycosphingolipid (GSL) antigens by the semi-invariant TCR remains to be determined.

Glycosphingolipid antigens derived from *Sphingomonas* bacteria

The first chemically well-defined microbial antigens to activate *i*NKT cells were obtained from *Sphingomonas ssp.*, which are Gram-negative bacteria widely distributed in nature, especially in water and soil. In addition, certain *Sphingomonas* species may be commensal organisms in the intestines of mice and some humans,[83,84] and although these *Sphingomonas* usually are not human pathogens, they have been implicated in causing a number of nosocomial and community-acquired infections,

for example bacteremia, meningitis, and peritonitis.[83,85–87] It also was proposed that these bacteria could contribute to the cause of primary biliary cirrhosis, an autoimmune disease leading to small bile duct damage. According to studies, *Sphingomonas* infection can induce antibodies against the microbe's pyruvate dehydrogenase complex E2 enzyme that are cross-reactive with host mitochondrial counterparts, in addition to inducing chronic T cell–mediated autoimmunity.[83,88] In another mouse disease model, administration of a *Sphingomonas*-derived synthetic glycolipid antigen rapidly induced airway hyperreactivity by activating *i*NKT cells,[89] although there is no evidence yet that bacterial infection might lead to the same outcome.

Within the group of Gram-negative bacteria, *Sphingomonas* are different because they do not express lipopolysaccharide (LPS), and instead use GSLs to construct their cell walls. The main GSL components of these cell walls are heterogenous GSLs that include α-glucuronosyl-ceramide, α-galacturonosylceramide, and GSLs with more complex carbohydrates containing up to four saccharide units, in addition to variations in the ceramide lipid.[90–92] Mattner *et al.*, using both purified material from *Sphingomonas capsulata* and synthetic material based on this, and Kinjo *et al.*, taking a similar path with materials from *Sphingomonas paucimobilis* and *Sphingomonas yanoikuyae*, identified the main glycolipid antigens that stimulate *i*NKT cells.[93,94] Almost at the same time Sriram *et al.* found that crude glycolipid extracts from *S. paucimobilis* were able to activate several *i*NKT cell hybridomas.[95] The GSLs purified from these *Sphingomonas* strains have minimal variations in structure, with a glucuronic acid (GlcA) sugar in *S. paucimobilis* to form GlcA-GSL-1 and a mixture of glucuronic and galacturonic acids (GalA) in *S. yanoikuyae* to form GalA-GSL-1 (Fig. 1). All of these antigens, however, have an α-linkage to connect the carbohydrate to the ceramide.[93] CD1d tetramers loaded with the *Sphingomonas* compounds identified the *i*NKT cell population in wild-type splenocytes or liver mononuclear cells, but *i*NKT cells from Jα18[−/−] mice were not identified (i.e., stained) because these mice do not express the invariant TCRα chain due to a deficiency of the required Jα segment. Furthermore, CD1d-deficient or Jα18[−/−] mice had a significantly higher burden of bacteria at early times after

Sphingomonas infection, pointing to a defect in bacterial clearance due to the absence of *i*NKT cells.[93,94]

Although, monosaccharide-containing GSLs are the main glycolipid components of the cell walls of *Sphingomonas* species, they are not the only ones. Not only do different strains express distinct GSL profiles according to the environmental conditions, these bacteria are also able to synthesize various GSLs with mono- or oligosaccharides linked to ceramides consisting of different sphingosine bases.[90–92] *Sphingomonas* GSLs with more complex carbohydrate head groups include trisaccharide and tetrasaccharide glycosylceramides, originally designated "GSL-3" and "GSL-4," respectively (Fig. 1). Both of these compounds contain glucuronic acid linked to the lipid. The antigenic activity of the GSL-4 variants for *i*NKT cells was found to be either very weak or undetectable, probably because mammalian APCs do not have glycosidases that can degrade the carbohydrate head groups to a monosaccharide form containing glucuronic acid.[96,97] However, it is notable that a variant designated "GSL-4A" (Fig. 1), composed of a tandem tetrasaccharide of α-mannose(1–2)α-galactose(1–6)α-glucosamine(1–4)α-glucuronosyl linked to a ceramide with an α-anomeric conformation, not only stimulated *i*NKT cells—though to a lesser extent than a synthetic *Sphingomonas*-based antigen GlcA–GSL—but it did so without the requirement of lysosomal processing to trim the relatively complex oligosaccharide head group.

Besides variations in the sugar, the sphingosine chains of *Sphingomonas* bacterial GSLs also vary in the length and unsaturation, including a C18:0, C20:1, and a cyclopropyl-C21:0.[90,92,98,99] A synthetic GSL with a single galactosyl group and a cyclopropyl-C21:0 sphingosine chain, designated "GalGSL-C21cycl" (Fig. 1), showed antigenic activity that was intermediate between GalA–GSL and αGalCer, which further confirms the conclusion that the hydrophobic glycolipid backbone also contributes to the ability of glycolipid antigens to activate *i*NKT cells.

The synthetic GalA–GSL compound (Fig. 1) was analyzed in an X-ray crystallographic structural study for its interactions with CD1d alone and as part of a trimolecular complex with CD1d and the semi-invariant TCR. One important difference distinguishing GalA–GSL from αGalCer is the sugar moiety, as the 6′-OH in galactose is substituted with

a carboxylate in GalA–GSL. This compound also has a much shorter fatty acyl chain composed of only 14 carbons, versus 26 carbons for αGalCer, and the GalA–GSL sphingosine lacks the 4′-OH of the phytosphingosine of αGalCer. The hydrophobic CD1d binding groove has two large pockets, designated as A′ and F′.[100] Analysis of the crystal structure of the GalA–GSL–CD1d complex showed that similar to the αGalCer complex, the fatty acyl chain of the antigen inserted into the A′ pocket of CD1d, with the sphingosine chain in the F′ pocket and the sugar head group protruded from the center of the binding groove at the CD1d surface.[101] However, lack of the 4′-OH led to a lower orientation of sphingosine in the pocket in order to make optimal hydrogen bonds with Asp at CD1d position 80, which resulted in GalA–GSL more deeply inserted into the CD1d groove. Moreover, the hydrogen bond between Arg-79 and the 3′-OH in the αGalCer-CD1d complex was lost, and therefore Arg-79 orients differently and may interfere partially with the interaction between CD1d and the TCR. A shortened fatty acyl chain was not able to fully occupy the A′ pocket, and a so-called *spacer lipid*, a fatty acid with 16 carbons, also bound in this pocket and presumably stabilized the GalA–GSL–CD1d complex. Overall, these structural alternations caused the galacturonosyl group of GalA–GSL to shift laterally, compared with the galactosyl group of αGalCer, in the complex with CD1d.[101] There were no hydrogen bonds linking the 6′-COOH in the galacturonic head group to CD1d, consistent with the ability of this sugar position to tolerate variation in chemical structure.[101]

Glycophospholipid antigens derived from *Borrelia* species

Borrelia spp. are Gram-negative spirochetes that do not express LPS in their cell walls. Twelve of the thirty-six known species are able to cause Lyme disease, also called *borreliosis*—a zoonotic, vector-borne illness transmitted by the bite of infected ticks. According to a report from the United States Centers for Disease Control and Prevention, the rate of this disease is 7.9 cases out of 100,000 persons and the number of reported cases has been increasing, making it the most common vector-borne disease in the United States. In North America, *B. burgdorferi* is the predominant one of the three major *Borrelia* species that cause Lyme disease. Several *Borrelia* strains, including *B. hermsii, B. parkeri,* and

B. recurrentis, can cause relapsing fever, often together with severe bacteremia.

CD1d deficiency rendered mice that are highly resistant to *Borrelia* susceptible to arthritis, a principal disease manifestation in the mouse model of Lyme disease,[102] suggesting a role for *i*NKT cells in host protection. Compared with wild-type mice, after infection CD1d$^{-/-}$ mice had a dramatically increased level of spirochete DNA in tissues and a higher titer of *Borrelia*-specific IgG2a in serum; the increased IgG2a might reflect an increased bacterial load in the absence of *i*NKT cells. A subsequent study analyzed the role of splenic MZB cells that express the highest amount of CD1d[103] on the clearance of *Borrelia* bacteria. The study found that MZB cells secreted *Borrelia*-specific IgM and helped to reduce bacterial load in wild-type mice by 24 hours post-infection, whereas by 96 hours the MZB cells in CD1d knockout mice produced only minimal pathogen-specific IgM and the bacterial burden remained elevated.[104] CD1d has multiple functions, however, as it activates both type II and type I NKT cells, and thus experiments looking at CD1d knockout mice may suffer from confounding effects. Tupin *et al.* therefore analyzed Jα18$^{-/-}$ mice to demonstrate the involvement of *i*NKT cells in the clearance of *Borrelia* bacteria more directly. They showed increased joint inflammation and effects on spirochete clearance in BALB/c Jα18$^{-/-}$ mice.[105] They also demonstrated *i*NKT cell activation *in vivo*, including the secretion of cytokines, at one week after infection. In a subsequent study, bacterial load in joints was increased 25-fold at day three in the absence of *i*NKT cells, and following infection, alteration in *i*NKT cell migration in the liver sinusoids and interactions with Kuppfer cells (liver macrophages) were visualized by intravital microscopy.[106] In C57BL/6 mice, however, there was increased carditis more prominently than arthritis in the absence of *i*NKT cells, as well as evidence for a local accumulation of *i*NKT cells in the heart and increased local IFN-γ.[107]

Analytical chemistry was required in order to determine if *B. burgdorferi* has a glycolipid antigen that can activate *i*NKT cells. There are two major glycolipids in this pathogen. One of these, *B. burgdorferi* glycolipid 1 (BbGL-I), is a cholesteryl 6-O-acyl-β-galactoside, accounting for about 23% of the total glycolipids. A second one, BbGL-II (Fig. 1 and Table 1), is a 1,2-diacyl-3-O-α-galactosyl-*sn*-

Table 1. Fatty acid components of synthetic BbGL-II compounds

BbGL-II compounds	R1	R2
BbGL-IIa	C16:0	C18:1
BbGL-IIb	C16:0	C18:2
BbGL-IIc	C18:1	C16:0
BbGL-IId	C18:1	C18:2
BbGL-IIe	C18:2	C16:0
BbGL-IIf	C18:2	C18:1
BbGL-IIg	C16:0	C18:0
BbGL-IIh	C16:0	C16:0

R1 and R2 represent *sn*-1 and *sn*-2 positions of glycerol, respectively (Fig. 1). The length of fatty acyl chains and the number of unsaturated bonds are indicated.

glycerol, accounting for approximately 12% of the total glycolipids.[108] When plated on the microwells precoated with purified mouse CD1d protein, the diacylglycerol (DAG)-containing BbGL-II, but not BbGL-I, was able to stimulate *i*NKT cell hybridomas to secrete IL-2, indicative of TCR engagement. Natural BbGL-II is a mixture of glyco-DAG compounds with C14:0, C16:0, C18:0, C18:1, and C18:2 fatty acid chains, with C16:0 and C18:1 as the most abundant ones (Table 1). Based on this, eight BbGL-II compounds comprised of different combination of C16:0, C18:1, and C18:2 fatty acids were synthesized and their antigenic activity was determined. A compound designated "BbGL-IIc," with a C18:1 oleic acid in the *sn*–1 position and C16:0 in the *sn*-2 position of the glycerol backbone, was the optimal one for stimulating mouse *i*NKT cells. BbGL-IIf, by contrast, with the C18:1 fatty acid in the *sn*-2 position, could bind to mouse CD1d but was not antigenic. Structural analysis showed that the two compounds bound to CD1d differently. BbGL-IIc could be accommodated by mouse CD1d with the kinked *sn*-1 oleoyl chain, caused by a *cis* double bond at the nine position, inserted into A′ pocket, while the saturated C16:0 palmitoyl chain resided in the F′ pocket (Fig. 2). This permitted the formation of polar interactions between the galactosyl head group of BbGL-IIc and the amino acids Asp153 and Thr156 in the CD1d α2-helix, similar to that of αGalCer. However, different from αGalCer, there were no hydrogen bonds between Arg79 and Asp80 with BbGL-IIc because of the different structures of the DAG and ceramide backbones.[109] In the case of

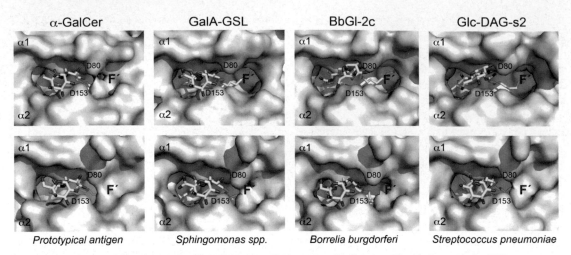

α-GalCer	GalA-GSL	BbGl-2c	Glc-DAG-s2
Prototypical antigen	Sphingomonas spp.	Borrelia burgdorferi	Streptococcus pneumoniae

Figure 2. Comparison of αGalCer, GalA-GSL, BbGL-IIc, and Glc-DAG-s2 binding to mCD1d before and after TCR engagement. (Top) Glycolipid ligand presentation shown in "TCR view" before TCR binding. Note how the galactose of BbGL-IIc and the glucose of Glc-DAG-s2 lose intimate contacts with α2-helix of mouse CD1d and instead rotate counter-clockwise (∼60°) toward Asp80 (D80) of α1-helix, in contrast to αGalCer. (Bottom) Glycolipid bound in the TCR-containing ternary complexes shown in the same view as the top, but with the bound TCR removed needed for clarity. Note that the TCR forces both the galactose of BbGL-IIc as well as the glucose of Glc-DAG-s2 into a position similar to that of αGalCer and GalA-GSL before TCR engagement. This structural change induced upon TCR binding is necessary for the conserved TCR binding footprint on CD1d. Also note how the TCR is able to induce the formation of the F′ roof in all CD1d–glycolipid complexes that do not have a preformed F′ roof (e.g. αGalCer, top panel).

BbGL-IIf, the C18:1 oleic acid also was bound to the A′ pocket of CD1d, but since in BbGL-IIf this fatty acid is in the *sn*-2 rather than the *sn*-1 position, the galactose sugar protruding from the top of CD1d was oriented differently and in a suboptimal position compared to the BbGL-IIc–CD1d complex. This illustrates a more general property of the two major classes of microbial glycolipid antigens. Glycosphingolipids are bound tightly in a single orientation with the sphingoid base in the F′ pocket due to a network of hydrogen bonds that hold this category of glycolipids rigidly in place. The DAG containing antigens, by contrast, have reduced hydrogen binding and therefore can bind more flexibly in two orientations.[110]

Glycolipids from Gram-positive pathogens also engage the semi-invariant TCR

Streptococcus pneumoniae (*S. pneumoniae*, also known as pneumococcus) is an agent of pneumonia, bloodstream infections, and meningitis in children and the elderly. Neurologic sequelae and/or learning disabilities can occur in meningitis patients. Hearing impairment can result from recurrent otitis media, now estimated to cause 11% of all deaths in children from 1 month to 5 years of age.[111] Group B *Strepto-coccus* (GBS) is a cause of life-threatening bacterial infections such as sepsis and meningitis in human newborns. Both of these Gram-positive pathogens have antigens that activate *i*NKT cells. We focused on *S. pneumoniae* because mice lacking *i*NKT cells previously were shown to be highly sensitive to intratracheal infection with these bacteria.[112,113] The previous work showed that during an infection with *S. pneumoniae*, IFN-γ derived from liver mononuclear cells, likely *i*NKT cells, plays a role in clearance through stimulating increased production of TNF-α and MIP-2, which recruit neutrophils to the lung by 12 hours.[112] Subsequent research showed more directly that *i*NKT were activated in an antigen-dependent fashion nearly immediately following infection. *i*NKT cells in the lungs of mice infected with *S. pneumonia* expressed IFN-γ and IL-17 by 13 hours, and treatment with a CD1d-blocking antibody substantially reduced IFN-γ production and caused a higher bacterial burden in the lung. Furthermore, DCs taken directly from infected mice could activate *i*NKT cell hybridomas *in vitro* to secrete IL-2, indicating that the DC presented an antigen that could activate *i*NKT cells.

Biochemical evidence for a glycolipid antigen in *S. pneumoniae* was first obtained from *i*NKT cell hybridomas that were activated by CD1d-coated plates exposed to total bacterial sonicates in a cell-free antigen presentation assay. Subsequent analyses demonstrated that the major antigen in *S. pneumoniae* and GBS contained a glucose as opposed to a galactose sugar α-linked to the *sn-3* position of a DAG antigen with palmitic acid (C16:0) in *sn-1* and *cis*-vaccenic acid (C18:1, n-7) linked at *sn-2* (Fig. 2).[114] This single species, which we refer to as *S. pneumoniae* glucosylated diacylglycerol (SPN Glc-DAG), consisted of up to 43% of the total glycolipid of some *S. pneumoniae* strains. Analysis of variants created by chemical synthesis indicated that the uncommon *cis*-vaccenic acid was required solely at the *sn-2* position, a variant we refer to as SPN Glc-DAG-s2. Four other variants tested, including a glycolipid lacking *cis*-vaccenic acid, one with *cis*-vaccenic acid at the *sn-1* position, and one with *cis*-vaccenic acid at both glycerol positions, all failed to activate mouse or human *i*NKT cells.[114]

As noted earlier, GSL bacterial antigens were shown to be more potent with a galactose or galacturonic acid compared to glucose or glucuronic acid, the difference between the two types of sugars being the orientation of the hydroxyl in the 4′ position of the hexose sugar. A recent structural analysis illustrates why the glucose sugar is preferred with DAG antigens containing *cis*-vaccenic acid, providing a vivid illustration of the interplay between the sugar and the CD1d-buried lipid in forming an epitope for the semi-invariant TCR. The glucose in Glc-DAG-s2 leads to a loss of a hydrogen bond with Asn at position 30 of CDR1 of the TCR α chain compared to binding the galactose from BbGL-IIc, but once the TCR is bound, a new contact is made between the equatorial 4′ hydroxyl of the glucose with Gly155 on the α2 helix of mouse CD1d. This change does not lead to a large difference in the affinity of the semi-invariant TCR for glycolipid mouse CD1d complexes, which is 4.4 μM for SPN Glc-DAG-s2 and 6.2 μM for BbGL-IIc.[110,115] However, the glucose is important in the context of the Glc-DAG-s2 lipid because the axial hydroxyl at the 4′ position of galactose—meaning the hydroxyl is facing up from, rather than parallel to, the hexose ring—would not permit contact with position 155 of mouse CD1d. In fact, when galactose is combined in a synthetic antigen in α linkage with the *S. pneumoniae* DAG containing vaccenic acid, it does not stimulate *i*NKT cells.[114]

Interestingly, although the equilibrium binding constants are similar, when binding the *S. pneumoniae* antigen the TCR has both a slower association rate and a slower dissociation rate compared to binding to *Borrelia* antigen. This fits our model that the TCR first contacts the exposed sugar, with the ability to contact the sugar affecting primarily the association rate. Subsequent contacts with, and accommodation of, CD1d affect primarily the dissociation rate, with glucose interaction with position 155 proposed to be responsible for more stable binding to complexes of Glc-DAG-s2 by the semi-invariant TCR (compared to BbGL-IIc) once the trimolecular complex has formed. Also consistent with this hypothesis are the changes observed over the F′ roof of the CD1d groove. Figure 2 shows that in the absence of TCR engagement only αGalCer binding causes a closing of the roof over the F′ pocket. Even for the GalA–GSL antigen, this F′ roof is open, exposing part of the lipid tail in the groove. Although the association rates of the *i*NKT cell TCR for the two GSL antigens are similar, the TCR dissociation rate for GalA–GSL–CD1d complexes is faster. We attribute this faster rate to the opening of the F′ pocket and a possible flexing of the CD1d molecule even after the TCR is bound. Therefore, the ability of αGalCer to lock CD1d into the optimal conformation accounts in part for its extraordinary ability of αGalCer-CD1d complexes bind to the semi-invariant TCR.

Summary, conclusion, and perspectives

From the discovery of the first antigen in a marine sponge, which was most likely associated with a *Sphingomonas*-like microorganism that produced the GSL antigen originally isolated, much has been learned in recent years about the diverse microbial sources of the glycolipid antigens that activate *i*NKT cells. Although such antigens may not be found in all bacteria, they can be derived from either Gram-negative or Gram-positive organisms, as well as from pathogens, commensals, or environmental bacteria. They have an α-linked hexose sugar and primarily contain one of two types of lipids: GSLs or DAGs. Recently, however, a cholesterol-containing glycolipid that activates *i*NKT cells was obtained from *Helicobacter pylori*, the organism that causes stomach ulcers.[116] It is likely, therefore, that the diversity of antigens recognized by the semi-invariant TCR has not been exhausted.

Much has been learned as well about the biochemical basis for the specificity of the semi-invariant TCR, including the factors determining the high affinity reactivity for complexes of αGalCer bound to CD1d, how the binding of GSLs and DAG antigens to CD1d differs, and how the kinetic parameters of TCR binding are determined by the position of the sugar for the association rate, and the flexibility of CD1d for the dissociation rate. Although ultimately most of the critical features of antigen recognition by the semi-invariant TCR relate to the specific binding to monosaccharides displayed on the surface of the CD1d antigen-presenting protein, critically important as well is the role that the CD1d-bound lipid plays in positioning this sugar.

Years ago, Janeway proposed that pattern recognition receptors would allow cells of the innate immune system to recognize critical or conserved molecular features of microbes that are not present in the host.[117] Considering how abundant the microbial glycolipid antigens are, the likelihood they are critical for microbe survival, and the absence of α-linked GSL or DAG antigens in mammalian hosts, when bound to CD1d microbial glycolipids constitute a kind of microbial-associated molecular pattern that stimulates a subpopulation of T cells that carry out innate-like responses. The analogy to pattern recognition receptors may not be perfect, as self-GSL compounds are able to activate *i*NKT cells to some degree, although this requires a TCR-induced accommodation of the β-linked sugar to a position similar to the α-linked saccharides. However, this does not invalidate the comparison, as Toll-like receptors—the paradigm of innate cell pattern recognition receptors—are also known to bind to self as well as microbial components. It should be noted that *i*NKT cells also can be activated by cytokines, particularly IL-12,[44] or by self-antigens presented by CD1d plus IL-12. In a recent study, the authors found that the responses to bacteria that are known to have antigens, such *S. pneumoniae*, and those not known to have antigens, such as *E. coli*, are similar in intensity and the requirement for CD1d expression for IL-12.[118] From these data, the authors concluded that even for microbes expressing a foreign antigen, the responses to self-antigens are predominant. However, this conclusion remains speculative, in part because removal of either the

self antigen(s) or the bacterial antigens is not yet possible for technical reasons, making it difficult to weigh their relative importance for *i*NKT cell stimulation. In our view, it seems unlikely that a glycolipid antigen comprising more than 40% of the total glycolipid, as is the case for *S. pneumoniae*, does not contribute to *i*NKT cell activation, although this does not exclude a role for self-antigen.

In addition to further exploration of microbial antigens, future work likely will explore in depth the structural basis for the recognition of different glycolipids by human *i*NKT cells. Furthermore, as αGalCer was first developed as an anticancer agent, research undoubtedly will focus on the design of antigens that might be even more effective in clinical settings, such as in the treatment of cancer.

Acknowledgments

Supported by NIH Grants AI45053, AI71922, AI69296, and AI74952.

Conflicts of interest

The authors declare no conflicts of interest.

References

1. Kronenberg, M. 2005. Toward an understanding of NKT cell biology: progress and paradoxes. *Annu. Rev. Immunol.* **23:** 877–900.
2. Bendelac, A., P.B. Savage & L. Teyton. 2007. The biology of NKT cells. *Annu. Rev. Immunol.* **25:** 297–336.
3. Parekh, V.V., M.T. Wilson & L. Van Kaer. 2005. iNKT-cell responses to glycolipids. *Crit. Rev. Immunol.* **25:** 183–213.
4. Matsuda, J.L. *et al.* 2008. CD1d-restricted iNKT cells, the 'Swiss-Army knife' of the immune system. *Curr. Opin. Immunol.* **20:** 358–368.
5. Cohen, N.R., S. Garg & M.B. Brenner. 2009. Antigen presentation by CD1 Lipids, T Cells, and NKT Cells in microbial immunity. *Adv. Immunol.* **102:** 1–94.
6. Godfrey, D.I. *et al.* 2004. NKT cells: what's in a name? *Nat Rev Immunol.* **4:** 231–237.
7. Brigl, M. & M.B. Brenner. 2004. CD1: antigen presentation and T cell function. *Annu. Rev. Immunol.* **22:** 817–890.
8. Silk, J.D. *et al.* 2008. Structural and functional aspects of lipid binding by CD1 molecules. *Annu. Rev. Cell Dev. Biol.* **24:** 369–395.
9. Tupin, E., Y. Kinjo & M. Kronenberg. 2007. The unique role of natural killer T cells in the response to microorganisms. *Nat. Rev. Microbiol.* **5:** 405–417.
10. Behar, S.M. & S.A. Porcelli. 2007. CD1-restricted T cells in host defense to infectious diseases. *Curr. Top Microbiol. Immunol.* **314:** 215–250.
11. Scott-Browne, J.P. *et al.* 2007. Germline-encoded recognition of diverse glycolipids by natural killer T cells. *Nat. Immunol.* **8:** 1105–1113.

12. Germain, R.N. & D.H. Margulies. 1993. The biochemistry and cell biology of antigen processing and presentation. *Annu. Rev. Immunol.* **11:** 403–450.

13. Cresswell, P. 1994. Assembly, transport, and function of MHC class II molecules. *Annu. Rev. Immunol.* **12:** 259–293.

14. Heemels, M.T. & H. Ploegh. 1995. Generation, translocation, and presentation of MHC class I-restricted peptides. *Annu. Rev. Biochem.* **64:** 463–491.

15. McMichael, A.J. *et al.* 1979. A human thymocyte antigen defined by a hybrid myeloma monoclonal antibody. *Eur. J. Immunol.* **9:** 205–210.

16. Bernard, A. & L. Boumsell. 1984. The clusters of differentiation (CD) defined by the First International Workshop on Human Leucocyte Differentiation Antigens. *Hum. Immunol.* **11:** 1–10.

17. Kasmar, A., I. Van Rhijn & D.B. Moody. 2009. The evolved functions of CD1 during infection. *Curr. Opin. Immunol.* **21:** 397–403.

18. Calabi, F. & C. Milstein. 2000. The molecular biology of CD1. *Semin. Immunol.* **12:** 503–509.

19. Delia, D. *et al.* 1988. CD1c but neither CD1a nor CD1b molecules are expressed on normal, activated, and malignant human B cells: identification of a new B-cell subset. *Blood.* **72:** 241–247.

20. Smith, M.E., J.A. Thomas & W.F. Bodmer. 1988. CD1c antigens are present in normal and neoplastic B-cells. *J Pathol.* **156:** 169–177.

21. Res, P. *et al.* 1997. Downregulation of CD1 marks acquisition of functional maturation of human thymocytes and defines a control point in late stages of human T cell development. *J. Exp. Med.* **185:** 141–151.

22. Exley, M. *et al.* 2000. CD1d structure and regulation on human thymocytes, peripheral blood T cells, B cells and monocytes. *Immunology* **100:** 37–47.

23. Pena-Cruz, V. *et al.* 2003. Epidermal Langerhans cells efficiently mediate CD1a-dependent presentation of microbial lipid antigens to T cells. *J. Invest. Dermatol.* **121:** 517–521.

24. Dougan, S.K., A. Kaser & R.S. Blumberg. 2007. CD1 expression on antigen-presenting cells. *Curr. Top Microbiol. Immunol.* **314:** 113–141.

25. Knowles, R.W. & W.F. Bodmer. 1982. A monoclonal antibody recognizing a human thymus leukemia-like antigen associated with beta 2-microglobulin. *Eur. J. Immunol.* **12:** 676–681.

26. Martin, L.H. *et al.* 1987. Structure and expression of the human thymocyte antigens CD1a, CD1b, and CD1c. *Proc. Natl. Acad. Sci. U. S. A.* **84:** 9189–9193.

27. Longley, J. *et al.* 1989. Molecular cloning of CD1a (T6), a human epidermal dendritic cell marker related to class I MHC molecules. *J. Invest. Dermatol.* **92:** 628–631.

28. Bilsland, C.A. & C. Milstein. 1991. The identification of the beta 2-microglobulin binding antigen encoded by the human CD1D gene. *Eur. J. Immunol.* **21:** 71–78.

29. de la Salle, H. *et al.* 2005. Assistance of microbial glycolipid antigen processing by CD1e. *Science* **310:** 1321–1324.

30. Prigozy, T.I. *et al.* 2001. Glycolipid antigen processing for presentation by CD1d molecules. *Science* **291:** 664–667.

31. Kang, S.J. & P. Cresswell. 2004. Saposins facilitate CD1d-restricted presentation of an exogenous lipid antigen to T cells. *Nat. Immunol.* **5:** 175–181.

32. Winau, F. *et al.* 2004. Saposin C is required for lipid presentation by human CD1b. *Nat. Immunol.* **5:** 169–174.

33. Zhou, D. *et al.* 2004. Editing of CD1d-bound lipid antigens by endosomal lipid transfer proteins. *Science* **303:** 523–527.

34. Yuan, W. *et al.* 2007. Saposin B is the dominant saposin that facilitates lipid binding to human CD1d molecules. *Proc. Natl. Acad. Sci. U. S. A.* **104:** 5551–5556.

35. Chiu, Y.H. *et al.* 2002. Multiple defects in antigen presentation and T cell development by mice expressing cytoplasmic tail-truncated CD1d. *Nat. Immunol.* **3:** 55–60.

36. de Jong, A. *et al.* 2010. CD1a-autoreactive T cells are a normal component of the human alphabeta T cell repertoire. *Nat. Immunol.* **11:** 1102–1109.

37. Spada, F.M. *et al.* 2000. Self-recognition of CD1 by gamma/delta T cells: implications for innate immunity. *J. Exp. Med.* **191:** 937–948.

38. Bendelac, A. 1995. CD1: presenting unusual antigens to unusual T lymphocytes. *Science* **269:** 185–186.

39. Benlagha, K. *et al.* 2000. In vivo identification of glycolipid antigen-specific T cells using fluorescent CD1d tetramers. *J. Exp. Med.* **191:** 1895–1903.

40. Matsuda, J.L. *et al.* 2000. Tracking the response of natural killer T cells to a glycolipid antigen using CD1d tetramers. *J Exp. Med.* **192:** 741–754.

41. Kawano, T. *et al.* 1997. CD1d-restricted and TCR-mediated activation of valpha14 NKT cells by glycosylceramides. *Science* **278:** 1626–1629.

42. Godfrey, D.I. & S.P. Berzins. 2007. Control points in NKT-cell development. *Nat. Rev. Immunol.* **7:** 505–518.

43. Brigl, M. *et al.* 2003. Mechanism of CD1d-restricted natural killer T cell activation during microbial infection. *Nat. Immunol.* **4:** 1230–1237.

44. Nagarajan, N.A. & M. Kronenberg. 2007. Invariant NKT cells amplify the innate immune response to lipopolysaccharide. *J. Immunol.* **178:** 2706–2713.

45. Sada-Ovalle, I. *et al.* 2008. Innate invariant NKT cells recognize Mycobacterium tuberculosis-infected macrophages, produce interferon-gamma, and kill intracellular bacteria. *PLoS Pathog.* **4:** e1000239.

46. Tyznik, A.J. *et al.* 2008. Cutting edge: the mechanism of invariant NKT cell responses to viral danger signals. *J. Immunol.* **181:** 4452–4456.

47. Kitamura, H. *et al.* 1999. The natural killer T (NKT) cell ligand alpha-galactosylceramide demonstrates its immunopotentiating effect by inducing interleukin (IL)-12 production by dendritic cells and IL-12 receptor expression on NKT cells. *J. Exp. Med.* **189:** 1121–1128.

48. Vincent, M.S. *et al.* 2002. CD1-dependent dendritic cell instruction. *Nat. Immunol.* **3:** 1163–1168.

49. Galli, G. *et al.* 2003. CD1d-restricted help to B cells by human invariant natural killer T lymphocytes. *J. Exp. Med.* **197:** 1051–1057.

50. Leadbetter, E.A. *et al.* 2008. NK T cells provide lipid antigen-specific cognate help for B cells. *Proc. Natl. Acad. Sci. U. S. A.* **105:** 8339–8344.

51. Liu, T.Y. *et al.* 2008. Distinct subsets of human invariant NKT cells differentially regulate T helper responses via dendritic cells. *Eur. J. Immunol.* **38:** 1012–1023.

52. Cui, J. *et al.* 1997. Requirement for Valpha14 NKT cells in IL-12-mediated rejection of tumors. *Science* **278:** 1623–1626.

53. Hong, S. *et al.* 2001. The natural killer T-cell ligand alpha-galactosylceramide prevents autoimmune diabetes in non-obese diabetic mice. *Nat. Med.* **7:** 1052–1056.

54. van der Vliet, H.J. *et al.* 2004. The immunoregulatory role of CD1d-restricted natural killer T cells in disease. *Clin. Immunol.* **112:** 8–23.

55. Crowe, N.Y. *et al.* 2005. Differential antitumor immunity mediated by NKT cell subsets in vivo. *J. Exp. Med.* **202:** 1279–1288.

56. Lappas, C.M. *et al.* 2006. Adenosine A2A receptor activation reduces hepatic ischemia reperfusion injury by inhibiting CD1d-dependent NKT cell activation. *J. Exp. Med.* **203:** 2639–2648.

57. VanderLaan, P.A. *et al.* 2007. Characterization of the natural killer T-cell response in an adoptive transfer model of atherosclerosis. *Am. J. Pathol.* **170:** 1100–1107.

58. Pichavant, M. *et al.* 2008. Ozone exposure in a mouse model induces airway hyperreactivity that requires the presence of natural killer T cells and IL-17. *J. Exper. Med.* **205:** 385–393.

59. Getz, G.S., P.A. Vanderlaan & C.A. Reardon. 2011. Natural killer T cells in lipoprotein metabolism and atherosclerosis. *Thromb. Haemost.* **106:** 814–819.

60. Kenna, T. *et al.* 2003. NKT cells from normal and tumor-bearing human livers are phenotypically and functionally distinct from murine NKT cells. *J. Immunol.* **171:** 1775–1779.

61. Molling, J.W. *et al.* 2005. Peripheral blood IFN-gamma-secreting Valpha24+Vbeta11+ NKT cell numbers are decreased in cancer patients independent of tumor type or tumor load. *Int. J. Cancer.* **116:** 87–93.

62. Natori, T. *et al.* 1994. Agelasphins, novel antitumor and immunostimulatory cerebrosides from the marine sponge Agelas mauritianus. *Tetrahedron.* **50:** 2771–2784.

63. Sidobre, S. *et al.* 2002. The V alpha 14 NKT cell TCR exhibits high-affinity binding to a glycolipid/CD1d complex. *J. Immunol.* **169:** 1340–1348.

64. Cantu, C., 3rd *et al.* 2003. The paradox of immune molecular recognition of alpha-galactosylceramide: low affinity, low specificity for CD1d, high affinity for alpha beta TCRs. *J. Immunol.* **170:** 4673–4682.

65. Sidobre, S. *et al.* 2004. The T cell antigen receptor expressed by Valpha14i NKT cells has a unique mode of glycosphingolipid antigen recognition. *Proc. Natl. Acad. Sci. U. S. A.* **101:** 12254–12259.

66. Zajonc, D.M. *et al.* 2008. Crystal structures of mouse CD1d-iGb3 complex and its cognate Valpha14 T cell receptor suggest a model for dual recognition of foreign and self glycolipids. *J. Mol. Biol.* **377:** 1104–1116.

67. Patel, O. *et al.* 2011. NKT TCR Recognition of CD1d-{alpha}-C-Galactosylceramide. *J. Immunol.* **187:** 4705–4713.

68. Wun, K.S. *et al.* 2011. A molecular basis for the exquisite CD1d-restricted antigen specificity and functional responses of natural killer T cells. *Immunity* **34:** 327–339.

69. Ortaldo, J.R. *et al.* 2004. Dissociation of NKT stimulation, cytokine induction, and NK activation in vivo by the use of distinct TCR-binding ceramides. *J. Immunol.* **172:** 943–953.

70. Xing, G.W. *et al.* 2005. Synthesis and human NKT cell stimulating properties of 3-O-sulfo-alpha/beta-galactosylceramides. *Bioorg. Med. Chem.* **13:** 2907–2916.

71. Wu, D. *et al.* 2005. Bacterial glycolipids and analogs as antigens for CD1d-restricted NKT cells. *Proc. Natl. Acad. Sci. U. S. A.* **102:** 1351–1356.

72. Raju, R. *et al.* 2009. Synthesis and evaluation of 3''- and 4''-deoxy and -fluoro analogs of the immunostimulatory glycolipid, KRN7000. *Bioorg. Med. Chem. Lett.* **19:** 4122–4125.

73. Zhou, D. *et al.* 2004. Lysosomal glycosphingolipid recognition by NKT cells. *Science* **306:** 1786–1789.

74. Morita, M. *et al.* 1995. Structure-activity relationship of alpha-galactosylceramides against B16-bearing mice. *J. Med. Chem.* **38:** 2176–2187.

75. Brossay, L. *et al.* 1998. Structural requirements for galactosylceramide recognition by CD1-restricted NK T cells. *J. Immunol.* **161:** 5124–5128.

76. Sakai, T. *et al.* 1999. Syntheses of biotinylated alpha-galactosylceramides and their effects on the immune system and CD1 molecules. *J. Med. Chem.* **42:** 1836–1841.

77. Miyamoto, K., S. Miyake & T. Yamamura. 2001. A synthetic glycolipid prevents autoimmune encephalomyelitis by inducing TH2 bias of natural killer T cells. *Nature* **413:** 531–534.

78. Ndonye, R.M. *et al.* 2005. Synthesis and evaluation of sphinganine analogues of KRN7000 and OCH. *J. Org. Chem.* **70:** 10260–10270.

79. Pellicci, D.G. *et al.* 2011. Recognition of beta-linked self glycolipids mediated by natural killer T cell antigen receptors. *Nat. Immunol.* **12:** 827–833.

80. Yu, E.D. *et al.* 2011. Cutting Edge: structural basis for the recognition of beta-linked glycolipid antigens by invariant NKT cells. *J. Immunol.* **187:** 2079–2083.

81. O'Konek, J.J. *et al.* 2011. Mouse and human iNKT cell agonist beta-mannosylceramide reveals a distinct mechanism of tumor immunity. *J. Clin. Invest.* **121:** 683–694.

82. Brennan, P.J. *et al.* 2011. Invariant natural killer T cells recognize lipid self antigen induced by microbial danger signals. *Nat. Immunol.* **12:** 1202–1211.

83. Selmi, C. *et al.* 2003. Patients with primary biliary cirrhosis react against a ubiquitous xenobiotic-metabolizing bacterium. *Hepatology* **38:** 1250–1257.

84. Wei, B. *et al.* 2010. Commensal microbiota and CD8+ T cells shape the formation of invariant NKT cells. *J. Immunol.* **184:** 1218–1226.

85. Hajiroussou, V. *et al.* 1979. Meningitis caused by Pseudomonas paucimobilis. *J. Clin. Pathol.* **32:** 953–955.

86. Southern, P.M., Jr. & A.E. Kutscher. 1981. Pseudomonas paucimobilis bacteremia. *J. Clin. Microbiol.* **13:** 1070–1073.

87. Glupczynski, Y. *et al.* 1984. Pseudomonas paucimobilis peritonitis in patients treated by peritoneal dialysis. *J. Clin. Microbiol.* **20:** 1225–1226.

88. Mattner, J. *et al.* 2008. Liver autoimmunity triggered by microbial activation of natural killer T cells. *Cell Host Microbe.* **3:** 304–315.

89. Meyer, E.H. *et al.* 2006. Glycolipid activation of invariant T cell receptor+ NK T cells is sufficient to induce airway hyperreactivity independent of conventional CD4+ T cells. *Proc. Natl. Acad. Sci. U. S. A.* **103:** 2782–2787.

90. Kawahara, K. *et al.* 2000. Structural analysis of two glycosphingolipids from the lipopolysaccharide-lacking bacterium Sphingomonas capsulata. *Eur. J. Biochem.* **267:** 1837–1846.

91. Kawahara, K. *et al.* 2001. Structural analysis of a new glycosphingolipid from the lipopolysaccharide-lacking bacterium Sphingomonas adhaesiva. *Carbohydr. Res.* **333:** 87–93.

92. Kawahara, K. *et al.* 2006. Confirmation of the anomeric structure of galacturonic acid in the galacturonosyl-ceramide of Sphingomonas yanoikuyae. *Microbiol Immunol.* **50:** 67–71.

93. Kinjo, Y. *et al.* 2005. Recognition of bacterial glycosphingolipids by natural killer T cells. *Nature* **434:** 520–525.

94. Mattner, J. *et al.* 2005. Exogenous and endogenous glycolipid antigens activate NKT cells during microbial infections. *Nature* **434:** 525–529.

95. Sriram, V. *et al.* 2005. Cell wall glycosphingolipids of Sphingomonas paucimobilis are CD1d-specific ligands for NKT cells. *Eur. J. Immunol.* **35:** 1692–1701.

96. Long, X. *et al.* 2007. Synthesis and evaluation of stimulatory properties of Sphingomonadaceae glycolipids. *Nat. Chem. Biol.* **3:** 559–564.

97. Kinjo, Y. *et al.* 2008. Natural Sphingomonas glycolipids vary greatly in their ability to activate natural killer T cells. *Chem. Biol.* **15:** 654–664.

98. Kawahara, K. *et al.* 1991. Chemical structure of glycosphingolipids isolated from Sphingomonas paucimobilis. *FEBS Lett.* **292:** 107–110.

99. Kawahara, K. *et al.* 2002. Occurrence of an alpha-galacturonosyl-ceramide in the dioxin-degrading bacterium Sphingomonas wittichii. *FEMS Microbiol. Lett.* **214:** 289–294.

100. Zeng, Z. *et al.* 1997. Crystal structure of mouse CD1: an MHC-like fold with a large hydrophobic binding groove. *Science* **277:** 339–345.

101. Wu, D. *et al.* 2006. Design of natural killer T cell activators: structure and function of a microbial glycosphingolipid bound to mouse CD1d. *Proc. Natl. Acad. Sci. U. S. A.* **103:** 3972–3977.

102. Kumar, H. *et al.* 2000. Cutting edge: CD1d deficiency impairs murine host defense against the spirochete, Borrelia burgdorferi. *J. Immunol.* **165:** 4797–4801.

103. Roark, J.H. *et al.* 1998. CD1.1 expression by mouse antigen-presenting cells and marginal zone B cells. *J. Immunol.* **160:** 3121–3127.

104. Belperron, A.A., C.M. Dailey & L.K. Bockenstedt. 2005. Infection-induced marginal zone B cell production of Borrelia hermsii-specific antibody is impaired in the absence of CD1d. *J. Immunol.* **174:** 5681–5686.

105. Tupin, E. *et al.* 2008. NKT cells prevent chronic joint inflammation after infection with Borrelia burgdorferi. *Proc. Natl. Acad. Sci. U. S. A.* **105:** 19863–19868.

106. Lee, W.Y. *et al.* 2010. An intravascular immune response to Borrelia burgdorferi involves Kupffer cells and iNKT cells. *Nat. Immunol.* **11:** 295–302.

107. Olson, C.M., Jr. *et al.* 2009. Local production of IFN-gamma by invariant NKT cells modulates acute Lyme carditis. *J. Immunol.* **182:** 3728–3734.

108. Ben-Menachem, G. *et al.* 2003. A newly discovered cholesteryl galactoside from Borrelia burgdorferi. *Proc. Natl. Acad. Sci. U. S. A.* **100:** 7913–7918.

109. Kinjo, Y. *et al.* 2006. Natural killer T cells recognize diacylglycerol antigens from pathogenic bacteria. *Nat. Immunol.* **7:** 978–986.

110. Wang, J. *et al.* 2010. Lipid binding orientation within CD1d affects recognition of Borrelia burgorferi antigens by NKT cells. *Proc. Natl. Acad. Sci. U. S. A.* **107:** 1535–1540.

111. O'Brien, K.L. *et al.* 2009. Burden of disease caused by Streptococcus pneumoniae in children younger than 5 years: global estimates. *Lancet.* **374:** 893–902.

112. Kawakami, K. *et al.* 2003. Critical role of Valpha14+ natural killer T cells in the innate phase of host protection against Streptococcus pneumoniae infection. *Eur. J. Immunol.* **33:** 3322–3330.

113. Nakamatsu, M. *et al.* 2007. Role of interferon-gamma in Valpha14+ natural killer T cell-mediated host defense against Streptococcus pneumoniae infection in murine lungs. *Microbes Infect.* **9:** 364–374.

114. Kinjo, Y. *et al.* 2011. Invariant natural killer T cells recognize glycolipids from pathogenic Gram-positive bacteria. *Nat. Immunol.* **12:** 966–974.

115. Girardi, E. *et al.* 2011. Unique interplay between sugar and lipid in determining the antigenic potency of bacterial antigens for NKT cells. *PLoS Biol.* **9:** e1001189.

116. Chang, Y.J. *et al.* 2011. Influenza infection in suckling mice expands an NKT cell subset that protects against airway hyperreactivity. *J. Clin. Invest.* **121:** 57–69.

117. Medzhitov, R. & C.A. Janeway, Jr. 1997. Innate immunity: the virtues of a nonclonal system of recognition. *Cell* **91:** 295–298.

118. Brigl, M. *et al.* 2011. Innate and cytokine-driven signals, rather than microbial antigens, dominate in natural killer T cell activation during microbial infection. *J. Exper. Med.* **208:** 1163–1177.

Ann. N.Y. Acad. Sci. ISSN 0077-8923

ANNALS OF THE NEW YORK ACADEMY OF SCIENCES
Issue: *Glycobiology of the Immune Response*

Galectins in acute and chronic inflammation

Fu-Tong Liu,[1,2] Ri-Yao Yang,[1] and Daniel K. Hsu[1,2]

[1]Department of Dermatology, University of California Davis, School of Medicine, Sacramento, California. [2]Institute of Biomedical Sciences, Academia Sinica, Taipei, Taiwan

Address for correspondence: Fu-Tong Liu, M.D., Ph.D., Department of Dermatology, University of California, Davis School of Medicine, 3301 C Street, Suite 1400, Sacramento, CA 95816. fliu@ucdavis.edu

Galectins are animal lectins that bind to β-galactosides, such as lactose and *N*-acetyllactosamine, in free form or contained in glycoproteins or glycolipids. They are located intracellularly or extracellularly. In the latter they exhibit bivalent or multivalent interactions with glycans on cell surfaces and induce various cellular responses, including production of cytokines and other inflammatory mediators, cell adhesion, migration, and apoptosis. Furthermore, they can form lattices with membrane glycoprotein receptors and modulate receptor properties. Intracellular galectins can participate in signaling pathways and alter biological responses, including apoptosis, cell differentiation, and cell motility. Current evidence indicates that galectins play important roles in acute and chronic inflammatory responses, as well as other diverse pathological processes. Galectin involvement in some processes *in vivo* has been discovered, or confirmed, through studies of genetically engineered mouse strains, each deficient in a given galectin. Current evidence also suggests that galectins may be therapeutic targets or employed as therapeutic agents for these inflammatory responses.

Keywords: galectins; inflammation; allergic inflammation; autoimmune disease; atherosclerosis

Introduction

Prototypic galectins (galectin-1, -2, -5, -7, -10, -11, -13, -14, -15) consist entirely of one CRD containing 130 amino acids. Tandem repeat galectins (galectin-4, -6, -8, -9, and -12) contain two homologous CRDs, which exhibit differential preferences for carbohydrate binding, separated by a linker of up to 70 amino acids in a single polypeptide chain. Galectin-3 is uniquely chimeric and contains a CRD preceded by a nonlectin region of about 120 residues consisting of short proline/glycine-rich tandem repeats that participate in protein oligomerization and may be involved in interaction with other intracellular proteins (Fig. 1A).[1] The presence of two carbohydrate-binding sites in a single polypeptide chain confers tandem-repeat galectins with bivalency at minimum, with regard to carbohydrate-binding activity.[2] Other galectins may be bivalent or multivalent due to oligomerization, which results in increased avidity of carbohydrate binding.[3] Dimerization, for example, is observed for galectin-1 and

galectin-2 in solution, while galectin-3 forms pentamers upon binding to multivalent carbohydrates.[4] Different galectin members can interact to form hetero-dimers or hetero-oligomers.[5]

Galectin genes do not encode for proteins with a classical signal sequence. Accordingly, the proteins are primarily localized in the cytoplasm and nucleus, although they are present extracellularly. Some galectins have been identified in exosomes, which are 50–90 nm membrane vesicles shed by many mammalian cell types (e.g., galectin-5, see Ref. 6), suggesting another mechanism by which galectins are released from cells. Galectins may shuttle between cytoplasm and nucleus depending on cell types and proliferation states.[7] Many galectins are distributed widely in tissues, but a few demonstrate high tissue specificity. For example, galectin-4 is predominantly expressed in the mouse gastrointestinal system; galectin-7 is expressed in human stratified epithelia; and galectin-12 is preferentially expressed by human and mouse adipose tissues (reviewed in Ref. 3).

doi: 10.1111/j.1749-6632.2011.06386.x

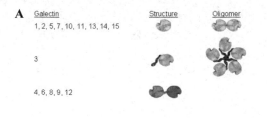

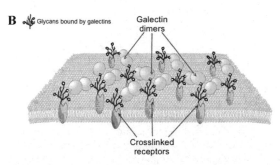

Figure 1. Structures of galectins. (A) Single-CRD galectins contain one domain, but may exist naturally as dimers, thus conferring bivalency. Galectin-3 is the sole member composed of a nonlectin N-terminal domain, through which the protein may oligomerize and form pentamers, upon binding to multivalent glycoproteins or glycolipids. Double-CRD galectins contain a short linker between CRDs, and are bivalent if each CRD retains full lectin activity. At present, it is uncertain if galectin-10 forms homodimers. (B) Formation of galectin-glycoprotein lattices on the cell surface. Multivalent galectins (dimerized one-CRD galectins are represented) are able to simultaneously ligate two glycans on neighboring glycoproteins. As each glycoprotein consists of multiple galectin ligands, an interconnected lattice is formed. As appropriate on each cell surface, this lattice could conceivably consist of heterogeneous glycoproteins. In addition, multiple galectins may participate in the formation of lattices.

Extracellular versus intracellular functions

Galectins are able to exert various *in vitro* activities by engaging cell surface glycans, as demonstrated by addition of recombinant proteins to cell lines or primary cells from humans or animals. The concept of galectin lattice formation was proposed to describe ordered arrays constituting galectin dimers or oligomers bound to selected glycans on cells (Fig. 1B).[8] The formation of such lattices can result in modulation of receptor responses and properties, such as lateral cell surface mobilities, rates of endocytosis, and signal transmission at the cell surface (reviewed in Refs. 9–11).

Galectins can regulate cellular responses by functioning intracellularly,[3] including participation in

some fundamental processes such as the pre-mRNA splicing in the nucleus—where galectins are components of spliceosomes[12]—and regulation of expression of certain genes.[13] Figure 2 summarizes some observed functions of galectins in intra- and extracellular compartments.

Galectin-3 and galectin-4 have been shown to control intracellular trafficking of glycoproteins.[14,15] In the case of galectin-3, this may be linked to its ability to translocate into the lumen of transport vesicles, such as post-Golgi vesicles. This implies that galectins recognize intracellular glycoproteins residing within these vesicles, though this has not been formally demonstrated.

Functions related to inflammation demonstrated *in vitro*

The roles of galectins in immunity and inflammation have been extensively reviewed.[3,16,17] The following sections summarize selected information, focusing primarily on more recent reports.

Effects of recombinant proteins on lymphoid cells

Galectin-1. The ability of recombinant galectin-1 to affect T cell homeostasis in humans and mice by inducing apoptosis has been extensively documented.[3,18,19] It should be mentioned that relatively high amounts (micromolar) of the protein are necessary to induce the response. However, lower levels may be efficacious when presented on a solid-phase matrix under appropriate conditions.[20] Galectin-1 induces Th2 polarization by selective killing of Th1- and Th17-differentiated cells, sparing Th2 cells due to the presence of extensive $\alpha2$–6 sialylation on cell-surface glycoproteins of these helper T cells.[21] Some investigators, however, contend that sensitivity of T cells to galectin-1 is conferred by the addition of reducing agents during apoptosis assays; for example, in their absence galectin-1 was able to bind to the cell surface but did not induce apoptosis in activated T cells,[22] although galectin-1 was still able to induce IL-10 production and attenuate IFN-γ production.[23] This provides an alternative mechanism of producing a Th2 phenotype. In addition, it has been shown that galectin-1 alters T cell trafficking by blocking adhesion to extracellular matrix, thus suppressing transendothelial migration.[24,25]

Galectin-1 is expressed in mouse regulatory T cells and may contribute to the inhibitory function

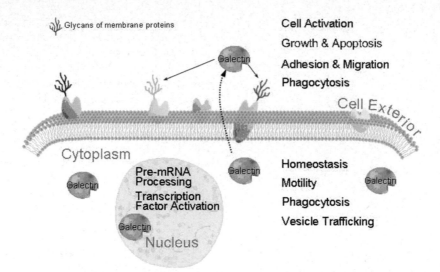

Figure 2. Overview of functions attributed to galectins in cell compartments. Galectins can function either extracellularly or intracellularly. This diagram lists examples of extracellular and intracellular functions. The majority of the extracellular actions of galectins can be attributed to lectin function mediated through ligation of glycosylated receptors. Many intracellular functions involve galectins' engagement in signaling pathways and do not appear to depend on lectin–carbohydrate interactions. Mechanisms by which galectins are secreted are yet to be defined.

of this cell type.[26–28] Galectin-1 can bind to pre-B cell receptor (BCR) in mouse and human B cells and participate in pre-BCR activation.[29] Galectin-1 can also regulate B cell proliferation and BCR-mediated signal transduction and promote immunoglobulin production during plasma cell differentiation.[30]

Galectin-2. Galectin-2 can also induce apoptosis in activated T cells,[31] including those residing in lamina propria of the intestines[32] and activated mouse CD8+ T cells.[33] Treatment with galectin-2 inhibited proinflammatory cytokine release from mucosal T cells and ameliorated acute and chronic experimental colitis.[32]

Galectin-3. Recombinant galectin-3 was shown to induce IL-2 production[34] and calcium influx[35] in Jurkat T cells and to induce apoptosis in activated T cells and T cell lines in a number of studies.[36,37] In MOLT-4 cells, galectin-3 was subsequently found to induce phosphatidylserine exposure—an early event in apoptosis—in the absence of cell death.[23] Galectin-3 induces apoptosis in both Th1 and Th2 cells, in contrast to galectin-1 (as mentioned above).[21]

T cells from mice deficient in the glycosyltransferase β1–6 *N*-acetylglucosaminyltransferase V (GlcNAc-TV or Mgat5) were found to have significantly decreased thresholds for T cell receptor

(TCR) activation induced by several stimuli, compared to wild-type cells. Since this enzyme regulates availability of suitable ligands for galectin-3 by generating branched glycans with *N*-acetyllactosamine groups, the authors suggested that one mechanism by which this phenotype could arise is through elevated lateral motility of the TCR.[38] They further proposed that formation of galectin-3–TCR lattices results in physical restriction and attenuation of TCR signal initiation. This phenomenon was reversed by treatment with lactose, which is capable of inhibiting galectin-3 binding to cell surfaces.[38]

Galectin-4. Some functional discrepancy exists for galectin-4, as galectin-4 was shown to stimulate IL-6 production by CD4+ T cells and to contribute to the development of inflammatory bowel disease;[39] yet, another study described the ability of galectin-4 to induce apoptosis in mucosal T cells, to reduce secretion of proinflammatory cytokines, including IL-17, and to promote resolution of inflammatory disease.[40]

Galectin-8. Both soluble and matrix-bound forms of galectin-8 were found to induce apoptosis in Jurkat T cells and in human peripheral mononuclear cells activated by anti-CD3 and anti-CD28 antibodies;[41] yet, galectin-8 has also been shown to induce proliferation of naive mouse CD4+ T cell and provide costimulatory signals.[42] In addition, galectin-8 can modulate the adhesive properties of T cells by specific interaction with α_4 integrins.[43]

Galectin-9. Galectin-9, like galectin-1, induces Th1 but not Th2 cell death through interaction with Tim-3, which is a cell-surface protein marker specifically expressed by Th1 cells.[44] However, galectin-9 has also been shown to induce production of pro-inflammatory cytokines from T helper cells independent of Tim-3.[45] More recently, it was shown that galectin-9 is expressed by mouse T regulatory cells and contributes to their suppressive activity.[46]

Effects of recombinant galectins on myeloid cells

Galectin-1. Recombinant galectin-1 was shown to impair chemotaxis of neutrophils and the interaction of these cells with endothelial cells.[47,48] Galectin-1 targets activated neutrophils for phagocytosis by causing exposure of phosphatidylserine, which can result in cell turnover independently of apoptosis.[22,49,50] In monocytes/macrophages, galectin-1 triggers alternative activation (M2 polarization) by inhibiting arachidonic acid release[51] and nitric oxide synthesis and increasing arginase activity.[52] It down-regulates IFN-γ–induced FcγRI-dependent phagocytosis and MHC class II expression[53] and IL-12 secretion from parasite-infected macrophages.[54] Galectin-1 also regulates the physiology of mouse antigen-presenting dendritic cells; it promotes the formation of CD43/CD45 coclusters on dendritic cells and induces cell activation and migration.[55] Galectin-1 also facilitates the maturation of IL-27–producing mouse tolerogenic DCs, promotes IL-10–mediated T cell tolerance, and suppresses autoimmune neuroinflammation.[56] Lung cancer cells produce galectin-1 that mediates IL-10 expression in monocytes and dendritic cells, inducing dendritic cell anergy and immune suppression.[57]

Galectin-3. Recombinant galectin-3 influences neutrophils, monocytes/macrophages, and dendritic cells through multiple mechanisms (reviewed in Ref. 16). In neutrophils, conflicting reports exist on the ability of galectin-3 to induce apoptosis; in one report, recombinant galectin-3 was protective of neutrophils from apoptosis,[57] but in the other it induced apoptosis.[58] In yet another study, galectin-3 induced phosphatidylserine expo-

sure in neutrophils in the absence of apoptosis.[23] Recombinant galectin-3 can induce mediator release from both IgE-sensitized and nonsensitized mast cells.[59,60] On the other hand, prolonged exposure (for example, 18–44 hours) of mast cells to recombinant galectin-3 results in apoptosis.[61]

Galectin-4. Like galectin-1, galectin-4 induces phosphatidylserine exposure in activated but not resting human neutrophils.[22]

Galectin-8. Recombinant galectin-8 bound to α_M integrin on neutrophils enhances adhesive properties and stimulates production of superoxide anion.[62] In the human promyelocytic cell line HL60, galectin-8 induced phosphatidylserine exposure but not apoptosis.[63]

Galectin-9. Human galectin-9 was originally identified as a potent eosinophil chemoattractant.[64] It was subsequently shown to significantly suppress apoptosis in eosinophils from eosinophilic patients but enhance apoptosis in those from healthy subjects; the mechanism for this differential response is unknown.[65] More recently, it was shown that galectin-9 in BAL fluid from patients with acute and chronic eosinophilic pneumonia demonstrated chemotactic properties on activated eosinophils, and thus might be responsible for lung eosinophilia in these patients.[66] Galectin-9 was found to suppress the production of TNF-α and IL-1β but enhance the production of IL-10 by macrophages stimulated with immune complexes (ICs).[67] In the same studies, the authors found that in galectin-9–deficient macrophages FcγRIIb expression was lower while FcγRIII expression was higher compared to wild-type macrophages. They concluded that galectin-9 functions by modulating FcγR expression in macrophages, though the underlying mechanism for this is not clear. Galectin-9 was found to suppress IgE-mediated degranulation of RBL-2H2 mast cells in a way that was dependent on its lectin function. This activity is related to its ability to disrupt antigen binding by interacting with IgE. Thus, galectin-9 secreted by activated MC/9 mast cells may in turn inhibit IgE-mediated activation of these cells.[68]

Endogenous functions of galectins

Galectin-1. Inhibitory effects of endogenous galectin-1 on the adhesion and trafficking of T cells and neutrophils were demonstrated by suppressing galectin-1 expression on endothelial cells. This resulted in a significant increase in the number of leukocytes exhibiting capture and rolling on a monolayer of endothelial cells.[48,69]

Galectin-3. Endogenous galectin-3 has been shown to have antiapoptotic activity in the Jurkat human T cell line.[70] In primary T cells, endogenous galectin-3 was found to be necessary for IL-2–dependent cell growth.[71] Overexpression of galectin-3 resulted in decreased apoptosis in B cell lymphoma cell lines,[72] but it is unknown if the endogenous protein also inhibits apoptosis in primary B cells. Suppression of galectin-3 expression in B cells caused differentiation into plasma cells rather than memory cells.[73]

Using macrophages from *Lgals3*[−/−] mice, galectin-3 was shown to be critical for phagocytosis of opsonized sheep red blood cells and apoptotic thymocytes through FcγR and phosphatidylserine receptors, respectively.[74] Macrophages from *Lgals3*[−/−] mice exhibited reduced IL-4/IL-13–induced alternative macrophage activation *in vitro* compared to macrophages from wild-type mice, but the two genotypes were comparable in IFN-γ/LPS-induced classical activation and IL-10–induced deactivation.[75] This defect was confirmed in wild-type cells treated with galectin-3–specific siRNA to suppress galectin-3 expression and was noted *in vivo* in *Lgals3*[−/−] mice. IL-4–induced alternative activation was blocked by the galectin-3 inhibitor bis-(3-deoxy-3-(3-methoxybenzamido)-beta-D-galactopyranosyl) sulfane.[75] On the other hand, another study reported that galectin-3 was positively associated with LPS-induced production of inflammatory cytokines IL-6, IL-12, and TNF-α; this was demonstrated with *Lgals3*[−/−] macrophages and with neutralizing antibody or the galectin-3 inhibitor lactose to inhibit galectin-3 in wild-type macrophages.[76] According to this study, galectin-3 also promotes classical macrophage activation. Additional studies are necessary to firmly establish the role of galectin-3 in classical macrophage activation versus alternative macrophage activation.

A suppressive role for galectin-3 in the production of IL-12 was demonstrated with the use of *Lgals3*[−/−] dendritic cells, which produced significantly larger amounts of IL-12 than wild-type cells when stimulated with LPS and IFN-γ.[77] Because IL-12 is a key factor that promotes Th1 responses, galectin-3 is capable of suppressing Th1 responses through

dendritic cells.[78] Galectin-3 also regulates the migratory pattern of dendritic cells, as demonstrated by comparing *Lgals3*[−/−] dendritic cells and their wild-type counterparts both *in vitro* and *in vivo*. This effect is via an intracellular mechanism because migration *in vitro* was unaffected by the presence of lactose, which would inhibit activity of extracellular galectins.[79]

Lgals3[−/−] mast cells exhibited impaired degranulation and decreased cytokine production compared to wild-type cells when activated by crosslinking IgE receptor.[84]

Galectin-3 was found to contribute to increased rolling and firm adhesion on immobilized VCAM-1 under conditions of flow via engagement of α_4 integrin on eosinophils.[80]

Galectin-9. Transfected galectin-9 was shown to transactivate inflammatory cytokine genes through direct physical interaction with the transcription factor NF-IL6 (C/EBPβ) in the monocytic cell line THP-1.[81]

Galectin-10. By proteomic analysis, human CD4[+]CD25[+] regulatory T (T_{reg}) cells were found to express galectin-10. Additional results indicate that this protein is primarily intracellular and essential for the suppressive function of T_{reg} cells.[82]

Functions demonstrated *in vivo*

There are a large number of studies on the functions of galectins *in vivo*. Many involved the administration of recombinant proteins to experimental animals, while others employed genetically engineered mice deficient in a given galectin.

Hypersensitivity and airway inflammation
Galectin-1. In a mouse model of contact hypersensitivity, galectin-1 alleviated T cell-dependent inflammation by promoting Th2 polarization[83] and suppressing T cell homing to mesenteric lymph nodes and sites of inflammation.[69]

Galectin-2. Galectin-2 was demonstrated to suppress contact hypersensitivity induced by the hapten 2,4-dinitro-1-fluorobenzene.[33] Treatment with galectin-2 *in vivo* effectively suppressed contact hypersensitivity by inducing apoptosis in activated CD8[+] T cells.[33]

Galectin-3. Consistent with the positive regulatory role of this protein in mast cell response observed *in vitro*, *Lgals3*[−/−] mice exhibited diminished IgE-mediated passive cutaneous anaphylactic reactions.[84] In a mouse model of asthma, *Lgals3*[−/−] mice developed less severe allergic airway inflammation and hyperresponsiveness compared to wild-type mice.[85] *Lgals3*[−/−] mice also developed lower levels of serum IgE and reduced IgE and IL-4, but higher levels of IFN-γ in bronchioalveolar lavage fluid. These data suggest that galectin-3 promotes a Th2 response. The role of galectin-3 in a mouse model of chronic allergic airway inflammation has also been investigated, in which mice were exposed to ovalubumin (OVA) for 12 weeks. *Lgals3*[−/−] mice developed significantly lower amounts of airway remodeling than did wild-type mice, and significantly lower airway inflammatory responses.[86]

However, other investigators showed reduction of eosinophil infiltration in rats and mice treated by intranasal delivery of cDNA encoding galectin-3 following airway antigen challenge.[87,88] These authors demonstrated that this treatment resulted in the inhibition of expression of suppressors of cytokine signaling (SOCS), which regulate the Th1–Th2 balance.[89] It thus appears that exogenously added galectin-3 may not exactly reproduce the function of endogenous galectin-3, possibly due to differences in the tissues and cells in which the protein is expressed and in the intra- versus extra-cellular modes of action in mice expressing a transgene compared to wild-type mice. Nevertheless, these studies demonstrate that recombinant galectin-3 or galectin-3–expressing plasmid may be used for treatment of allergic inflammation.

However, under certain conditions galectin-3 may promote atopic dermatitis; this was observed in a mouse model of atopic dermatitis induced by chronic epicutaneous exposure to ovalbumin. *Lgals3*[−/−] mice developed reduced disease severity, as measured by lower epidermal thickening, and reduced eosinophilic infiltrations. Relative to wild-type, skin IL-4 and serum IgE were reduced and IFN-γ was increased,[78] but the IgG2a/IgG1 ratio (a Th1 index) was elevated in *Lgals3*[−/−] mice. Endogenous galectin-3 promotes a characteristic Th2 response in this disease model. This polarization involves galectin-3 effects on T cells, as wild-type recipients of transgenic antigen-specific TCR (OT-II) CD4[+] cells from *Lgals3*[−/−] mice recapitulated the above results. Finally, *Lgals3*[−/−] mice developed a less severe response to hapten in skin in a model of contact hypersensitivity,[79] which was attributable to its positive regulation of dendritic cell migration, as mentioned earlier.

Galectin-9. In patients with food allergy galectin-9 expression in intestinal epithelial cells (IEC) was noted to be increased. The expression of galectin-9 in IEC was induced when the cells were co-cultured with mast cells. In a mouse model of food allergy, inhibition of galectin-9 with a neutralizing antibody ameliorated allergic hypersensitivity and antigen-specific Th2 responses in the intestine.[90] The results suggest that galectin-9 may play an important role in the development of food allergy.

Galectin-9 was shown to be immunosuppressive in a Th2 mouse model of allergic asthma, and it reduced cytokine levels and both eosinophilic and T cell infiltrations in the airways.[91] In an endotoxin-induced model of acute lung injury, mice pretreated by subcutaneous injection of recombinant galectin-9 showed reduced levels of several pro-inflammatory cytokines and chemokines and reduced numbers of neutrophils in bronchoalveolar lavage fluids, but increased numbers of macrophages that appeared to possess immunoprotective properties.[92] Lung injury in *Lgals9*[−/−] mice was intensified relative to wild-type mice, suggesting that properties of the endogenous lectin are similar to the exogenously added galectin-9.

Autoimmunity and other chronic inflammatory disorders

Galectin-1. Galectin-1 has attracted broad interest as a potential immunosuppressive agent in restoring immune cell homeostasis in autoimmunity and inflammation. In several animal models of experimentally-induced or spontaneous autoimmune diseases, galectin-1 suppressed clinical and histopathological signs of inflammation by blunting proinflammatory Th1 and Th17 responses and by skewing the cytokine balance toward a Th2 profile.[17] Recombinant galectin-1 ameliorated retinal inflammation disease by promoting T$_{reg}$ cell–mediated anti-inflammatory responses.[93] Consistent with these results, *Lgals1*[−/−] mice exhibited greater antigen-specific Th1 and Th17 responses, more severe autoimmune inflammation, and augmented trafficking of T cells to mesenteric lymphoid organs and inflamed tissues, compared to their wild-type counterparts.[21,69,94]

Administration of recombinant galectin-1 protected NZB × NZW F1 mice from developing disease resembling human systemic lupus erythematosus. Reduced levels of activated lymphocytes were present in galectin-1–treated mice, and the frequency of peripheral T_{reg} cells was increased.[95] In contrast, galectin-1–deficient mice developed higher titers of antibodies to double-stranded DNA, consistent with an immunosuppressive function of endogenous galectin-1.[95]

Galectin-3. In a model of antigen-induced arthritis by immunization with methylated bovine serum albumin, it was found that $Lgals3^{-/-}$ mice developed significantly less joint inflammation and bone erosion compared to wild-type mice. The former also exhibited lower amounts of IL-17–producing cells in the spleen. Exogenously added recombinant galectin-3 augmented the inflammatory and arthritic responses in $Lgals3^{-/-}$ mice, suggesting that galectin-3 plays a crucial role in the development of arthritis.[96]

In an animal model of autoimmune encephalomyelitis (EAE) induced by immunization with myelin oligodendrocyte glycoprotein peptide, $Lgals3^{-/-}$ mice were found to develop significantly less disease compared to wild-type mice, including lower infiltrations of macrophages and dendritic cells in neural tissue. Moreover, $Lgals3^{-/-}$ mice produced lower levels of proinflammatory cytokines, both in isolated T cells and neural tissue, compared to wild-type littermates. Significantly higher frequencies of Foxp3$^+$ T_{reg} cells were found in spleens and central nervous system of $Lgals3^{-/-}$ mice, suggesting a significant role for T_{reg} in ameliorating disease in these mice.[97] Finally, there are a large number of studies documenting the role of galectin-3 in the development of diabetes[98–102] (reviewed in Refs. 3 and 17).

Galectin-9. Administration of galectin-9 to mice reduced the severity of IC-induced arthritis, proinflammatory cytokine levels in inflamed joints, and serum C5a. The immunosuppressive function of galectin-9 is also suggested by the fact that $Lgals9^{-/-}$ mice exhibited more severe disease. The authors provided evidence suggesting that galectin-9 suppressed IC-induced inflammation partly by regulating Fcγ R expression on macrophages.[67]

Galectin-9 expression in bronchial epithelia of patients with cystic fibrosis (CF) was elevated and the protein was found to activate TIM-3 on airway epithelial cells, resulting in secretion of neutrophil attractant IL-8. As TIM-3 was found to be constitutively overexpressed in the airways of CF patients, the authors concluded that galectin-9 contributes to the neutrophil-dominated immune response in the CF airways.[103]

Intestinal and peritoneal inflammation

Galectin-1. Galectin-1 can contribute to suppression of intestinal inflammation via induction of effector T cell apoptosis.[104] In rodent models of peritonitis, recombinant galectin-1 inhibited leukocyte recruitment into the peritoneal cavity.[47,105] The finding that both MHC II expression and allostimulatory capacity were increased in peritoneal macrophages from $Lgals1^{-/-}$ mice supports an immunosuppressive function of galectin-1.[53]

Galectin-2. Administration of recombinant galectin-2 has protective effects on epithelial cell integrity in experimental colitis in mice, resulting in suppression of acute and chronic disease.[32]

Galectin-3. When intraperitoneally injected with thioglycollate broth, $Lgals3^{-/-}$ mice exhibited significantly reduced numbers of neutrophils[106] and macrophages[107] in the peritoneal cavity compared to wild-type mice, indicating proinflammatory properties for galectin-3.

Galectin-4. Administration of a galectin-4 antibody into mice with intestinal inflammation suppressed the severity of the disease.[39] However, in a model of experimental colitis, administration of recombinant galectin-4 resulted in the suppression of proinflammatory cytokine secretion and mucosal inflammation, and caused apoptosis of mucosal T cells.[40] These conflicting reports may result from differences in the experimental methods and remain to be reconciled.

Inflammation related to atherosclerosis and other cardiovascular diseases

Galectin-1. Galectin-1 binds to lipoprotein(a) (a modified LDL molecule implicated in atherogenesis) *in situ*, suggesting a mechanism by which this lipoprotein accumulates within arterial walls in atherogenesis.[108]

Galectin-3. Galectin-3 was found to be expressed in foam cells and macrophages in atherosclerotic lesions,[109] and a role for galectin-3 in development of atherosclerosis was subsequently demonstrated by comparing apolipoprotein (Apo)E-deficient and ApoE/galectin-3 double-knockout mice. Compared to the former, the latter developed a significantly lower number of atherosclerotic lesions and atheromatous plaques. This was associated with lower numbers of perivascular inflammatory infiltrates in the latter.[110] It has been shown that in this animal model early atherosclerotic lesions were infiltrated by M2 macrophages that favor the proliferation of smooth muscle cells, whereas lesion progression was correlated with the dominance of M1 over M2 phenotype.[111] Further analysis of disease progression in ApoE/galectin-3 double-knockout mice will be helpful to clarify the role of galectin-3 in macrophage polarization, as experiments with *in vitro* cultured macrophages seemed to produce conflicting results as to whether galectin-3 favors the M1 or M2 phenotype.[74–76]

Recent studies have implicated galectin-3 as a serum marker for heart failure and unfavorable patient prognosis.[112–115] Mechanisms of action of galectin-3 in this disease are not fully understood but may involve roles in inflammation and fibrosis.[116,117]

Conclusions

Although many studies examining the roles of galectins in inflammation were most readily performed *in vitro* with recombinant proteins, the described outcomes typically required high concentrations (micromolar). It is unknown if such high levels of protein are present under physiological or pathological conditions, leaving some functions *in vivo* yet to be recapitulated. However, as mentioned earlier, lower levels of galectins may be just as efficacious when presented on a solid-phase matrix under appropriate conditions *in vivo*.[20] Although some properties were confirmed by exploiting galectin-deficient conditions in gene-targeted mice or knock-down conditions with siRNA, a challenge still remains in our ability to differentiate the roles of the endogenous protein functioning intracellularly or extracellularly. In this regard, support for an extracellular lectin function may be suggested in studies employing neutralizing antibodies to reverse galectin functions in conditions in which galectin-producer cells and target cells coexist.

Administration of recombinant proteins to experimental animals has revealed many galectin functions. Although animal studies have yielded results that seem to support some of the functions of galectins suggested by *in vitro* studies, additional work is still needed to definitively establish the *in vivo* functions. This is due to the abilities of galectins to bind and act on multiple cell types, and to induce a variety of responses in these cells. This is exemplified by galectin-3 in allergic airway inflammation: administration of transgenic galectin-3 was efficacious in suppressing disease activity in one animal model, in contrast to other studies in *Lgal3*$^{-/-}$ mice in which endogenous galectin-3 promoted airway inflammation. These results suggest that the functions of exogenously delivered galectin may differ from the endogenous protein. Despite this potential conundrum, studies have established the use of recombinant galectins as therapeutics for a number of different diseases.

An increasing number of studies described *in vivo* functions in galectin-deficient mice. The remaining challenge here is the ability to link observed phenotypes to the specific functions demonstrated *in vitro*. This is especially challenging for galectins that exhibit widespread expression in several different cell types and tissues. The remaining task is to distinguish extracellular and intracellular functions of individual galectins. Future studies employing mutant mice with conditionally regulated galectin expression in a tissue- and cell-type specific manner are required to more firmly establish the *in vivo* functions of galectins. In addition, this approach would be strengthened by recapitulation of the phenotypes present in mutant mice through the use of specific neutralizing antibodies or other specific inhibitors administered to wild-type mice.

Acknowledgments

Work in the authors' laboratories is supported by grants from NIH to F.T.L.

Conflicts of Interest

The authors declare no conflicts of interest.

References

1. Cummings, R.D. & F.-T. Liu. 2009. Galectins. In *Essentials of Glycobiology*. A. Varki, R.D. Cummings, J.D. Esko, H.H. Freeze, P. Stanley, C.R. Bertozzi, G.W. Hart & M.E. Etzler, Eds.: 475–487. CSHL Press. Woodbury, New York.

2. Hirabayashi, J. *et al.* 2002. Oligosaccharide specificity of galectins: a search by frontal affinity chromatography. *Biochim. Biophys. Acta* **1572:** 232–254.

3. Yang, R.Y., G.A. Rabinovich & F.T. Liu. 2008. Galectins: structure, function and therapeutic potential. *Expert Rev. Mol. Med.* **10:** e17.

4. Ahmad, N. *et al.* 2004. Galectin-3 precipitates as a pentamer with synthetic multivalent carbohydrates and forms heterogeneous cross-linked complexes. *J. Biol. Chem.* **279:** 10841–10847.

5. Miyanishi, N. *et al.* 2007. Carbohydrate-recognition domains of galectin-9 are involved in intermolecular interaction with galectin-9 itself and other members of the galectin family. *Glycobiology* **17:** 423–432.

6. Barres, C. *et al.* 2010. Galectin-5 is bound onto the surface of rat reticulocyte exosomes and modulates vesicle uptake by macrophages. *Blood* **115:** 696–705.

7. Moutsatsos, I.K. *et al.* 1987. Endogenous lectins from cultured cells: Nuclear localization of carbohydrate-binding protein 35 in proliferating 3T3 fibroblasts. *Proc. Natl. Acad. Sci. USA.* **84:** 6452–6456.

8. Sacchettini, J.C., L.G. Baum & C.F. Brewer. 2001. Multivalent protein-carbohydrate interactions. A new paradigm for supermolecular assembly and signal transduction. *Biochemistry* **40:** 3009–3015.

9. Dennis, J.W., I.R. Nabi & M. Demetriou. 2009. Metabolism, cell surface organization, and disease. *Cell* **139:** 1229–1241.

10. Grigorian, A., S. Torossian & M. Demetriou. 2009. T-cell growth, cell surface organization, and the galectin-glycoprotein lattice. *Immunol. Rev.* **230:** 232–246.

11. Lajoie, P. *et al.* 2009. Lattices, rafts, and scaffolds: domain regulation of receptor signaling at the plasma membrane. *J. Cell Biol.* **185:** 381–385.

12. Haudek, K.C. *et al.* 2010. Dynamics of galectin-3 in the nucleus and cytoplasm. *Biochim. Biophys. Acta* **1800:** 181–189.

13. Nakahara, S. & A. Raz. 2007. Regulation of cancer-related gene expression by galectin-3 and the molecular mechanism of its nuclear import pathway. *Cancer Metastasis Rev.* **26:** 605–610.

14. Delacour, D., A. Koch & R. Jacob. 2009. The role of galectins in protein trafficking. *Traffic* **10:** 1405–1413.

15. Stechly, L. *et al.* 2009. Galectin-4-regulated delivery of glycoproteins to the brush border membrane of enterocyte-like cells. *Traffic* **10:** 438–450.

16. Liu, F.T. & G.A. Rabinovich. 2010. Galectins: regulators of acute and chronic inflammation. *Ann. N. Y. Acad. Sci.* **1183:** 158–182.

17. Rabinovich, G.A. & M.A. Toscano. 2009. Turning 'sweet' on immunity: galectin-glycan interactions in immune tolerance and inflammation. *Nat. Rev. Immunol.* **9:** 338–352.

18. Perillo, N.L. *et al.* 1995. Apoptosis of T cells mediated by galectin-1. *Nature* **378:** 736–739.

19. Brandt, B. *et al.* 2010. Role of the JNK/c-Jun/AP-1 signaling pathway in galectin-1-induced T-cell death. *Cell Death Dis* **1:** e23.

20. He, J. & L.G. Baum. 2004. Presentation of galectin-1 by extracellular matrix triggers T cell death. *J. Biol. Chem.* **279:** 4705–4712.

21. Toscano, M.A. *et al.* 2007. Differential glycosylation of T(H)1, T(H)2 and T(H)-17 effector cells selectively regulates susceptibility to cell death. *Nat. Immunol.* **8:** 825–834.

22. Stowell, S.R. *et al.* 2007. Human galectin-1, -2, and -4 induce surface exposure of phosphatidylserine in activated human neutrophils but not in activated T cells. *Blood* **109:** 219–227.

23. Stowell, S.R. *et al.* 2008. Differential roles of galectin-1 and galectin-3 in regulating leukocyte viability and cytokine secretion. *J. Immunol.* **180:** 3091–3102.

24. Rabinovich, G.A. *et al.* 1999. Specific inhibition of T-cell adhesion to extracellular matrix and proinflammatory cytokine secretion by human recombinant galectin-1. *Immunology* **97:** 100–106.

25. He, J. & L.G. Baum. 2006. Endothelial cell expression of galectin-1 induced by prostate cancer cells inhibits T-cell transendothelial migration. *Lab. Invest.* **86:** 578–590.

26. Sugimoto, N. *et al.* 2006. Foxp3-dependent and -independent molecules specific for CD25$^+$CD4$^+$ natural regulatory T cells revealed by DNA microarray analysis. *Int. Immunol.* **18:** 1197–1209.

27. Garin, M.I. *et al.* 2007. Galectin-1: a key effector of regulation mediated by CD4$^+$CD25$^+$ T cells. *Blood* **109:** 2058–2065.

28. Wang, J. *et al.* 2009. Cross-linking of GM1 ganglioside by galectin-1 mediates regulatory T cell activity involving TRPC5 channel activation: possible role in suppressing experimental autoimmune encephalomyelitis. *J. Immunol.* **182:** 4036–4045.

29. Gauthier, L. *et al.* 2002. Galectin-1 is a stromal cell ligand of the pre-B cell receptor (BCR) implicated in synapse formation between pre-B and stromal cells and in pre-BCR triggering. *Proc. Natl. Acad. Sci. USA* **99:** 13014–13019.

30. Tsai, C.M. *et al.* 2008. Galectin-1 promotes immunoglobulin production during plasma cell differentiation. *J. Immunol.* **181:** 4570–4579.

31. Sturm, A. *et al.* 2004. Human galectin-2: novel inducer of T cell apoptosis with distinct profile of caspase activation. *J. Immunol.* **173:** 3825–3837.

32. Paclik, D. *et al.* 2008. Galectin-2 induces apoptosis of lamina propria T lymphocytes and ameliorates acute and chronic experimental colitis in mice. *J. Mol. Med.* **86:** 1395–1406.

33. Loser, K. *et al.* 2009. Galectin-2 suppresses contact allergy by inducing apoptosis in activated CD8$^+$ T cells. *J. Immunol.* **182:** 5419–5429.

34. Hsu, D.K. *et al.* 1996. Human T lymphotropic virus-1 infection of human T lymphocytes induces expression of the b-galactose-binding lectin, galectin-3. *Am. J. Pathol.* **148:** 1661–1670.

35. Dong, S. & R.C. Hughes. 1996. Galectin-3 stimulates uptake of extracellular Ca2+ in human Jurkat T-cells. *FEBS Lett.* **395:** 165–169.

36. Stillman, B.N. *et al.* 2006. Galectin-3 and galectin-1 bind distinct cell surface glycoprotein receptors to induce T cell death. *J. Immunol.* **176:** 778–789.

37. Fukumori, T. *et al.* 2003. CD29 and CD7 mediate galectin-3-induced type II T-cell apoptosis. *Cancer Res.* **63:** 8302–8311.

38. Demetriou, M. *et al.* 2001. Negative regulation of T-cell activation and autoimmunity by Mgat5 N-glycosylation. *Nature* **409:** 733–779.

39. Hokama, A. *et al.* 2004. Induced reactivity of intestinal CD4+ T cells with an epithelial cell lectin, galectin-4, contributes to exacerbation of intestinal inflammation. *Immunity* **20**: 681–693.

40. Paclik, D. *et al.* 2008. Galectin-4 controls intestinal inflammation by selective regulation of peripheral and mucosal T cell apoptosis and cell cycle. *PLoS ONE* **3**: e2629.

41. Norambuena, A. *et al.* 2009. Galectin-8 induces apoptosis in jurkat T cells by phosphatidic acid-mediated ERK 1/2 activation supported by PKA down regulation. *J. Biol. Chem.* **284**: 12670–12679.

42. Tribulatti, M.V. *et al.* 2009. Galectin-8 provides costimulatory and proliferative signals to T lymphocytes. *J. Leukoc. Biol.* **86**: 371–380.

43. Yamamoto, H. *et al.* 2008. Induction of cell adhesion by galectin-8 and its target molecules in Jurkat T-cells. *J. Biochem.* **143**: 311–324.

44. Zhu, C. *et al.* 2005. The Tim-3 ligand galectin-9 negatively regulates T helper type 1 immunity. *Nat. Immunol.* **6**: 1245–1252.

45. Su, E.W., S. Bi & L.P. Kane. 2011. Galectin-9 regulates T helper cell function independently of Tim-3. *Glycobiology* **21**: 1258–1265.

46. Jayaraman, P. *et al.* 2010. Tim3 binding to galectin-9 stimulates antimicrobial immunity. *J. Exp. Med.* **207**: 2343–2354.

47. La, M. *et al.* 2003. A novel biological activity for galectin-1: inhibition of leukocyte-endothelial cell interactions in experimental inflammation. *Am. J. Pathol.* **163**: 1505–1515.

48. Cooper, D., L.V. Norling & M. Perretti. 2008. Novel insights into the inhibitory effects of Galectin-1 on neutrophil recruitment under flow. *J. Leukoc. Biol.* **83**: 1459–1466.

49. Karmakar, S. *et al.* 2008. Galectin-1 signaling in leukocytes requires expression of complex-type N-glycans. *Glycobiology* **18**: 770–778.

50. Stowell, S.R. *et al.* 2009. Galectin-1 induces reversible phosphatidylserine exposure at the plasma membrane. *Mol. Biol. Cell* **20**: 1408–1418.

51. Rabinovich, G.A. *et al.* 2000. Evidence of a role for galectin-1 in acute inflammation. *Eur. J. Immunol.* **30**: 1331–1339.

52. Correa, S.G. *et al.* 2003. Opposite effects of galectin-1 on alternative metabolic pathways of L-arginine in resident, inflammatory, and activated macrophages. *Glycobiology* **13**: 119–128.

53. Barrionuevo, P. *et al.* 2007. A novel function for galectin-1 at the crossroad of innate and adaptive immunity: galectin-1 regulates monocyte/macrophage physiology through a nonapoptotic ERK-dependent pathway. *J. Immunol.* **178**: 436–445.

54. Zuniga, E. *et al.* 2001. Regulated expression and effect of galectin-1 on Trypanosoma cruzi-infected macrophages: modulation of microbicidal activity and survival. *Infect. Immun.* **69**: 6804–6812.

55. Fulcher, J.A. *et al.* 2009. Galectin-1 co-clusters CD43/CD45 on dendritic cells and induces cell activation and migration through Syk and protein kinase C signaling. *J. Biol. Chem.* **284**: 26860–26870.

56. Ilarregui, J.M. *et al.* 2009. Tolerogenic signals delivered by dendritic cells to T cells through a galectin-1-driven immunoregulatory circuit involving interleukin 27 and interleukin 10. *Nat. Immunol.* **10**: 981–991.

57. Farnworth, S.L. *et al.* 2008. Galectin-3 reduces the severity of pneumococcal pneumonia by augmenting neutrophil function. *Am. J. Pathol.* **172**: 395–405.

58. Fernandez, G.C. *et al.* 2005. Galectin-3 and soluble fibrinogen act in concert to modulate neutrophil activation and survival: involvement of alternative MAPK pathways. *Glycobiology* **15**: 519–527.

59. Frigeri, L.G., R.I. Zuberi & F.T. Liu. 1993. eBP, a b-galactoside-binding animal lectin, recognizes IgE receptor (FceRI) and activates mast cells. *Biochemistry* **32**: 7644–7649.

60. Zuberi, R.I., L.G. Frigeri & F.T. Liu. 1994. Activation of rat basophilic leukemia cells by eBP, an IgE- binding endogenous lectin. *Cell Immunol.* **156**: 1–12.

61. Suzuki, Y. *et al.* 2008. Galectin-3 but not galectin-1 induces mast cell death by oxidative stress and mitochondrial permeability transition. *Biochim. Biophys. Acta* **1783**: 924–934.

62. Nishi, N. *et al.* 2003. Galectin-8 modulates neutrophil function via interaction with integrin alphaM. *Glycobiology* **13**: 755–763.

63. Stowell, S.R. *et al.* 2008. Dimeric Galectin-8 induces phosphatidylserine exposure in leukocytes through polylactosamine recognition by the C-terminal domain. *J. Biol. Chem.* **283**: 20547–20559.

64. Matsumoto, R. *et al.* 1998. Human ecalectin, a variant of human galectin-9, is a novel eosinophil chemoattractant produced by T lymphocytes. *J. Biol. Chem.* **273**: 16976–16984.

65. Saita, N. *et al.* 2002. Association of galectin-9 with eosinophil apoptosis. *Int. Arch. Allergy Immunol.* **128**: 42–50.

66. Katoh, S. *et al.* 2010. Involvement of galectin-9 in lung eosinophilia in patients with eosinophilic pneumonia. *Int. Arch. Allergy Immunol.* **153**: 294–302.

67. Arikawa, T. *et al.* 2009. Galectin-9 ameliorates immune complex-induced arthritis by regulating Fc gamma R expression on macrophages. *Clin. Immunol.* **133**: 382–392.

68. Niki, T. *et al.* 2009. Galectin-9 is a high affinity IgE-binding lectin with anti-allergic effect by blocking IgE-antigen complex formation. *J. Biol. Chem.* **284**: 32344–32352.

69. Norling, L.V. *et al.* 2008. Inhibitory control of endothelial galectin-1 on in vitro and in vivo lymphocyte trafficking. *FASEB J.* **22**: 682–690.

70. Yang, R.Y., D.K. Hsu & F.T. Liu. 1996. Expression of galectin-3 modulates T cell growth and apoptosis. *Proc. Natl. Acad. Sci. USA* **93**: 6737–6742.

71. Joo, H.G. *et al.* 2001. Expression and function of galectin-3, a beta-galactoside-binding protein in activated T lymphocytes. *J. Leukoc. Biol.* **69**: 555–564.

72. Hoyer, K.K. *et al.* 2004. An anti-apoptotic role for galectin-3 in diffuse large B-cell lymphomas. *Am. J. Pathol.* **164**: 893–902.

73. Acosta-Rodriguez, E.V. *et al.* 2004. Galectin-3 mediates IL-4-induced survival and differentiation of B cells: functional cross-talk and implications during Trypanosoma cruzi infection. *J. Immunol.* **172**: 493–502.

74. Sano, H. *et al.* 2003. Critical role of galectin-3 in phagocytosis by macrophages. *J. Clin. Invest.* **112**: 389–397.

75. MacKinnon, A.C. *et al.* 2008. Regulation of alternative macrophage activation by galectin-3. *J. Immunol.* **180:** 2650–2658.

76. Li, Y. *et al.* 2008. Galectin-3 is a negative regulator of lipopolysaccharide-mediated inflammation. *J. Immunol.* **181:** 2781–2789.

77. Bernardes, E.S. *et al.* 2006. Toxoplasma gondii infection reveals a novel regulatory role for galectin-3 in the interface of innate and adaptive immunity. *Am. J. Pathol.* **168:** 1910–1920.

78. Saegusa, J. *et al.* 2009. Galectin-3 is critical for the development of the allergic inflammatory response in a mouse model of atopic dermatitis. *Am. J. Pathol.* **174:** 922–931.

79. Hsu, D.K. *et al.* 2009. Endogenous galectin-3 is localized in membrane lipid rafts and regulates migration of dendritic cells. *J. Invest. Dermatol.* **129:** 573–583.

80. Rao, S.P. *et al.* 2007. Galectin-3 functions as an adhesion molecule to support eosinophil rolling and adhesion under conditions of flow. *J. Immunol.* **179:** 7800–7807.

81. Matsuura, A. *et al.* 2009. Intracellular galectin-9 activates inflammatory cytokines in monocytes. *Genes Cells* **14:** 511–521.

82. Kubach, J. *et al.* 2007. Human CD4$^+$CD25$^+$ regulatory T cells: proteome analysis identifies galectin-10 as a novel marker essential for their anergy and suppressive function. *Blood* **110:** 1550–1558.

83. Cedeno-Laurent, F. *et al.* 2010. Development of a nascent galectin-1 chimeric molecule for studying the role of leukocyte galectin-1 ligands and immune disease modulation. *J. Immunol.* **185:** 4659–4672.

84. Chen, H.Y. *et al.* 2006. Role of galectin-3 in mast cell functions: galectin-3-deficient mast cells exhibit impaired mediator release and defective JNK expression. *J. Immunol.* **177:** 4991–4997.

85. Zuberi, R.I. *et al.* 2004. Critical role for galectin-3 in airway inflammation and bronchial hyperresponsiveness in a murine model of asthma. *Am. J. Pathol.* **165:** 2045–2053.

86. Ge, X.N. *et al.* 2010. Allergen-induced airway remodeling is impaired in galectin-3-deficient mice. *J. Immunol.* **185:** 1205–1214.

87. del Pozo, V. *et al.* 2002. Gene therapy with galectin-3 inhibits bronchial obstruction and inflammation in antigen-challenged rats through interleukin-5 gene downregulation. *Am. J. Respir. Crit. Care Med.* **166:** 732–737.

88. Lopez, E. *et al.* 2006. Inhibition of chronic airway inflammation and remodeling by galectin-3 gene therapy in a murine model. *J. Immunol.* **176:** 1943–1950.

89. Lopez, E. *et al.* 2011. Gene expression profiling in lungs of chronic asthmatic mice treated with galectin-3: downregulation of inflammatory and regulatory genes. *Mediators Inflamm* **2011:** 823279.

90. Chen, X. *et al.* 2011. Intestinal epithelial cells express galectin-9 in patients with food allergy that plays a critical role in sustaining allergic status in mouse intestine. *Allergy* **66:** 1038–1046.

91. Katoh, S. *et al.* 2007. Galectin-9 inhibits CD44-Hyaluronan interaction and suppresses a Murine model of allergic asthma. *Am. J. Respir. Crit. Care Med.* **176:** 27–35.

92. Kojima, K. *et al.* 2011. Galectin-9 attenuates acute lung injury by expanding CD14- plasmacytoid dendritic cell-like macrophages. *Am. J. Respir. Crit. Care Med.* **184:** 328–339.

93. Toscano, M.A. *et al.* 2006. Galectin-1 suppresses autoimmune retinal disease by promoting concomitant Th2- and T regulatory-mediated anti-inflammatory responses. *J. Immunol.* **176:** 6323–6332.

94. Blois, S.M. *et al.* 2007. A pivotal role for galectin-1 in feto-maternal tolerance. *Nat. Med.* **13:** 1450–1457.

95. Liu, S.D. *et al.* 2011. Galectin-1-induced down-regulation of T lymphocyte activation protects (NZB x NZW) F1 mice from lupus-like disease. *Lupus* **20:** 473–484.

96. Forsman, H. *et al.* 2011. Galectin 3 aggravates joint inflammation and destruction in antigen-induced arthritis. *Arthritis Rheum.* **63:** 445–454.

97. Jiang, H.R. *et al.* 2009. Galectin-3 deficiency reduces the severity of experimental autoimmune encephalomyelitis. *J. Immunol.* **182:** 1167–1173.

98. Pugliese, G. *et al.* 2001. Accelerated diabetic glomerulopathy in galectin-3/AGE receptor 3 knockout mice. *FASEB J.* **15:** 2471–2479.

99. Iacobini, C. *et al.* 2003. Role of galectin-3 in diabetic nephropathy. *J. Am. Soc. Nephrol.* **14:** S264–S270.

100. Iacobini, C. *et al.* 2005. Development of age-dependent glomerular lesions in galectin-3/AGE-receptor-3 knockout mice. *Am. J. Physiol. Renal Physiol.* **289:** F611–F621.

101. Stitt, A.W. *et al.* 2005. Impaired retinal angiogenesis in diabetes: role of advanced glycation end products and galectin-3. *Diabetes* **54:** 785–794.

102. Mensah-Brown, E.P. *et al.* 2009. Targeted disruption of the galectin-3 gene results in decreased susceptibility to multiple low dose streptozotocin-induced diabetes in mice. *Clin. Immunol.* **130:** 83–88.

103. Vega-Carrascal, I. *et al.* 2011. Dysregulation of TIM-3-galectin-9 pathway in the cystic fibrosis airways. *J. Immunol.* **186:** 2897–2909.

104. Mizoguchi, E. & A. Mizoguchi. 2007. Is the sugar always sweet in intestinal inflammation? *Immunol. Res.* **37:** 47–60.

105. Gil, C.D. *et al.* 2006. Inflammation-induced modulation of cellular galectin-1 and -3 expression in a model of rat peritonitis. *Inflamm. Res.* **55:** 99–107.

106. Colnot, C. *et al.* 1998. Maintenance of granulocyte numbers during acute peritonitis is defective in galectin-3-null mutant mice. *Immunology* **94:** 290–296.

107. Hsu, D.K. *et al.* 2000. Targeted disruption of the galectin-3 gene results in attenuated peritoneal inflammatory responses. *Am. J. Pathol.* **156:** 1073–1083.

108. Chellan, B., J. Narayani & P.S. Appukuttan. 2007. Galectin-1, an endogenous lectin produced by arterial cells, binds lipoprotein(a) [Lp(a)] in situ: relevance to atherogenesis. *Exper. Mol. Pathol.* **83:** 399–404.

109. Nachtigal, M. *et al.* 1998. Galectin-3 expression in human atherosclerotic lesions. *Am. J. Pathol.* **152:** 1199–1208.

110. Nachtigal, M., A. Ghaffar & E.P. Mayer. 2008. Galectin-3 gene inactivation reduces atherosclerotic lesions and adventitial inflammation in ApoE-deficient mice. *Am. J. Pathol.* **172:** 247–255.

111. Khallou-Laschet, J. *et al.* 2010. Macrophage plasticity in experimental atherosclerosis. *PLoS ONE* **5:** e8852.

112. van Kimmenade, R.R. *et al.* 2006. Utility of amino-terminal pro-brain natriuretic peptide, galectin-3, and apelin for the evaluation of patients with acute heart failure. *J. Am. Coll. Cardiol.* **48:** 1217–1224.

113. Lok, D.J. *et al.* 2010. Prognostic value of galectin-3, a novel marker of fibrosis, in patients with chronic heart failure: data from the DEAL-HF study. *Clin. Res. Cardiol. (official journal of the German Cardiac Society)* **99:** 323–328.

114. Lainscak, M. *et al.* 2010. Clinical trials update from the Heart Failure Society of America Meeting 2009: FAST, IMPROVE-HF, COACH galectin-3 substudy, HF-ACTION nuclear substudy, DAD-HF, and MARVEL-1. *Eur. J. Heart Fail.* **12:** 193–196.

115. Ueland, T. *et al.* 2011. Galectin-3 in heart failure: high levels are associated with all-cause mortality. *Int. J. Cardiol.* **150:** 361–364.

116. Sharma, U.C. *et al.* 2004. Galectin-3 marks activated macrophages in failure-prone hypertrophied hearts and contributes to cardiac dysfunction. *Circulation* **110:** 3121–3128.

117. Liu, Y.H. *et al.* 2009. N-acetyl-seryl-aspartyl-lysyl-proline prevents cardiac remodeling and dysfunction induced by galectin-3, a mammalian adhesion/growth-regulatory lectin. *Am. J. Physiol. Heart Circ. Physiol.* **296:** H404–H412.

Ann. N.Y. Acad. Sci. ISSN 0077-8923

ANNALS OF THE NEW YORK ACADEMY OF SCIENCES
Issue: *Glycobiology of the Immune Response*

Mechanisms underlying *in vivo* polysaccharide-specific immunoglobulin responses to intact extracellular bacteria

Clifford M. Snapper

Department of Pathology, Uniformed Services University of the Health Sciences, Bethesda, Maryland

Address for correspondence: Clifford M. Snapper, Department of Pathology, Uniformed Services University of the Health Sciences, 4301 Jones Bridge Road, Bethesda, MD 20814. csnapper@usuhs.mil

Polysaccharides (PS) have historically been viewed as T cell–independent antigens. However, in this paper I propose a new concept, based on substantial data from my laboratory over the previous 10 years, that during infections with PS-encapsulated extracellular bacteria, PS-specific IgG responses are largely CD4[+] T cell dependent. Thus, capsular PS is typically encountered by the immune system covalently attached to the underlying subcapsular bacterial domain. I speculate that noncovalent association of PS with immunogenic proteins within the bacterial particle leads to recruitment of protein-specific CD4[+] T cell help for the anti-PS response. However, differences in the composition and/or structure of the subcapsular domain of different extracellular bacteria may result in distinct anti-PS responses as well as differential effects on the immune response to coimmunizing soluble antigens. CD4[+] T cell help for IgG anti-PS responses during infections with extracellular bacteria is likely to promote opsonic clearance of these rapidly growing pathogens. However, the expression of immunosuppressive components by certain bacteria may also serve to dampen such responses.

Keywords: polysaccharide; bacteria; antibody; vaccine; immunity; T cell

Regulation of Ig response to *isolated* T cell–independent (TI) versus T cell–dependent (TD) antigens

Antibody specific for both bacterial polysaccharide (PS) and protein can confer protection against extracellular bacteria.[1] Much of our current knowledge of the regulation of anti-PS and antiprotein responses has come from studies using isolated, soluble PS and proteins.[2] These studies revealed that Ig anti-PS, in contrast to antiprotein responses are more rapid, fail to induce robust germinal center reactions, show limited generation of memory or affinity maturation, and have a more restricted Ig isotype profile. PS (except zwitterionic PS),[3,4] in contrast to proteins, are not enzymatically processed within endosomes to generate MHC class II-PS complexes on the surface of APCs for presentation to CD4[+] T cells.[5,6] Hence, nonzwitterionic PS, representing the majority of capsular PS, fail to recruit classical cognate CD4[+] T cell help, likely ac-

counting for their inability to induce sustained GC reactions.[2] However, a regulatory role for T cells,[7,8] and CD40L-dependent help,[9] has been reported for some anti-PS responses, but the mechanisms of action and physiologic relevance of these observations have not been adequately clarified. More recently, a role for CD1-responsive CD8[+] T cells in stimulating an IgG response to isolated pneumococcal PS was demonstrated,[10] although a subsequent study failed to confirm this observation.[11] PS, in contrast to proteins, display repeating antigenic epitopes that mediate potent signaling through the B cell Ig receptor (BCR),[12] resulting in distinct functional outcomes relative to protein antigens.[13]

The mechanism by which isolated PS elicits Ig responses *in vivo* is not yet adequately clarified. A series of earlier studies addressed this issue in an *in vitro* model system using dextran-conjugated anti-IgD antibodies (αδ-dex) to mimic, in a polyclonal fashion, the multivalent Ig crosslinking mediated by PS. Highly purified B cells activated with αδ-dex

doi: 10.1111/j.1749-6632.2011.06329.x

Ann. N.Y. Acad. Sci. 1253 (2012) 92–101 © 2012 New York Academy of Sciences.

underwent vigorous proliferation with quantities of anti-IgD 1000-fold lower than that observed using unconjugated anti-IgD.[12] However, no Ig secretion or class switching was observed unless a second signal, mediated by certain cytokines or Toll-like receptor (TLR) ligands were also provided.[14,15] On this basis, a model was proposed whereby intact PS-encapsulated bacteria, through direct expression of TLR ligands, as well as TLR ligand–induced cytokines could provide the necessary second signals for a physiological TI anti-PS response.[13] This model presupposed that intrinsically, most PS do not possess critical costimulating activity for Ig induction, and thus left unresolved how *isolated* PS induce Ig responses *in vivo*. In this regard, a key role for a contaminating TLR2 ligand(s), present in isolated pneumococcal PS preparations, was demonstrated for TI IgM and IgG induction *in vivo*.[16] More recently (see below), we demonstrate that the primary TI IgM and IgG2a anti-PS response to intact *Neisseria meningitidis* requires endogenous TLR4 signaling.[17]

This costimulation model for TI anti-PS responses can also be extended to induction of TD PS-specific IgG in response to PS–protein conjugate vaccines and intact PS-encapsulated bacteria. Thus, a conjugate vaccine of capsular PS from *Hemophilus influenza*, type b linked to a Neisserial outer membrane protein complex (OMPC) that naturally expresses TLR2-activating and cytokine-inducing porin proteins,[18] demonstrated a higher potency than conventional conjugate vaccines in eliciting an anti-PS IgG response, that was dependent on TLR2 signaling.[19] Indeed, the presence of contaminating TLR2 ligands in commercial preparations of pneumococcal conjugate vaccine results in enhanced TD IgG anti-PS responses in mice[16] and perhaps has a similar stimulatory role in humans. In addition, mice genetically deficient in TLR2 elicit lower TD anti-PS IgG responses to intact *Streptococcus pneumoniae* (see below) relative to wild-type (WT) mice.[20]

More recent evidence suggests that TLR induction of BAFF and/or APRIL, particularly from DC and macrophages, is also important for induction of TI PS-specific Ig responses;[21–23] BAFF and APRIL bind to both TACI and BCMA, whereas BAFF additionally binds to BAFFR.[24] Indeed, mice genetically deficient in TACI have a selective defect in Ig responses to isolated TI-II Ags (i.e., NP-Ficoll and isolated

pneumococcal PS).[25] BAFF is also a critical survival factor for B cells[26] and thus may promote anti-PS responses emerging from the apoptosis-prone extrafollicular plasma cell response.[27] BAFF produced by FDC may additionally promote TI B cell activation in the follicle[28] as well as directly mediate T cell costimulation,[29] providing one potential mechanism for TD anti-PS responses to intact bacteria (see below). Thus, TLR ligands derived from intact bacteria can potentially provide critical second signals for BCR-activated PS-specific B cells via direct TLR-mediated activation of B cells, and TLR induction of both cytokines and BAFF/APRIL that promote TI, and perhaps TD, anti-PS responses.

Intact PS-encapsulated extracellular bacteria are potentially unique immunogens for elicitation of an anti-PS response

The historic use of isolated, soluble PS in most immunologic studies makes the current dogma that PS are TI antigens, potentially divorced from the physiologic context of the intact microbe that coexpresses these antigens. Thus, the association of PS with protein and TLR,[30] NOD-like (NLR),[31] and scavenger receptor ligands[32] within the particulate bacterial structure[33] may confer unique immunologic properties to the coexpressed PS on the intact bacterium. In particular, particulate antigens exhibit distinct immunologic characteristics relative to their soluble counterparts. For example, particulates, relative to soluble antigens, tend to concentrate in the marginal zone of the spleen,[34] are more efficiently internalized by APC for enhancement in antigen presentation,[35,36] although requiring more antigen processing time,[37] are presented poorly by B cells, and upon internalization by APC lead to quantitative and qualitative differences in the antigenic epitopes generated relative to soluble antigen.[38] Indeed, strikingly enhanced antitype III and antitype VIII pneumococcal capsular PS responses were observed when repeated doses of intact, formalinized pneumococcus was used as an immunogen as opposed to isolated PS.[39] Of note, *covalent* linkage of protein and PS to create a soluble conjugate vaccine converts the PS into a classic TD antigen, including the ability to generate PS-specific memory.[40,41] T cell clones generated from mice immunized with a conjugate vaccine were mostly MHC-II–restricted CD4$^+$ T cells specific for the protein carrier, with some specific for the protein-PS combination.[42] However, several MHC-II–nonrestricted CD4[+] T cells specific for the PS alone were also isolated.

The distinct structural and compositional features of Gram-positive (GP) versus Gram-negative (GN) bacteria might further distinguish the anti-PS responses to these two classes of bacteria. Thus, capsular PS expressed by GP bacteria are covalently linked to a thick, underlying cell wall peptidoglycan to which a number of proteins are also covalently attached.[43,44] Capsular PS expressed by GN bacteria, which express a thin peptidoglycan cell wall, is covalently attached to the acyl glycerol moiety of the outer membrane, which contains highly immunogenic proteins, including porins, and lipopolysaccharide (LPS), the latter a potentially potent stimulator of the innate immune system, depending on its biochemical composition.[45,46] Shedding of the outer membrane/capsular PS complex, to form vesicles, is a unique property of GN bacteria that may have distinct immunologic consequences for the anti-PS response.[19,47] Collectively, these considerations raise questions of whether (1) most isolated PS by themselves are intrinsically capable of inducing Ig responses in vivo, (2) PS expressed by intact bacteria, in contrast to isolated PS, behave as TD antigens, (3) PS associated with GP versus GN elicit different PS-specific Ig responses, and (4) PS expressed by the intact particulate bacterium is immunologically different relative to PS covalently attached to an immunogenic protein in the form of a soluble conjugate vaccine.

That PS expressed by intact bacteria may indeed behave as a TD antigen has been suggested by sequencing of variable regions of natural human PS-specific Ig. These studies demonstrated a significant degree of somatic hypermutation (SH),[48,49] suggesting that host contact with colonizing encapsulated bacteria may have induced TD GC reactions where SH is known to occur. Similarly, repeated immunization of rabbits with formalinized capsular type III and type VIII pneumococci, followed by sequence analysis of the induced antibodies, strongly suggested a high degree of somatic diversification.[50] These observations are consistent with an early study demonstrating the ability of isolated type III pneumococcal PS to elicit a secondary (anamnestic) response following priming with intact type III pneumococci.[51]

Potential immunologic consequences of biochemically distinct PS

Potential differences in PS-specific Ig responses elicited by any two bacteria might also reflect distinct biochemical differences of the expressed PS. For example, isolated type C capsular PS from *Neisseria meningitidis* and type V capsular PS from *Streptococcus agalactiae*, but not NP-Ficoll inhibit a number of B cell functions that are mediated by BAFF and APRIL[52] (see below). Other bacterial PS, such as C-polysaccharide and PPS1 of *Streptococcus pneumoniae* (Pn), *Staphylococcus aureus* capsular polysaccharides types 5 and 8, and PS-A of *Bacteroides fragilis*, are zwitterionic on the basis of expression of both positive and negative charges, and thus may stimulate CD4[+] T cells through association with MHC-II.[3,4] The ability of certain PS to bind to SIGN-R1, a scavenger receptor on splenic marginal zone macrophages,[52–54] to fix complement,[55] or express terminal sialic acid residues that interact with the inhibitory CD22 receptor on B cells,[56] may also have potential effects on PS-specific Ig responses to intact bacteria. Capsular PS may additionally exhibit different degrees of immunogenicity based on molecular weight.[57] Further, distinct pneumococcal PS serotypes exhibit significant differences in immunogenicity even when covalently attached to the same carrier protein in a conjugate vaccine, when comparable degrees of carrier-specific T cell memory are observed.[58] Finally, bacteria may express PS linked to phosphorylcholine (PC; e.g., cell wall teichoic acid/C-polysaccharide of Pn), which may inhibit immunity.[59] Thus, a determination of how differences in bacterial composition and structure may potentially impact the specific Ig response to an expressed PS may need to take into account the possible immunologic consequences of the PS itself.

The murine IgG anticapsular PS response to intact *Streptococcus pneumoniae* uniquely combines both TD and TI features

A series of studies were initially conducted to determine key immunologic parameters that underlie anti-PS and antiprotein responses to the encapsulated GP extracellular bacterium, *Streptococcus pneumoniae* (Pn) and to compare these results with those obtained using a soluble pneumococcal conjugate vaccine or isolated pneumococcal PS. To simplify the analysis and to focus specifically on the bacterium as an intact particulate immunogen, heat-killed bacteria and a systemic immunization approach were utilized. Pn, capsular type 14 (Pn14) was chosen since the type 14 capsular PS (PPS14) uniquely demonstrated very high specificity when used in ELISA assays to measure serum titers of anti-PPS14. The highly immunogenic cell wall, choline-binding protein, pneumococcal surface protein A (PspA)[60] was chosen as the main target antigen for measuring an antiprotein Ig response to Pn.

Whereas the IgM anti-PPS14 response to intact Pn14 was TI, the IgG anti-PPS14 response was dependent on CD4[+] T cells, CD28 costimulation, and CD40L, and comprised all four IgG isotypes.[61,62] These parameters were also required for the IgG anti-PspA response and suggested that PPS14 associated with intact Pn14 was behaving as a classic TD antigen. However, the IgG anti-PPS14 response exhibited a number of distinct features relative to PspA, in that the primary response peaked more rapidly, required a shorter period of dependence on CD4[+] T cell help,[61] was ICOS independent (although SAP dependent),[62] and failed to elicit a boosted IgG anti-PPS14 response following secondary immunization. In light of the critical role for ICOS in the development of the GC reaction,[63] this suggested that the primary IgG anti-PPS14, in contrast to the IgG anti-PspA, response was extrafollicular in nature. These latter features were more analogous to those exhibited by TI antigens. Indeed, mice expressing transgenic Bcl-2 or Bcl-X_L, two antiapoptotic proteins,[64] selectively within B cells demonstrated a significantly higher primary IgG anti-PPS14, but not IgG anti-PspA, response[65] supporting the notion that the former represents an apoptosis-prone extrafollicular plasma cell response.[27] Bcl-2 and Bcl-X_L transgenic mice did not elicit a boosted secondary IgG anti-PPS14 response to Pn14.[65]

The parameters regulating the anti-PPS14 response to a pneumococcal conjugate vaccine (PPS14-PspA) versus intact Pn14 are distinct

Covalent linkage of PS to an immunogenic protein creates a "conjugate" vaccine[66] that results in the elicitation of TD high-titer, protective IgG anti-PS responses and the generation of immunologic memory, including immunogenicity in the infant host.[67–69] Using a conjugate of type 3 capsular PS from *Streptococcus agalactiae* (group B Streptococcus) and tetanus toxoid (TT) (i.e., GBSIII-TT) it was demonstrated that the IgG anti-GBSIII response depended on CD4[+] T cells, CD28 costimulation and CD40L. A boosted IgG anti-GBSIII response was observed upon secondary immunization.[40,41] Subsequent studies utilizing a soluble pneumococcal conjugate (PPS14-PspA) confirmed these findings for the IgG anti-PPS14 response and further demonstrated a dependence on ICOS.[62,70] Thus, although the IgG anti-PPS14 responses to Pn14 and PPS14-PspA both depend on CD4[+] T cells, CD28 costimulation, and CD40L, only the latter demonstrated secondary ICOS-dependent boosting. Of note, immunization of transgenic mice exhibiting a selective reduction in B cell expression of Bruton's tyrosine kinase (Btk[low] mice),[71] a molecule critical for BCR signaling,[72] demonstrated strongly reduced IgG anti-PS, but normal antiprotein responses to both intact Pn14 and PPS14-PspA conjugate.[73] Btk[low] mice demonstrated normal conventional (FB and MZB) B cell development, although reduced B-1 cell numbers, but continued to exhibit significant defects in BCR signaling. This study strongly suggested that although PS and protein within a conjugate vaccine, both appear to behave as "classical" TD Ags, they nevertheless exhibit different requirements for BCR signaling.

The propensity of particulate antigens to be sequestered within the marginal zone (MZ),[34] and the observation that MZ B cells (MZB) in relation to follicular B cells (FB) are programmed to favor plasma cell over memory cell differentiation,[74] led to the hypothesis that the failure of intact Pn to generate PPS14-specific IgG memory was secondary to its selective usage of MZB cells to generate this response. Mice genetically deficient in Lsc ($Lsc^{-/-}$) were thus utilized to determine the B cell subset responsible for the TD IgG anti-PPS14 response to intact Pn14 versus PPS14-PspA. $Lsc^{-/-}$ mice exhibit a marked defect in MZB migration from the marginal zone following immunization, precluding MZB interaction with CD4[+] T cells.[75] Lsc acts selectively on MZB cells.[75,76] Thus, the TD IgM anti-NP response to NP-CGG, which is dependent on MZB, is markedly reduced in $Lsc^{-/-}$ mice, whereas the TI anti-NP response to NP-Ficoll is not.[75] Of note, the IgG anti-PPS14 response to Pn14 was markedly inhibited in $Lsc^{-/-}$ mice, whereas no difference was observed between

$Lsc^{-/-}$ and WT mice in response to PPS14-PspA.[35,65] These data strongly suggested that the intact bacteria and soluble conjugate vaccine utilized predominantly MZB and FB cells, respectively for the IgG anti-PPS14 response. Our unpublished data supporting this notion demonstrate that the IgG anti-PPS14 antibody elicited in response to Pn14 versus PPS14-PspA expressed different idiotypes (J. Colino et al., submitted). However, additional experiments demonstrated that MZB cells could indeed generate a boosted secondary IgG anti-PPS14 response under appropriate experimental conditions. Thus, immunization of mice with PspA-depleted, unencapsulated Pn to which PPS14-PspA conjugate was stably adsorbed to the bacterial surface, resulted in a boosted IgG anti-PPS14 response that was markedly inhibited in $Lsc^{-/-}$ mice.[35] These data indicated that the failure of intact Pn14 to elicit a boosted PPS14-specific IgG secondary response could not be simply explained by its preferential usage of MZB cells. In addition, no role for pre-existing PPS14-specific Ig, the bacterial surface or particulation, or the ability to activate and recruit Pn14 protein-specific CD4$^+$ T cells by Pn14 could account for the lack of secondary boosting, in contrast to that observed in response to PPS14-PspA.[35]

The murine *in vivo* anticapsular PS response to intact *Neisseria meningitidis* is distinct relative to that elicited by Pn14

Studies on the IgG anti-PPS14 response to intact Pn14 indicated that the bacterium could markedly influence the humoral immune response to the expressed PS Ag. These studies, however, left unresolved whether the nature of this response was characteristic of intact PS-expressing extracellular bacteria in general, or perhaps represented either a characteristic feature of PPS14, the underlying structure and/or composition of intact Pn itself, or perhaps a more general dichotomy between GP and GN bacteria. To begin to address this issue, a series of studies analogous to those conducted utilizing intact heat-killed Pn14, were performed using intact, heat-killed *Neisseria meningitidis*, serogroup C (MenC), a GN extracellular bacterium, with analysis of the anticapsular PS (MCPS) Ig response following systemic immunization.[17] In contrast to Pn14, our data indicated that the IgG anti-MCPS response to MenC exhibited delayed primary kinetics and a highly boosted secondary IgG anti-MCPS

response, whereas the IgG anti-MCPS response to isolated MCPS was rapid, failed to show secondary boosting, and consisted of only IgG1 and IgG3, as opposed to all four IgG isotypes in response to intact MenC. The secondary, but not primary, IgG anti-MCPS response to MenC was dependent on CD4$^+$ T cells, CD40L, CD28 and ICOS. These data are consistent with a previous study using live MenC that demonstrated a boosted IgG anti-MCPS response following secondary MenC exposure.[77] Collectively, these data indicate that although *isolated* soluble PPS14 and MCPS both behave as TI antigens, their presence on the surface of intact bacteria markedly alters their immunologic properties, including a newly acquired capacity to recruit CD4$^+$ T cell help, but in ways that are strikingly distinct from each other. Specifically, Pn14 induces an ICOS-independent TD primary anti-PS response with rapid kinetics of induction, but no secondary boosting, whereas MenC induces a TI primary response with delayed kinetics of induction, but a significant TD secondary boosting response that is ICOS dependent.

The murine PPS14-specific IgG responses to *Streptococcus agalactiae* (Group B Streptococcus) and Pn14 are distinct

The studies cited above demonstrated that different intact bacteria (Pn14 and MenC) could elicit fundamentally distinct anti-PS responses that further differed from those observed in response to their isolated PS counterparts. However, the differences in the anti-PPS14 and anti-MCPS responses to these distinct intact bacteria could, in theory, reflect intrinsic differences in the two PS themselves. It has been demonstrated that PPS14 expressed naturally by Pn14 is identical biochemically with the core type III PS of GBS (III-PS), the latter differing from the native type III PS by lacking the terminal sialic acid residue.[78] In this regard, women immunized with either a type III PS conjugate vaccine or isolated (unconjugated) type III PS of GBS elicited IgG antibodies that cross-reacted with PPS14.[79]

The biochemical identities of PPS14 and core type III-PS were exploited to determine whether two different GP bacteria could elicit distinct Ig responses to the same expressed capsular PS. Thus, mice were immunized systemically with heat-killed intact GB-SIII bacteria and serum titers of both IgG anti-III PS and anti-PPS14 were measured (S. Arjunaraja et al., submitted). Of interest, GBSIII induced a rapid primary *and* highly boosted secondary IgG anti-III PS *and* IgG anti-PPS14 response. Both the primary and secondary IgG anti-PS responses to GBSIII were dependent upon CD4$^+$ T cells and CD40L, and the boosted secondary response was also dependent upon ICOS. These data clearly establish that the anti-PS response to an intact bacterium can be markedly influenced by the composition and/or architecture of the underlying bacterial subcapsular domain (i.e., GBSIII versus Pn14). The mechanism underlying these distinct anti-PPS14 responses remains to be determined, as well as the additional differences observed in the primary responses between Pn14 and GBSIII versus MenC. In this regard, endogenous CD4$^+$CD25$^+$ regulatory T cells do not appear to account for the lack of secondary boosting of the IgG anti-PPS14 response to Pn14.[80]

Defining the essential features for induction of boostable anti-PS responses to intact extracellular bacteria

TD IgG anti-PS responses to soluble conjugate vaccines occur on the basis of the covalent linkage of the PS to an immunogenic protein. It was recently demonstrated that, in response to intact vaccinia virus, CD4$^+$ T cell help for an *in vivo* IgG response to a specific viral protein was mediated strictly by CD4$^+$ T cells with the same protein specificity as the B cell, despite there being a number of distinct immunogenic proteins expressed by the intact virion.[81] This dependence on "intramolecular" help suggests that covalent attachment of the PS to the protein, and not merely association within a multimolecular physical structure would be required for protein-specific CD4$^+$ T cells to deliver help to PS-specific B cells. This raised the question of how intact Pn14, GBSIII or MenC induce TD anticapsular PS responses, in light of the *noncovalent* association of the capsular PS with the subcapsular bacterial proteins, combined with the inability of the PS itself to recruit cognate T cell help.[5,6] A requirement for specificity for CD4$^+$ T cells to mediate help for both the IgG anti-PPS14 and IgG anti-PspA responses to Pn14, was previously suggested using a nonleaky TCR transgenic mouse model in which CD4$^+$ T cells lacked specificity for Pn Ags.[61] These mice exhibited strongly reduced IgG (TD), but normal IgM (TI) anti-PPS14, and undetectable IgG anti-PspA responses to intact Pn14. More recently, purified

recombinant PspA and isolated PPS14 were noncovalently attached to 1 μm polystyrene beads to model intact Pn. Mice were immunized with these Ag-coated beads in the presence of alum + CpG-ODN as adjuvant. Our unpublished data show that both the primary IgG anti-PPS14 and anti-PspA responses are dependent on CD4$^+$ T cells (J. Colino, *et al.*, submitted). Beads coated only with PS and blocked with glycine to prevent binding of serum proteins following immunization, elicited only a TI response similar to that observed for isolated soluble PS.

These data indicated that noncovalent attachment of PS and an immunogenic protein on the surface of a particle can elicit a TD anti-PS response similar to that observed for intact bacteria. Of note, in contrast to intact Pn, but similar to GBSIII and MenC, secondary immunization with PPS14 + PspA-coated beads elicited a boosted, ICOS-dependent PS-specific IgG response. The basis for the distinct secondary IgG anti-PPS14 responses to intact Pn versus Ag-coated beads remain unexplained.

Different intact, heat-killed extracellular bacteria exert distinct effects on the antibody response to coimmunized heterologous soluble protein antigens

During infections with extracellular bacteria, the immune system likely encounters a variety of microbial components in soluble form, as well as those associated with the intact bacterium.[44,82] Thus, secreted hydrolases such as hyaluronidases, neuraminidases, and endoglycosidases can mediate bacterial spread and destruction of host tissue through degradation of hyaluronan, mucins, and glycolipids. In addition, during the stationary growth phase, Pn expresses a major autolysin (LytA amidase) that degrades its own peptidoglycan cell wall, resulting in release of cytoplasmic proteins.[83,84] In addition, since both capsular polysaccharide (PS) and a number of proteins are covalently attached to the bacterial cell wall peptidoglycan,[43,85] the release of soluble fragments containing both PS and protein upon bacterial lysis is also likely. Although soluble and particulate antigens may exhibit distinct immunologic properties,[34–38] their potential cross-regulatory effects on the humoral immune response, following concomitant immunization, as might occur during bacterial infections, is unknown.

In this regard, coimmunization of mice with Pn and a soluble PS–protein conjugate resulted in marked inhibition of conjugate-induced PS-specific IgG memory, and primary and memory antiprotein IgG responses.[86] Inhibition occurred with unencapsulated Pn, encapsulated Pn expressing different capsular types of PS from those present in the conjugate, and with conjugate containing protein not expressed by Pn. In contrast, coimmunization of mice with conjugate and 1 μm latex beads in adjuvant resulted in no inhibition of the Ig response. Inhibition was long lasting, was induced only during the early phase of the immune response (within 24 h), but was not associated with tolerance. More recently, using soluble chicken ovalbumin (OVA) as the immunizing protein, our unpublished data confirm that coimmunization with soluble OVA in adjuvant and an unencapsulated variant of Pn is inhibitory for the IgG anti-OVA response. In contrast, coimmunization with GBSIII or MenC (both unencapsulated variants) is highly stimulatory (Saumyaa *et al.*, submitted). Of note, the expression by Pn of phosphorylcholine (PC), which is not expressed by the strains of GBSIII or MenC used in this study, appears to be critical for this inhibitory effect. Indeed, Pn lacking PC was stimulatory for the IgG anti-OVA response, similar to GBSIII and MenC. These effects of Pn do not appear to be dependent on the presence or absence of PC-binding proteins. These unpublished data are consistent with previous studies demonstrating the ability of a secreted filarial protein (ES-62) to mediate immunosuppression through a mechanism that is dependent on ES-62 expression of PC.[87] Thus, intact bacteria have the capacity to either inhibit or enhance humoral immune responses to coimmunizing soluble heterologous Ags, dependent on their structural composition.

Conclusions and future directions

Understanding the parameters that regulate Ig responses to isolated PS and PS–protein conjugates has helped to guide the design of PS-based vaccines against extracellular bacteria. However, during *infections* with these pathogens, the immune system encounters PS in the context of the bacteria that express them on their surface. The evidence provided herein suggests that this radically changes the immunobiology of PS-specific Ig induction. In particular, intact bacteria recruit CD4[+] T cell help for PS-specific IgG responses. Nevertheless, the humoral immune response to PS on the surface of bacteria exhibit certain features that make these antigens distinct from proteins expressed by the same bacterium, as well as PS that are covalently attached to protein as part of a soluble conjugate vaccine. An added level of complexity is the differential effect on anti-PS responses of the subcapsular domains of distinct bacteria.

We believe that a number of unanswered questions are worthy of future study. (1) Do CD4[+] T cells indeed access bacterial protein to deliver cognate help to PS-specific B cells in response to intact bacteria, and if so, how do PS-specific B cells obtain this protein for MHC-II/peptide display necessary for receiving T cell help? (2) Are some of the basic lessons learned from studying the peripheral immune system in response to systemic injection of inactivated bacteria apply to the mucosal immune system, as well as to live bacteria where added levels of complexity are no doubt operative? (3) What are the implications for immune homeostasis by PC-expressing bacteria, especially commensals or potentially pathogenic bacteria that colonize the mucosal surface in light of the PC-dependent, Pn-mediated suppression of Ig responses to soluble antigens? In this regard, it has been demonstrated that other extracellular bacteria can express PC (e.g., on pili or LPS) in a regulated fashion.[88,89] (4) Do anti-PS responses to GP versus GN extracellular bacteria differ, when studying the same expressed capsular PS? (5) What is the mechanism underlying the apparent selective usage of MZB cells for PS-specific Ig responses to bacteria, and FB cells for responses to a soluble conjugate vaccine containing the same PS and (6) are there distinct pathways of cellular regulation that underlie Ig responses to these two types of PS-containing immunogens?

Acknowledgments

I thank the following past and present members of my laboratory who contributed to the studies discussed in this review: Zheng-Qi Wu, Abdul Q. Khan, Yi Shen, Jesus Colino, Goutam Sen, Quanyi Chen, Katherine Lee, Gouri Chattopadhyay, Sam Vasilevsky, Swadhinya Arjunaraja, and Saumyaa, as well as my many collaborators.

Conflicts of interest

The author declares no conflicts of interest.

References

1. AlonsoDeVelasco, E., A.F. Verheul, J. Verhoef & H. Snippe. 1995. Streptococcus pneumoniae: virulence factors, pathogenesis, and vaccines. *Microbiol. Rev.* **59:** 591–603.

2. Mond, J.J., A. Lees & C.M. Snapper. 1995. T cell-independent antigens type 2. *Annu. Rev. Immunol.* **13:** 655–692.

3. Kalka-Moll, W.M., A.O. Tzianabos, P.W. Bryant, *et al.* 2002. Zwitterionic polysaccharides stimulate T cells by MHC class II-dependent interactions. *J. Immunol.* **169:** 6149–6153.

4. Cobb, B.A., Q. Wang, A.O. Tzianabos & D.L. Kasper. 2004. Polysaccharide processing and presentation by the MHCII pathway. *Cell* **117:** 677–687.

5. Harding, C.V., R.W. Roof, P.M. Allen & E.R. Unanue. 1991. Effects of pH and plysaccharides on peptide binding to class II major histocompatiblity complex molecules. *Proc. Natl. Acad. Sci. USA* **88:** 2740–2744.

6. Ishioka, G.Y., A.G. Lamont, D. Thomson, *et al.* 1992. MHC interaction and T cell recognition of carbohydrates and glycopeptides. *J. Immunol.* **148:** 2446–2451.

7. Baker, P.J. 1992. T cell regulation of the antibody response to bacterial polysaccharide antigens: an examination of some general characteristics and their implications. *J. Infect. Dis.* **165**(Suppl 1): S44–S48.

8. Mongini, P.K., K.E. Stein & W.E. Paul. 1981. T cell regulation of IgG subclass antibody production in response to T-independent antigens. *J. Exp. Med.* **153:** 1–12.

9. Jeurissen, A., M. Wuyts, A. Kasran, *et al.* 2002. Essential role for CD40 ligand interactions in T lymphocyte-mediated modulation of the murine immune response to pneumococcal capsular polysaccharides. *J. Immunol.* **168:** 2773–2781.

10. Kobrynski, L.J., A.O. Sousa, A.J. Nahmias & F.K. Lee. 2005. Cutting edge: antibody production to pneumococcal polysaccharides requires CD1 molecules and CD8+ T Cells. *J. Immunol.* **174:** 1787–1790.

11. Moens, L., A. Jeurissen, S. Nierkens, *et al.* 2009. Generation of antibody responses to pneumococcal capsular polysaccharides is independent of CD1 expression in mice. *Infect. Immun.* **77:** 1976–1980.

12. Brunswick, M., F.D. Finkelman, P.F. Highet, *et al.* 1988. Picogram quantities of anti-Ig antibodies coupled to dextran induce B cell proliferation. *J. Immunol.* **140:** 3364–3372.

13. Snapper, C.M. & J.J. Mond. 1996. A model for induction of T cell-independent humoral immunity in response to polysaccharide antigens. *J. Immunol.* **157:** 2229–2233.

14. Pecanha, L.M., C.M. Snapper, F.D. Finkelman, *et al.* 1991. Dextran-conjugated anti-Ig antibodies as a model for T cell-independent type 2 antigen-mediated stimulation of Ig secretion in vitro. I. Lymphokine dependence. *J. Immunol.* **146:** 833–839.

15. Snapper, C.M. & J.J. Mond. 1993. Towards a comprehensive view of immunoglobulin class switching. *Immunol. Today* **14:** 15–17.

16. Sen, G., A.Q. Khan, Q. Chen & C.M. Snapper. 2005. In vivo humoral immune responses to isolated pneumococcal polysaccharides are dependent on the presence of associated TLR ligands. *J. Immunol.* **175:** 3084–3091.

17. Arjunaraja, S., P. Massari, L.M. Wetzler, A. Lees, J. Colino & C.M. Snapper. 2011. The nature of an in vivo anti-capsular polysaccharide response is markedly influenced by the composition and/or architecture of the bacterial subcapsular domain. *J. Immunol.* In press.

18. Massari, P., P. Henneke, Y. Ho, *et al.* 2002. Cutting edge: immune stimulation by neisserial porins is toll-like receptor 2 and MyD88 dependent. *J. Immunol.* **168:** 1533–1537.

19. Latz, E., J. Franko, D.T. Golenbock & J.R. Schreiber. 2004. Haemophilus influenzae type b-outer membrane protein complex glycoconjugate vaccine induces cytokine production by engaging human toll-like receptor 2 (TLR2) and requires the presence of TLR2 for optimal immunogenicity. *J. Immunol.* **172:** 2431–2438.

20. Khan, A.Q., Q. Chen, Z.Q. Wu, *et al.* 2005. Both innate immunity and type 1 humoral immunity to Streptococcus pneumoniae are mediated by MyD88 but differ in their relative levels of dependence on toll-like receptor 2. *Infect. Immun.* **73:** 298–307.

21. Balazs, M., F. Martin, T. Zhou & J.F. Kearney. 2002. Blood dendritic cells interact with splenic marginal zone B cells to initiate T-independent immune responses. *Immunity* **17:** 341–352.

22. Litinskiy, M.B., B. Nardelli, D.M. Hilbert, *et al.* 2002. DCs induce CD40-independent immunoglobulin class switching through BLyS and APRIL. *Nat. Immunol.* **3:** 822–829.

23. Craxton, A., D. Magaletti, E.J. Ryan & E.A. Clark. 2003. Macrophage- and dendritic cell-dependent regulation of human B-cell proliferation requires the TNF family ligand BAFF. *Blood* **101:** 4464–4471.

24. Schneider, P. 2005. The role of APRIL and BAFF in lymphocyte activation. *Curr. Opin. Immunol.* **17:** 282–289.

25. von Bulow, G.U., J.M. van Deursen & R.J. Bram. 2001. Regulation of the T-independent humoral response by TACI. *Immunity* **14:** 573–582.

26. Mackay, F. & J.L. Browning. 2002. BAFF: a fundamental survival factor for B cells. *Nat. Rev. Immunol.* **2:** 465–475.

27. MacLennan, I.C., K.M. Toellner, A.F. Cunningham, *et al.* 2003. Extrafollicular antibody responses. *Immunol. Rev.* **194:** 8–18.

28. El Shikh, M.E., R.M. El Sayed, A.K. Szakal & J.G. Tew. 2009. T-independent antibody responses to T-dependent antigens: a novel follicular dendritic cell-dependent activity. *J. Immunol.* **182:** 3482–3491.

29. Huard, B., L. Arlettaz, C. Ambrose, *et al.* 2004. BAFF production by antigen-presenting cells provides T cell costimulation. *Int. Immunol.* **16:** 467–475.

30. Trinchieri, G. & A. Sher. 2007. Cooperation of Toll-like receptor signals in innate immune defence. *Nat. Rev. Immunol.* **7:** 179–190.

31. Franchi, L., C. McDonald, T.D. Kanneganti, *et al.* 2006. Nucleotide-binding oligomerization domain-like receptors: intracellular pattern recognition molecules for pathogen detection and host defense. *J. Immunol.* **177:** 3507–3513.

32. Robinson, M.J., D. Sancho, E.C. Slack, *et al.* 2006. Myeloid C-type lectins in innate immunity. *Nat. Immunol.* **7:** 1258–1265.

33. Bachmann, M.F. & G.T. Jennings. 2010. Vaccine delivery: a matter of size, geometry, kinetics and molecular patterns. *Nat. Rev. Immunol.* **10:** 787–796.

34. Kraal, G., H. Ter Hart, C. Meelhuizen, *et al.* 1989. Marginal zone macrophages and their role in the immune response

against T-independent type 2 antigens: modulation of the cells with specific antibody. *Eur. J. Immunol.* **19:** 675–680.

35. Colino, J., G. Chattopadhyay, G. Sen, *et al.* 2009. Parameters underlying distinct T cell-dependent polysaccharide-specific IgG responses to an intact gram-positive bacterium versus a soluble conjugate vaccine. *J. Immunol.* **183:** 1551–1559.

36. Kovacsovics-Bankowski, M., K. Clark, B. Benacerraf & K.L. Rock. 1993. Efficient major histocompatibility complex class I presentation of exogenous antigen upon phagocytosis by macrophages. *Proc. Natl. Acad. Sci. USA.* **90:** 4942–4946.

37. Ziegler, H.K., C.A. Orlin & C.W. Cluff. 1987. Differential requirements for the processing and presentation of soluble and particulate bacterial antigens by macrophages. *Eur. J. Immunol.* **17:** 1287–1296.

38. Vidard, L., M. Kovacsovics-Bankowski, S.K. Kraeft, *et al.* 1996. Analysis of MHC class II presentation of particulate antigens of B lymphocytes. *J. Immunol.* **156:** 2809–2818.

39. Pincus, J.H., E. Haber, M. Katz & A.M. Pappenheimer, Jr. 1968. Antibodies to pneumococcal polysaccharides: relation between binding and electrophoretic heterogeneity. *Science* **162:** 667–668.

40. Guttormsen, H.K., L.M. Wetzler, R.W. Finberg & D.L. Kasper. 1998. Immunologic memory induced by a glycoconjugate vaccine in a murine adoptive lymphocyte transfer model. *Infect. Immun.* **66:** 2026–2032.

41. Guttormsen, H.K., A.H. Sharpe, A.K. Chandraker, *et al.* 1999. Cognate stimulatory B-cell-T-cell interactions are critical for T-cell help recruited by glycoconjugate vaccines. *Infect. Immun.* **67:** 6375–6384.

42. Muthukkumar, S. & K.E. Stein. 2004. Immunization with meningococcal polysaccharide-tetanus toxoid conjugate induces polysaccharide-reactive T cells in mice. *Vaccine* **22:** 1290–1299.

43. Sorensen, U.B., J. Henrichsen, H.C. Chen & S.C. Szu. 1990. Covalent linkage between the capsular polysaccharide and the cell wall peptidoglycan of Streptococcus pneumoniae revealed by immunochemical methods. *Microb. Pathog.* **8:** 325–334.

44. Jedrzejas, M.J. 2004. Extracellular virulence factors of Streptococcus pneumoniae. *Front. Biosci.* **9:** 891–914.

45. Whitfield, C. 2006. Biosynthesis and assembly of capsular polysaccharides in Escherichia coli. *Annu. Rev. Biochem.* **75:** 39–68.

46. Rietschel, E.T., H. Brade, O. Holst, *et al.* 1996. Bacterial endotoxin: chemical constitution, biological recognition, host response, and immunological detoxification. *Curr. Top. Microbiol. Immunol.* **216:** 39–81.

47. Beuvery, E.C., F. Miedema, R.W. van Delft, *et al.* 1983. Preparation and physicochemical and immunological characterization of polysaccharide-outer membrane protein complexes of Neisseria meningitidis. *Infect. Immun.* **40:** 369–380.

48. Lucas, A.H., K.D. Moulton, V.R. Tang & D.C. Reason. 2001. Combinatorial library cloning of human antibodies to Streptococcus pneumoniae capsular polysaccharides: variable region primary structures and evidence for somatic mutation of Fab fragments specific for capsular serotypes 6B, 14, and 23F. *Infect. Immun.* **69:** 853–864.

49. Zhou, J., K.R. Lottenbach, S.J. Barenkamp, *et al.* 2002. Recurrent variable region gene usage and somatic mutation in the human antibody response to the capsular polysaccharide

of Streptococcus pneumoniae type 23F. *Infect. Immun.* **70:** 4083–4091.

50. Haber, E., M.N. Margolies & L.E. Cannon. 1977. Origins of antibody diversity: insight gained from amino acid sequence studies of elicited antibodies. *Cold. Spring. Harb. Symp. Quant. Biol.* **41:** Pt 2 647–659.

51. Paul, W.E., G.W. Siskind, B. Benacerraf & Z. Ovary. 1967. Secondary antibody responses in haptenic systems: cell population selection by antigen. *J. Immunol.* **99:** 760–770.

52. Kanswal, S., N. Katsenelson, W. Allman, *et al.* 2011. Suppressive effect of bacterial polysaccharides on BAFF system is responsible for their poor immunogenicity. *J. Immunol.* **186:** 2430–2443.

53. Lanoue, A., M. R. Clatworthy, P. Smith, *et al.* 2004. SIGN-R1 contributes to protection against lethal pneumococcal infection in mice. *J. Exp. Med.* **200:** 1383–1393.

54. Koppel, E.A., C.W. Wieland, V.C. van den Berg, *et al.* 2005. Specific ICAM-3 grabbing nonintegrin-related 1 (SIGNR1) expressed by marginal zone macrophages is essential for defense against pulmonary Streptococcus pneumoniae infection. *Eur. J. Immunol.* **35:** 2962–2969.

55. Chen, Z., S.B. Koralov & G. Kelsoe. 2000. Regulation of humoral immune responses by CD21/CD35. *Immunol. Rev.* **176:** 194–204.

56. Courtney, A.H., E.B. Puffer, J.K. Pontrello, *et al.* 2009. Sialylated multivalent antigens engage CD22 in trans and inhibit B cell activation. *Proc. Natl. Acad. Sci. USA.* **106:** 2500–2505.

57. Dintzis, H.M., R.Z. Dintzis & B. Vogelstein. 1976. Molecular determinants of immunogenicity: the immunon model of immune response. *Proc. Natl. Acad. Sci. USA.* **73:** 3671–3675.

58. Kamboj, K.K., H.L. Kirchner, R. Kimmel, *et al.* 2003. Significant variation in serotype-specific immunogenicity of the seven-valent Streptococcus pneumoniae capsular polysaccharide-CRM197 conjugate vaccine occurs despite vigorous T cell help induced by the carrier protein. *J. Infect. Dis.* **187:** 1629–1638.

59. Grabitzki, J. & G. Lochnit. 2009. Immunomodulation by phosphocholine–biosynthesis, structures and immunological implications of parasitic PC-epitopes. *Mol. Immunol.* **47:** 149–163.

60. Briles, D.E., J.D. King, M.A. Gray, *et al.* 1996. PspA, a protection-eliciting pneumococcal protein: immunogenicity of isolated native PspA in mice. *Vaccine* **14:** 858–867.

61. Khan, A.Q., A. Lees & C.M. Snapper. 2004. Differential regulation of IgG anti-capsular polysaccharide and antiprotein responses to intact Streptococcus pneumoniae in the presence of cognate CD4+ T cell help. *J. Immunol.* **172:** 532–539.

62. Chen, Q., J.L. Cannons, J.C. Paton, *et al.* 2008. A novel ICOS-independent, but CD28- and SAP-dependent, pathway of T cell-dependent, polysaccharide-specific humoral immunity in response to intact Streptococcus pneumoniae versus pneumococcal conjugate vaccine. *J. Immunol.* **181:** 8258–8266.

63. Dong, C., U.A. Temann & R.A. Flavell. 2001. Cutting edge: critical role of inducible costimulator in germinal center reactions. *J. Immunol.* **166:** 3659–3662.

64. Huang, D.C., S. Cory & A. Strasser. 1997. Bcl-2, Bcl-XL and adenovirus protein E1B19kD are functionally

equivalent in their ability to inhibit cell death. *Oncogene* **14:** 405–414.

65. Chattopadhyay, G., A.Q. Khan, G. Sen, *et al.* 2007. Transgenic expression of Bcl-xL or Bcl-2 by Murine B Cells enhances the In vivo antipolysaccharide, but Not antiprotein, response to intact streptococcus pneumoniae. *J. Immunol.* **179:** 7523–7534.

66. Avery, O.T. & W.F. Goebel. 1931. Chemo-immunological studies on conjugated carbohydrate-proteins: V. The immunological specificity of an antigen prepared by combining the capsular polysaccharide of type III pneumococcus with foreign protein. *J. Exp. Med.* **54:** 437–447.

67. Ahmad, H. & E.K. Chapnick. 1999. Conjugated polysaccharide vaccines. *Infect. Dis. Clin. North. Am.* **13:** 113–133, vii.

68. Beuvery, E.C., F. van Rossum & J. Nagel. 1982. Comparison of the induction of immunoglobulin M and G antibodies in mice with purified pneumococcal type 3 and meningococcal group C polysaccharides and their protein conjugates. *Infect. Immun.* **37:** 15–22.

69. Schneerson, R., O. Barrera, A. Sutton & J.B. Robbins. 1980. Preparation, characterization, and immunogenicity of Haemophilus influenzae type b polysaccharide-protein conjugates. *J. Exp. Med.* **152:** 361–376.

70. Sen, G., Q. Chen & C.M. Snapper. 2006. Immunization of aged mice with a pneumococcal conjugate vaccine combined with an unmethylated CpG-containing oligodeoxynucleotide restores defective immunoglobulin G antipolysaccharide responses and specific CD4+-T-cell priming to young adult levels. *Infect. Immun.* **74:** 2177–2186.

71. Satterthwaite, A.B., H. Cheroutre, W.N. Khan, *et al.* 1997. Btk dosage determines sensitivity to B cell antigen receptor cross-linking. *Proc. Natl. Acad. Sci. USA.* **94:** 13152–13157.

72. Satterthwaite, A.B., Z. Li & O.N. Witte. 1998. Btk function in B cell development and response. *Semin. Immunol.* **10:** 309–316.

73. Khan, A.Q., G. Sen, S. Guo, *et al.* 2006. Induction of in vivo antipolysaccharide immunoglobulin responses to intact Streptococcus pneumoniae is more heavily dependent on Btk-mediated B-cell receptor signaling than antiprotein responses. *Infect. Immun.* **74:** 1419–1424.

74. Song, H. & J. Cerny. 2003. Functional heterogeneity of marginal zone B cells revealed by their ability to generate both early antibody-forming cells and germinal centers with hypermutation and memory in response to a T-dependent antigen. *J. Exp. Med.* **198:** 1923–1935.

75. Rubtsov, A., P. Strauch, A. Digiacomo, *et al.* 2005. Lsc regulates marginal-zone B cell migration and adhesion and is required for the IgM T-dependent antibody response. *Immunity* **23:** 527–538.

76. Girkontaite, I., K. Missy, V. Sakk, *et al.* 2001. Lsc is required for marginal zone B cells, regulation of lymphocyte motility and immune responses. *Nat. Immunol.* **2:** 855–862.

77. Colino, J. & I. Outschoorn. 1998. Dynamics of the murine humoral immune respose to neisseria meningitidis group B capsular polysaccharide. *Infect. Immun.* **66:** 505–513.

78. Kasper, D.L., C.J. Baker, R.S. Baltimore, *et al.* 1979. Immunodeterminant specificity of human immunity to type III group B streptococcus. *J. Exp. Med.* **149:** 327–339.

79. Guttormsen, H.K., C.J. Baker, M.H. Nahm, *et al.* 2002. Type III group B streptococcal polysaccharide induces antibodies that cross-react with Streptococcus pneumoniae type 14. *Infect. Immun.* **70:** 1724–1738.

80. Lee, K.S., G. Sen & C.M. Snapper. 2005. Endogenous CD4+ CD25+ regulatory T cells play no apparent role in the acute humoral response to intact Streptococcus pneumoniae. *Infect. Immun.* **73:** 4427–4431.

81. Sette, A., M. Moutaftsi, J. Moyron-Quiroz, *et al.* 2008. Selective CD4+ T cell help for antibody responses to a large viral pathogen: deterministic linkage of specificities. *Immunity* **28:** 847–858.

82. Rigden, D.J., M.Y. Galperin & M.J. Jedrzejas. 2003. Analysis of structure and function of putative surface-exposed proteins encoded in the Streptococcus pneumoniae genome: a bioinformatics-based approach to vaccine and drug design. *Crit. Rev. Biochem. Mol. Biol.* **38:** 143–168.

83. Lopez, R., J.L. Garcia, E. Garcia, *et al.* 1992. Structural analysis and biological significance of the cell wall lytic enzymes of Streptococcus pneumoniae and its bacteriophage. *FEMS Microbiol. Lett.* **79:** 439–447.

84. Ronda, C., J.L. Garcia, E. Garcia, *et al.* 1987. Biological role of the pneumococcal amidase. Cloning of the lytA gene in Streptococcus pneumoniae. *Eur. J. Biochem.* **164:** 621–624.

85. Navarre, W.W. & O. Schneewind. 1999. Surface proteins of gram-positive bacteria and mechanisms of their targeting to the cell wall envelope. *Microbiol. Mol. Biol. Rev.* **63:** 174–229.

86. Chattopadhyay, G., Q. Chen, J. Colino, *et al.* 2009. Intact bacteria inhibit the induction of humoral immune responses to bacterial-derived and heterologous soluble T cell-dependent antigens. *J. Immunol.* **182:** 2011–2019.

87. Harnett, M.M., A.J. Melendez & W. Harnett. 2009. The therapeutic potential of the filarial nematode-derived immunodulator, ES-62 in inflammatory disease. *Clin. Exp. Immunol.* **159:** 256–267.

88. Casey, R., J. Newcombe, J. McFadden & K.B. Bodman-Smith. 2008. The acute-phase reactant C-reactive protein binds to phosphorylcholine-expressing Neisseria meningitidis and increases uptake by human phagocytes. *Infect. Immun.* **76:** 1298–1304.

89. Lysenko, E.S., J. Gould, R. Bals, *et al.* 2000. Bacterial phosphorylcholine decreases susceptibility to the antimicrobial peptide LL-37/hCAP18 expressed in the upper respiratory tract. *Infect. Immun.* **68:** 1664–1671.

Ann. N.Y. Acad. Sci. ISSN 0077-8923

ANNALS OF THE NEW YORK ACADEMY OF SCIENCES
Issue: *Glycobiology of the Immune Response*

CD33-related siglecs as potential modulators of inflammatory responses

Paul R. Crocker, Sarah J. McMillan, and Hannah E. Richards

Division of Cell Signaling and Immunology, College of Life Sciences, University of Dundee, Dundee, United Kingdom

Address for correspondence: Paul R. Crocker, Division of Cell Signaling and Immunology, Wellcome Trust Biocentre, College of Life Sciences, University of Dundee, Dundee, DD1 5EH, UK. p.r.crocker@dundee.ac.uk

The immune system must be tightly regulated to prevent unwanted tissue damage caused by exaggerated immune and inflammatory reactions. Inhibitory and activating immune receptors play a crucial role in this function via phosphotyrosine-dependent signaling pathways. A significant body of evidence has accumulated suggesting that the siglec family of sialic acid binding Ig-like lectins makes an important contribution to this immunoregulation. The CD33-related siglecs are a distinct subset of inhibitory and activating receptors, expressed primarily on leukocytes in a cell type–specific manner. Here, we critically assess the *in vitro* and *in vivo* evidence on the functional role for CD33-related siglecs in modulation of inflammatory and immune responses.

Keywords: immunoglobulin superfamily; inhibitory receptors; innate immunity; siglecs; sialic acid; CD33

Introduction

Siglecs (sialic-acid-binding immunoglobulin-like lectins) are type I transmembrane proteins belonging to the immunoglobulin superfamily.[1] Their sialylated ligands are commonly found at terminal positions of cell surface and secreted glycoconjugates. Siglecs consist of a sialic acid (Sia)–binding N-terminal V-set immunoglobulin (Ig)–like domain and a variable number of C2-set Ig-like domains connected via a transmembrane domain to a cytoplasmic region. The siglec family can be grouped into two subsets based upon evolutionary conservation and sequence similarity. Four siglecs are conserved in mammalian species: sialoadhesin (siglec-1, Sn), CD22 (siglec-2), myelin-associated glycoprotein (MAG, siglec-4), and siglec-15.[2] The second group, the CD33-related siglecs, exhibit a high degree of sequence similarity with CD33. Ten have been characterized in humans (CD33, siglec-5 to siglec-11, siglec-14, and siglec-16) and five in mice (mCD33, siglec-E, siglec-F, siglec-G, and siglec-H).

CD33-related siglecs usually contain one or more immunoreceptor tyrosine-based inhibitory motifs (ITIMs) or ITIM-like motifs in their cytoplasmic

tails and can function as inhibitory receptors by dampening tyrosine kinase-driven signaling pathways. This occurs predominantly via recruitment and activation of the tyrosine phosphatases SHP-1 and SHP-2 although other SH2-containing signaling molecules can also contribute, as reviewed.[1] The exceptions are human siglec-14 and siglec-16 and murine siglec-H, which lack ITIMs and interact with DAP12, an immunoreceptor tyrosine-based activation motif (ITAM)–containing adaptor protein that can trigger both activating and inhibitory signaling.[3–6]

The CD33-related siglecs are expressed primarily on leukocyte subsets and thus are thought to be involved in regulation of leukocyte functions during inflammatory and immune responses.[7,8] Figure 1 illustrates the differential expression patterns of CD33-related siglecs on leukocytes in humans compared to mice. Some are highly restricted in their expression pattern; for example, siglec-8 is found predominantly on circulating eosinophils[9,10] and at very low levels on basophils,[10] and siglec-H is an excellent marker of rodent plasmacytoid dendritic cells (pDCs).[4,11] In contrast, other siglecs demonstrate broad expression patterns; for example,

doi: 10.1111/j.1749-6632.2011.06449.x

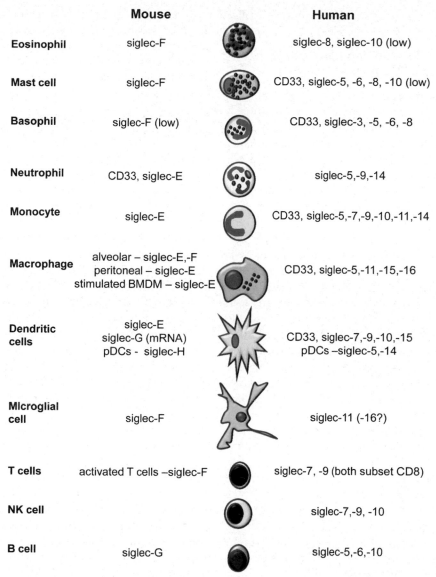

Figure 1. Comparison of mouse and human CD33-related siglecs on leukocyte populations. Expression of each siglec on the various cell populations has been validated using specific mAbs, except for siglec-G on dendritic cells, which has been shown at the mRNA level.

siglec-9 and siglec-E are found on monocytes, macrophages, neutrophils, and dendritic cells (DCs).[8,12] Several siglecs can be present on the same cell type. For example, human monocytes have been shown to express CD33 and siglec-5, siglec-7, siglec-9, and siglec-10, which could suggest functional redundancy among these receptors.[12–17] However, analysis of glycan-binding specificity has shown that each siglec exhibits a distinct binding pattern,[18,19] arguing that even when several siglecs are expressed

simultaneously, each may perform a unique function in fine-tuning cellular activation.

The molecular features and signaling properties of CD33-related siglecs point to a potentially important role in regulation of cellular activation in the innate immune system and the ensuing inflammatory responses. In this short review, we critically discuss the evidence accumulated to date in support of this notion and suggest areas where future research is needed. For general aspects of siglec structure, function, and glycan-binding properties, the reader is referred to several recent reviews.[20–23]

Approaches to studying cell signaling mediated by CD33-related siglecs

Although the structure, expression profile, and Sia recognition of CD33-related siglecs are well studied, the functional outcomes of physiological ligation and subsequent signaling are less well understood. Many of the studies have been carried out using cell lines overexpressing siglec constructs and/or using anti-siglec antibodies to ligate the receptors. A potential limitation of this strategy is that (1) overexpression may lead to aberrant signaling and (2) the use of antibodies as surrogate ligands may trigger non-physiological signaling pathways due to their high affinity, bivalent nature, and an Fc region that may be recognized by Fc receptors on leukocytes. The natural siglec–sialic acid interactions are much weaker than siglec–antibody interactions, typically in the affinity range of 100–1,000 μM. Siglec–sialic acid interactions are also highly complex since siglec recognition is strongly influenced by the glycosidic linkage to other sugars, the underlying glycan structure, glycan modifications such as sulfation, and the nature of the sialic acid moiety itself that, being a complex family of sugars, can vary across species. In addition, CD33-related siglecs are normally masked on the cell surface via *cis* interactions with cell-expressed sialic acids, which limits the ability of exogenous *trans* ligands to induce clustering at the cell surface (reviewed in Ref. 1). Alternative *in vitro* approaches include the use of synthetic sialylated carbohydrates to crosslink siglecs, use of naturally siglec-expressing cell types, and comparisons with cells from siglec-deficient mice.

To complicate matters, the sialic acid profile ("sialoglycome") of immune cells alters during both maturation and activation, and sialidases can be produced by both host and pathogen to modify this further and "unmask" siglecs, therefore potentially allowing interactions with *trans* ligands (reviewed in Ref. 1). Ultimately, *in vivo* studies using mouse models are required to identify the roles siglecs play in inflammation and immune modulation and to highlight potential therapeutic targets. However, the CD33-related siglec gene clusters are rapidly evolving, resulting in differential gene expansion and loss across species. It is therefore difficult to directly compare the biology of rodent and human

siglecs, despite the high conservation of Sia binding and signaling properties. Nevertheless, principles established using mouse models of disease can be extremely useful as a starting point and then translated into a clinical context using appropriate human systems.

Inhibitory siglecs as regulators of inflammatory responses

A number of studies have shown that antibody ligation of ITIM-bearing CD33-related siglecs induces rapid and transient src kinase-mediated phosphorylation of the tyrosine motifs and recruitment of SHP-1 and SHP-2 (Refs. 24–28). Orr *et al.*[29,30] showed that the SH2-domain containing suppressor of cytokine signaling-3 (SOCS-3) can compete with SHP1/2 for binding to human CD33 and siglec-7 ITIMs, which recruits ECS (Elongin C - CUL2 - SOCS-box) E3 ubiquitin ligase and results in concomitant proteosomal degradation of the siglec and SOCS3. Since SOCS3 expression is induced by cytokines and pathogen derivatives, this process potentially prevents siglec-induced inhibition of proliferation under inflammatory conditions, suggesting siglecs lie at the fulcrum of tolerance and activation.[29,30]

The interplay between siglecs and activating pattern recognition receptors, which primarily detect injured host or pathogen derivatives, is of significant interest. Toll-like receptors (TLRs) and NOD-like receptors (Nucleotide Oligomerization Domain, NLRs) are two families that can trigger inflammatory responses upon ligation and, in some cases, can modulate siglec expression. Murine siglec-E is expressed constitutively on neutrophils and other myeloid cells,[8] and is upregulated on bone marrow–derived macrophages following TLR stimulation with lipopolysaccharide (LPS) and other microbial components in a MyD88-dependent manner.[31] Antibody crosslinking of siglec-E on LPS-stimulated macrophages led to reduced production of proinflammatory cytokines, TNF-α and IL-6. Furthermore, siglec-E was tyrosine phosphorylated and associated with SHP-1 and SHP-2, suggesting that siglec-E upregulation on macrophages represents a negative feedback pathway that limits the inflammatory response to LPS signaling.

Similar findings were reported in macrophage cell lines. When cells overexpressed wild-type siglec-9 (Ref. 32), TLR ligation led to reduced TNF-α and

IL-6 production, and antiinflammatory cytokine IL-10 was concomitantly upregulated. This effect was not observed with cells overexpressing siglec-9 containing a mutated ITIM motif. These findings suggested that siglec-9 signaling dampens responses to TLR ligands. Although these studies failed to address the role of sialic acid recognition in siglec-dependent effects, another study suggested that mucin ligation of siglec-9 suppressed IL-12 production from LPS-stimulated monocyte-derived DCs.[33] Interestingly, and in contrast to siglec-E induction on macrophages, siglec-9 was downregulated upon LPS maturation of DCs. The authors also showed that the active mucins were present in the blood of two gastric cancer patients; however, they neither compared blood of healthy controls in this study, nor could they rule out the possibility that the binding of mucins to other receptors mediated this effect.[33] In apparent contrast to the inhibitory properties of these siglecs, siglec-5 ligation on neutrophils by either antibodies or α_1-acid glycoprotein (AGP) has been shown to prime N-formyl-methionine-leucine-phenylalanine (FMLP)-induced oxidative burst[34] and increases in cytosolic Ca^{2+} responses,[35] respectively. Further studies are required to investigate whether this siglec-dependent activation of neutrophils involves the activating siglec-14, which is coexpressed with siglec-5 on neutrophils.[36]

The differences in results seen in these studies could be due to a number of factors, such as the different siglecs and cell types under study. It is certainly possible that similar siglecs mediate distinct functions depending on their ligand specificity and/or host cell environment. Another possibility is that the results of siglec manipulation depend greatly on experimental context, and that their signaling functions may be masked in some cases by other more dominant signaling pathways. Siglec-dependent signaling may also be influenced by multiple experimental variables, such as the state of maturation of cells used and the precise nature and concentration of diverse activation stimuli.

Self versus nonself recognition?

A recent study has suggested that by forming a *cis* complex with the CD24 sialoglycoprotein, siglec-10/G is important for distinguishing host-derived *danger-associated* molecular patterns from *pathogen-associated* molecular patterns.[37,38] The authors showed that in a sublethal model of acetaminophen-induced liver necrosis, both CD24-deficient and siglec-G–deficient mice had elevated levels of proinflammatory cytokines and more severe liver necrosis, ultimately leading to enhanced mortality. High mobility group box 1 (HMGB1) was released during cellular injury, and was shown to form a complex with CD24/siglec-10/G. It was also demonstrated that addition of HMGB1 to wild-type (WT) bone marrow–derived DCs inhibited IL-6 and TNF-α production but not bone marrow–derived DCs from either CD24- or siglec-G–deficient animals. In contrast, no defect was shown in cytokine production upon LPS stimulation of DCs and no decrease was seen in mouse survival following a lethal dose of LPS. However, since the dose of LPS used was already lethal in WT mice, any increase in mortality in deficient animals would be difficult to resolve. Taken together with the high dose of LPS used to stimulate DCs, there may yet be an undetected role for siglec-G signaling during LPS stimulation.

Modulation of inflammatory responses by activating siglecs

Unlike most CD33-related siglecs, siglec-H, siglec-14, and siglec-16 contain only a short cytoplasmic tail, lacking ITIM and ITIM-like motifs. Instead, they contain a basic residue (lysine or arginine) in the transmembrane region, which is required for association with the ITAM-containing adaptor DAP12 and can trigger cellular activation. These properties suggest that these siglecs can function as activating receptors.

Siglec-H, expressed primarily on murine plasmacytoid DC (pDC), requires co-expression of DAP-12 for transport to the cell surface.[4,11] Although siglec-H displays structural features required for sialic acid recognition, binding to common sialoglycoconjugates has not been detected thus far.[11] In contrast to the predicted activating effects of siglec-H ligation, it was demonstrated that anti-siglec-H mAb inhibited IFN-α production by pDCs in response to the TLR9 ligand CpG.[4] In addition, at intermediate concentrations of CpG, pDC from DAP12-deficient mice had increased IFN-α production. In a separate study using different antibodies to siglec-H there was no impact on IL-6, IL-10, or TNF-α production. These seemingly contradictory findings could be explained by use of different antibodies and pDC populations, and may reflect the spatial and temporal effects of siglec signaling relative to TLR stimulation, as suggested for the "inhibitory" siglecs. In contrast to siglec-H, DAP-12–coupled siglec-14 produced activating effects when overexpressed in a monocytic cell line, leading to enhanced TNF-α production in response to LPS.[36] Interestingly, siglec-14 has high sequence similarity to the ITIM-containing siglec-5, and its identification warrants reinterpretation of some antibody-based studies with neutrophils, as they recognize both siglec-5 and siglec-14 that are coexpressed on these cells.[39] Identification of another receptor pair, siglec-11 (inhibitory) and siglec-16 (activating), suggests that siglec pairs have evolved to recognize similar ligands but have differential signaling outcomes (reviewed in Ref. 21).

Siglecs and inflammatory responses to sialic acid-bearing pathogens

Although enzymes of the Sia biosynthesis pathway are rare in lower organisms,[13,40] it is striking that a number of known human pathogens have independently evolved the capacity to synthesize or capture host Sia. This has led to the hypothesis that these pathogens have evolved to exploit inhibitory pathways and in turn drive the rapid evolution and expansion of this family of receptors to include activating forms (recently reviewed in Ref. 21).

However, sialoglycoconjugate incorporation into bacterial surfaces has been shown not only to hinder binding of complement components important for opsonization, phagocytosis, and killing by phagocytes but also to direct lysis by the complement membrane attack complex (MAC).[41] It has also been suggested that Sia integration passively disguises pathogens from the host immune system through a strategy of molecular mimicry. It is therefore difficult to resolve the direct impact of siglec binding to pathogens on inflammatory cells versus the other consequences of sialylation such as complement activation and negative charge.

Neisseria meningitidis was shown to interact selectively with Sn and siglec-5, leading to enhanced bacterial uptake by macrophages and cell transfectants, respectively.[42] However, only Sn was studied in context of primary cells, the functional consequences of enhanced phagocytosis mediated by these siglecs was not examined. In contrast to CD33-related siglecs, Sn is thought to function primarily as an adhesion molecule in immune cell interactions.[43] Several studies carried out with sia-lylated forms of *Campylobacter jejuni* demonstrated binding of bacteria and their lipooligosaccharides to siglec-7, siglec-9, and Sn.[25,44–47] Interestingly, siglec-7 recognition of bacterial sialic acid required sialidase pretreatment of the immune cells, suggesting either host- or bacteria-derived sialidase is important for determining the outcome of siglec–pathogen interaction. In a recent study,[47] targeting DC-expressed siglec-7 and Sn with α2,8-sialylated or α2,3-sialylated lipooligosaccharides led to differential modulation of cytokine production and Th1 or Th2 helper cell polarization, respectively. However, direct evidence that the effects on DCs were due to signaling via siglecs was not provided.

The best functional evidence that inhibitory siglecs can be targeted by sialylated pathogens has come from studies of group B *Streptococcus* (GBS).[48–50] Carlin *et al.* showed interactions of GBS with siglec-5, siglec-7, and siglec-9. Based on antibody inhibition experiments, engagement of siglec-9 on neutrophils by GBS or its derivatives impaired oxidative burst, granule protease release, neutrophil extracellular DNA traps, and extracellular bacterial killing. Interestingly, it was also shown that certain strains can engage siglec-5 in a sialic acid–independent manner with similar results.

While the above studies have shown that an interaction between pathogens and human siglecs can lead to alterations in the activity of immune cells *in vitro*, it remains to be shown how siglec targeting by pathogens might influence a productive infection *in vivo*. Since the repertoires of human and mouse CD33-related siglecs differ, and since the pathogens studied are principally human pathogens that do not readily infect mice to produce similar disease profiles, it is problematic to address the roles of siglecs during infection *in vivo*. However, in one attempt to address this, Erdmann *et al.* studied *trans* sialidase-expressing *Trypanosoma cruzi.*[51] They reported that strains expressing high levels of enzyme establish parasitemia in mice and suppress proinflammatory cytokine production and subsequent T cell activation by murine DC *in vitro*. Parasites also bound siglec-E–Fc fusion protein in a sialic acid–dependant manner, and in transfected CHO cell lines recruited siglec-E to the contact area, though without any apparent effect on infection rate. Although the authors showed that treatment of DC with anti-siglec-E Ig-coated beads down-regulated cytokine production, the interpretation of results using antibodies is problematic, as discussed earlier. The emerging use of siglec-deficient strains of mice will go further in addressing the significance of siglec ligation by pathogens and potential immunomodulation during infection.

Regulatory role of siglecs on granulocytes

Siglec-8 is selectively expressed on human eosinophils, basophils, and mast cells, which may permit selective targeting of these cells in eosinophil-mediated pathologies, for example, allergies, asthma, and parasitic disease. Indeed, it has been proposed that the siglec-8 gene may be a susceptibility locus for asthma.[52] *In vitro* ligation of siglec-8 through monoclonal antibody, alone or with crosslinking, or via autoantibodies in commercial gamma globulin (IVIG), have been reported to cause apoptosis in eosinophils.[53–55] Further analysis of the apoptotic pathways has revealed that siglec-8–induced apoptosis is mediated through the sequential production of reactive oxygen species, followed by induction of mitochondrial injury and caspase cleavage.[54] Moreover, in contrast to what was expected, eosinophil survival-promoting cytokines such as interleukin-5 (IL-5) and granulocyte-macrophage colony-stimulating factor (GM-CSF) enhanced the sensitivity of eosinophils to undergo apoptosis in response to siglec-8 antibody. Furthermore, *in vitro* crosslinking of siglec-F, the mouse paralog of siglec-8 on the surface of murine eosinophils, was also shown to induce apoptosis.[56] The potential clinical implication of these findings is that cytokine priming may render eosinophils susceptible to the proapoptotic effects of a siglec-8–engaging therapeutic agent.[57] However, use of a siglec-8 synthetic ligand was shown to be less effective than antibody at inducing IL-5 primed apoptosis,[58] suggesting that the responses observed *in vitro* through antibody crosslinking may be artificially enhanced.

Similarly, antibody crosslinking of siglec-9 was shown to induce apoptosis of neutrophils, which was further enhanced in the presence of proinflammatory cytokines.[59] Interestingly, a novel caspase-independent, nonapoptotic form of cell death was identified in GM-CSF-primed neutrophils prior to antibody crosslinking.[59] This phenomenon has also been observed *in vivo* in inflammatory settings. Since SHP-1 has been shown to be important in

neutrophil apoptosis,[60] siglec-9 could associate with SHP-1 in neutrophils to contribute to cell death, suggesting that in inflammatory situations, siglecs expressed on granulocytes are important in the regulation of the cellular accumulation. Indeed, this has been observed in siglec-F–deficient mice during a model of allergen-induced airway inflammation.[61]

Murine siglec-F is expressed on mature circulating eosinophils, mast cells, alveolar macrophages, and following allergen exposure on activated CD4+ T cells during a model of allergen-induced airway inflammation.[8,61,62] It is also a very useful marker for studies of eosinophil function and turnover.[63,64] Siglec-F shows weak binding to α-2,3 linked Sias, but has a strong preference for 6'-sulfo-sialyl Lewis X, the same ligand that selectively recognized by siglec-8.[62,65] Although siglec-F is not orthologous with siglec-8, their similarities in expression pattern[9] and ligand preference[62] suggest these paralogs play equivalent functional roles. Therefore, analysis of siglec-F–deficient mice is likely to give insights into siglec-8 functions in humans.

In models of allergen-induced airway inflammation, expression of both siglec-F on eosinophils and its sialic acid-containing ligands in the inflamed airways were shown to be elevated following allergen exposure of sensitized mice compared to sham controls.[61,66] Moreover, siglec-F–deficient mice demonstrated enhanced eosinophilic inflammation, possibly due to delayed resolution and diminished apoptosis.[61] Furthermore, using a model of chronic allergen-induced airway inflammation, increased eosinophilia in siglec-F–deficient mice was associated with enhanced features of airway remodeling.[66]

It is well documented that allergen-induced airway inflammation is driven by CD4+ Th2 cells.[67,68] In general, mouse T cells do not express siglecs, although CD22 has been reported at very low levels on primary mouse T cells.[69] Recently, work demonstrated that upon activation, CD4+ and CD8+ T cells *in vitro* and CD4+ T cells *in vivo* express siglec-F.[61] Since, siglec-F is also increased on CD4+ T cells, it would be interesting to investigate the effects on T cell function in siglec-F–deficient mice. Siglec-F is also expressed on alveolar macrophages and mast cells, and thus experiments should address the contribution of siglec-F on these populations *in vivo*. Indeed, in addition to promoting eosinophil apoptosis, targeting siglec-8 could also result in mast cell inhibitory biology *in vivo*.[70]

Parallel studies demonstrated that administration of anti-siglec-F antibody reduced eosinophilic inflammation and airway remodeling in a chronic model of allergen-induced airway inflammation and eosinophilic inflammation in the intestine.[71,72] However, since siglec-F is expressed on a number of cell populations, it is unknown what the antibodies are targeting *in vivo*. It has been documented that administration of anti-siglec-F antibody to IL-5 transgenic mice reduced eosinophil viability; but this has not been shown in more complex inflammatory situations where there are a multitude of signals involved.[56] Moreover, if targeting siglecs is going to progress as a potential therapy in the clinic, studies need to be performed using therapeutic dosing of the antibody since patients will already have established disease.

Together, these studies suggest that siglec-F expressed on eosinophils and activated T cells, in combination with an increased expression of sialylated ligands, provides a negative feedback loop in regulating allergic responses (in the described model). However, caution should be taken in interpreting the data for the mechanism of action *in vivo*. Further work is required before siglecs can be definitively ruled in as candidate therapeutics in controlling leukocyte expansion during inflammation of eosinophilic disorders.

CD33-related siglecs and neurodegenerative disease

In chronic neurodegenerative diseases it has been proposed that systemic inflammation may accelerate the onset and progression of clinical symptoms.[73] Microglia are the resident macrophages of the brain and thus act as the first and main form of active immune defense in the central nervous system.[74] Data suggest that while microglia priming is a general phenomenon of brain neuropathology, upon systemic infection priming can contribute to disease progression by inducing a switch in microglial phenotype. In response to injury, ischemia, and inflammatory stimuli, microglial cells assume an activated phenotype and can produce proinflammatory mediators such as IL-1, TNF-α, and nitric oxide, which are potent inducers of neuronal damage and cell death.[75] Conversely, microglia can also express antiinflammatory cytokines and initiate negative feedback pathways that result in repair, resolution of inflammation, and maintenance of tissue homeostasis. One receptor that may regulate the inhibitory signaling pathways in microglia is siglec-11, which has been shown to be expressed on microglia[76,77] to interact with SHP-1 and SHP-2 protein tyrosine phosphatases.[24]

Mice lack the ortholog of siglec-11; however, murine microglia have been used to study the role of ectopically expressed siglec-11 because the intracellular signaling pathways in mice and humans are thought to be highly conserved. Crosslinking of siglec-11 on microglia suppressed LPS-induced gene transcription of proinflammatory mediators.[77] Moreover, a neuroprotective effect of siglec-11 was observed in microglia–neuron coculture, as siglec-11–expressing microglial cells were less toxic to neurons and led to reduced neurite and neuronal cell body loss.[77] PSA, the long linear homopolymer of Neu5Ac-α2–8, has been shown to be a ligand of siglec-11. The PSA residues expressed on neurons, but not those on microglia, contributed to the neuroprotective function of microglial siglec-11. Recently, siglec-16 was found and described as being evolved from siglec-11 as a DAP12-associated receptor expressed in macrophages.[6] It can be hypothesized that opposing signals from these two paired receptors in the brain could be involved in the regulation of inflammatory responses important in the pathology of neurodegenerative disease.[76]

Interestingly, increased siglec-F expression in macrophages was observed in a mouse model of prion disease.[78] These data suggest that macrophage-expressed siglec-F may play an important role in the regulation of macrophage activation and signaling in the brain to prevent exaggerated inflammatory responses. Together, these data indicate possible new therapeutic interventions of microglial siglecs in neurodegenerative diseases.

Concluding comments

Although the various CD33-related siglecs contain ITIM motifs or associate with ITAM-containing adaptors, the functional outcome of siglec ligation on immune and inflammatory responses is complex and often unpredictable. Although simplified *in vitro* systems are necessary to understand molecular interactions and signaling pathways, their simplicity can hide the complex interplay between the leukocyte receptor repertoire and the inflammatory environment *in vivo*. Siglecs clearly have an important fine-tuning modulatory role to play in inflammation, and further study with appropriate model systems is likely to open new avenues toward exploiting these proteins as therapeutic targets in multiple diseases.

Conflicts of interest

The authors declare no conflicts of interest.

References

1. Crocker, P.R., J.C. Paulson & A. Varki. 2007. Siglecs and their roles in the immune system. *Nat. Rev. Immunol.* **7:** 255–266.
2. Angata, T., Y. Tabuchi, K. Nakamura & M. Nakamura. 2007. Siglec-15: an immune system Siglec conserved throughout vertebrate evolution. *Glycobiology* **17:** 838–846.
3. Attrill, H., A. Imamura, R.S. Sharma, *et al.* 2006. Siglec-7 undergoes a major conformational change when complexed with the alpha(2,8)-disialylganglioside GT1b. *J. Biol. Chem.* **281:** 32774–32783.
4. Blasius, A.L., M. Cella, J. Maldonado, *et al.* 2006. Siglec-H is an IPC-specific receptor that modulates type I IFN secretion through DAP12. *Blood* **107:** 2474–2476.
5. Hamerman, J.A., J.R. Jarjoura, M.B. Humphrey, *et al.* 2006. Cutting edge: inhibition of TLR and FcR responses in macrophages by triggering receptor expressed on myeloid cells (TREM)-2 and DAP12. *J. Immunol.* **177:** 2051–2055.
6. Cao, H., U. Lakner, B. de Bono, *et al.* 2008. SIGLEC16 encodes a DAP12-associated receptor expressed in macrophages that evolved from its inhibitory counterpart SIGLEC11 and has functional and non-functional alleles in humans. *Eur. J. Immunol.* **38:** 2303–2315.
7. Lock, K., J. Zhang, J. Lu, *et al.* 2004. Expression of CD33-related siglecs on human mononuclear phagocytes, monocyte-derived dendritic cells and plasmacytoid dendritic cells. *Immunobiology* **209:** 199–207.
8. Zhang, J.Q., B. Biedermann, L. Nitschke, *et al.* 2004. The murine inhibitory receptor mSiglec-E is expressed broadly on cells of the innate immune system whereas mSiglec-F is restricted to eosinophils. *Eur. J. Immunol.* **34:** 1175–1184.
9. Floyd, H., J. Ni, A.L. Cornish, *et al.* 2000. Siglec-8. A novel eosinophil-specific member of the immunoglobulin superfamily. *J. Biol. Chem.* **275:** 861–866.
10. Kikly, K.K., B.S. Bochner, S.D. Freeman, *et al.* 2000. Identification of SAF-2, a novel siglec expressed on eosinophils, mast cells, and basophils. *J. Allergy Clin. Immunol.* **105:** 1093–1100.
11. Zhang, J., A. Raper, N. Sugita, *et al.* 2006. Characterization of Siglec-H as a novel endocytic receptor expressed on murine plasmacytoid dendritic cell precursors. *Blood* **107:** 3600–3608.
12. Zhang, J.Q., G. Nicoll, C. Jones & P.R. Crocker. 2000. Siglec-9, a novel sialic acid binding member of the immunoglobulin superfamily expressed broadly on human blood leukocytes. *J. Biol. Chem.* **275:** 22121–22126.
13. Angata, T. & A. Varki. 2000. Cloning, characterization, and phylogenetic analysis of siglec-9, a new member of the CD33-related group of siglecs. Evidence for co-evolution with sialic acid synthesis pathways. *J. Biol. Chem.* **275:** 22127–22135.
14. Cornish, A.L., S. Freeman, G. Forbes, *et al.* 1998. Characterization of siglec-5, a novel glycoprotein expressed on myeloid cells related to CD33. *Blood* **92:** 2123–2132.
15. Munday, J., S. Kerr, J. Ni, *et al.* 2001. Identification, characterization and leucocyte expression of Siglec-10, a novel human sialic acid-binding receptor. *Biochem. J.* **355:** 489–497.
16. Nicoll, G., J. Ni, D. Liu, *et al.* 1999. Identification and characterization of a novel siglec, siglec-7, expressed by human natural killer cells and monocytes. *J. Biol. Chem.* **274:** 34089–34095.
17. Vitale, C., C. Romagnani, M. Falco, *et al.* 1999. Engagement of p75/AIRM1 or CD33 inhibits the proliferation of normal or leukemic myeloid cells. *Proc. Natl. Acad. Sci. U.S.A.* **96:** 15091–15096.
18. Blixt, O., B.E. Collins, I.M. van den Nieuwenhof, *et al.* 2003. Sialoside specificity of the siglec family assessed using novel multivalent probes: identification of potent inhibitors of myelin-associated glycoprotein. *J. Biol. Chem.* **278:** 31007–31019.
19. Campanero-Rhodes, M.A., R.A. Childs, M. Kiso, *et al.* 2006. Carbohydrate microarrays reveal sulphation as a modulator of siglec binding. *Biochem. Biophys. Res. Commun.* **344:** 1141–1146.
20. Magesh, S., H. Ando, T. Tsubata, *et al.* 2011. High-affinity ligands of Siglec receptors and their therapeutic potentials. *Curr. Med. Chem.* **18:** 3537–3550.
21. Cao, H. & P.R. Crocker. 2011. Evolution of CD33-related siglecs: regulating host immune functions and escaping pathogen exploitation? *Immunology* **132:** 18–26.
22. O'Reilly, M.K. & J.C. Paulson. 2009. Siglecs as targets for therapy in immune-cell-mediated disease. *Trends Pharmacol. Sci.* **30:** 240–248.
23. Varki, A. & T. Angata. 2006. Siglecs: the major subfamily of I-type lectins. *Glycobiology.* **16:** 1R–27R.
24. Angata, T., S.C. Kerr, D.R. Greaves, et al. 2002. Cloning and characterization of human Siglec-11. A recently evolved signaling that can interact with SHP-1 and SHP-2 and is expressed by tissue macrophages, including brain microglia. *J. Biol. Chem.* **277:** 24466–24474.
25. Avril, T., H. Floyd, F. Lopez, *et al.* 2004. The membrane-proximal immunoreceptor tyrosine-based inhibitory motif is critical for the inhibitory signaling mediated by Siglecs-7 and -9, CD33-related Siglecs expressed on human monocytes and NK cells. *J. Immunol.* **173:** 6841–6849.
26. Avril, T., S.D. Freeman, H. Attrill, *et al.* 2005. Siglec-5 (CD170) can mediate inhibitory signaling in the absence of immunoreceptor tyrosine-based inhibitory motif phosphorylation. *J. Biol. Chem.* **280:** 19843–19851.
27. Freeman, S.D., S. Kelm, E.K. Barber & P.R. Crocker. 1995. Characterization of CD33 as a new member of the sialoadhesin family of cellular interaction molecules. *Blood* **85:** 2005–2012.
28. Taylor, V.C., C.D. Buckley, M. Douglas, *et al.* 1999. The myeloid-specific sialic acid-binding receptor, CD33, associates with the protein-tyrosine phosphatases, SHP-1 and SHP-2. *J. Biol. Chem.* **274:** 11505–11512.
29. Orr, S.J., N.M. Morgan, R.J. Buick, *et al.* 2006. SOCS3 targets Siglec 7 for proteasomal degradation and blocks Siglec 7-mediated responses. *J. Biol. Chem.* **282:** 3418–3422.
30. Orr, S.J., N.M. Morgan, J. Elliott, *et al.* 2006. CD33 responses are blocked by SOCS3 through accelerated proteasomal-mediated turnover. *Blood* **109:** 1061–1068.

31. Boyd, C.R., S.J. Orr, S. Spence, *et al.* 2009. Siglec-E is up-regulated and phosphorylated following lipopolysaccharide stimulation in order to limit TLR-driven cytokine production. *J. Immunol.* **183:** 7703–7709.

32. Ando, M., W. Tu, K. Nishijima & S. Iijima. 2008. Siglec-9 enhances IL-10 production in macrophages via tyrosine-based motifs. *Biochem. Biophys. Res. Commun.* **369:** 878–883.

33. Ohta, M., A. Ishida, M. Toda, *et al.* 2010. Immunomodulation of monocyte-derived dendritic cells through ligation of tumor-produced mucins to Siglec-9. *Biochem. Biophys. Res. Commun.* **402:** 663–669.

34. Erickson-Miller, C.L., S.D. Freeman, C.B. Hopson, *et al.* 2003. Characterization of Siglec-5 (CD170) expression and functional activity of anti-Siglec-5 antibodies on human phagocytes. *Exp. Hematol.* **31:** 382–388.

35. Gunnarsson, P., L. Levander, P. Pahlsson & M. Grenegard. 2007. The acute-phase protein alpha 1-acid glycoprotein (AGP) induces rises in cytosolic Ca2+ in neutrophil granulocytes via sialic acid binding immunoglobulin-like lectins (siglecs). *FASEB J.* **21:** 4059–4069.

36. Yamanaka, M., Y. Kato, T. Angata & H. Narimatsu. 2009. Deletion polymorphism of SIGLEC14 and its functional implications. *Glycobiology* **19:** 841–846.

37. Chen, G.Y., J. Tang, P. Zheng & Y. Liu. 2009. CD24 and Siglec-10 selectively repress tissue damage-induced immune responses. *Science* **323:** 1722–1725.

38. Chen, G.Y., X. Chen, S. King, *et al.* 2011. Amelioration of sepsis by inhibiting sialidase-mediated disruption of the CD24-SiglecG interaction. *Nat. Biotechnol.* **29:** 428–435.

39. Angata, T., T. Hayakawa, M. Yamanaka, *et al.* 2006. Discovery of Siglec-14, a novel sialic acid receptor undergoing concerted evolution with Siglec-5 in primates. *FASEB J.* **20:** 1964–1973.

40. Schauer, R. & J. Kamerling. 1997. Chemistry, biochemistry and biology of silaic acids. In *Glycoproteins II.* J. Montreuil, J.F.G. Vliegenthart, & H. Schachter, Eds.: 243–402. Elsevier Science. Amsterdam.

41. Severi, E., D.W. Hood & G.H. Thomas. 2007. Sialic acid utilization by bacterial pathogens. *Microbiology* **153:** 2817–2822.

42. Jones, C., M. Virji & P.R. Crocker. 2003. Recognition of sialylated meningococcal lipopolysaccharide by siglecs expressed on myeloid cells leads to enhanced bacterial uptake. *Mol. Microbiol.* **49:** 1213–1225.

43. Wu, C., U. Rauch, E. Korpos, *et al.* 2009. Sialoadhesin-positive macrophages bind regulatory T cells, negatively controlling their expansion and autoimmune disease progression. *J. Immunol.* **182:** 6508–6516.

44. Avril, T., S.J. North, S.M. Haslam, *et al.* 2006. Probing the cis interactions of the inhibitory receptor Siglec-7 with alpha2,8-disialylated ligands on natural killer cells and other leukocytes using glycan-specific antibodies and by analysis of alpha2,8-sialyltransferase gene expression. *J. Leukoc. Biol.* **80:** 787–796.

45. Avril, T., E.R. Wagner, H.J. Willison & P.R. Crocker. 2006. Sialic acid-binding immunoglobulin-like lectin 7 mediates selective recognition of sialylated glycans expressed on *Campylobacter jejuni* lipooligosaccharides. *Infect. Immun.* **74:** 4133–4141.

46. Heikema, A.P., M.P. Bergman, H. Richards, *et al.* 2010. Characterization of the specific interaction between sialoadhesin and sialylated *Campylobacter jejuni* lipooligosaccharides. *Infect. Immun.* **78:** 3237–3246.

47. Bax, M., M.L. Kuijf, A.P. Heikema, *et al.* 2011. *Campylobacter jejuni* lipooligosaccharides modulate dendritic cell-mediated T cell polarization in a sialic acid linkage-dependent manner. *Infect. Immun.* **79:** 2681–2689.

48. Carlin, A.F., Y.C. Chang, T. Areschoug, *et al.* 2009. Group B *Streptococcus* suppression of phagocyte functions by protein-mediated engagement of human Siglec-5. *J. Exp. Med.* **206:** 1691–1699.

49. Carlin, A.F., A.L. Lewis, A. Varki & V. Nizet. 2007. Group B streptococcal capsular sialic acids interact with siglecs (immunoglobulin-like lectins) on human leukocytes. *J. Bacteriol.* **189:** 1231–1237.

50. Carlin, A.F., S. Uchiyama, Y.C. Chang, *et al.* 2009. Molecular mimicry of host sialylated glycans allows a bacterial pathogen to engage neutrophil Siglec-9 and dampen the innate immune response. *Blood* **113:** 3333–3336.

51. Erdmann, H., C. Steeg, F. Koch-Nolte, *et al.* 2009. Sialylated ligands on pathogenic Trypanosoma cruzi interact with Siglec-E (sialic acid-binding Ig-like lectin-E). *Cell Microbiol.* **11:** 1600–1611.

52. Gao, P.S., K. Shimizu, A.V. Grant, *et al.* 2010. Polymorphisms in the sialic acid-binding immunoglobulin-like lectin-8 (Siglec-8) gene are associated with susceptibility to asthma. *Eur. J. Hum. Genet.* **18:** 713–719.

53. Nutku, E., H. Aizawa, S.A. Hudson & B.S. Bochner. 2003. Ligation of Siglec-8: a selective mechanism for induction of human eosinophil apoptosis. *Blood* **101:** 5014–5020.

54. Nutku, E., S.A. Hudson & B.S. Bochner. 2005. Mechanism of Siglec-8-induced human eosinophil apoptosis: role of caspases and mitochondrial injury. *Biochem. Biophys. Res. Commun.* **336:** 918–924.

55. von Gunten, S., M. Vogel, A. Schaub, *et al.* 2007. Intravenous immunoglobulin preparations contain anti-Siglec-8 autoantibodies. *J. Allergy. Clin. Immunol.* **119:** 1005–1011.

56. Zimmermann, N., M.L. McBride, Y. Yamada, *et al.* 2008. Siglec-F antibody administration to mice selectively reduces blood and tissue eosinophils. *Allergy* **63:** 1156–1163.

57. Nutku-Bilir, E., S.A. Hudson & B.S. Bochner. 2008. Interleukin-5 priming of human eosinophils alters siglec-8 mediated apoptosis pathways. *Am. J. Respir. Cell Mol. Biol.* **38:** 121–124.

58. Hudson, S.A., N.V. Bovin, R.L. Schnaar, *et al.* 2009. Eosinophil-selective binding and proapoptotic effect in vitro of a synthetic Siglec-8 ligand, polymeric 6'-sulfated sialyl Lewis x. *J. Pharmacol. Exp. Ther.* **330:** 608–612.

59. von Gunten, S., S. Yousefi, M. Seitz, *et al.* 2005. Siglec-9 transduces apoptotic and nonapoptotic death signals into neutrophils depending on the proinflammatory cytokine environment. *Blood* **106:** 1423–1431.

60. Yousefi, S. & H.U. Simon. 2003. SHP-1: a regulator of neutrophil apoptosis. *Semin. Immunol.* **15:** 195–199.

61. Zhang, M., T. Angata, J.Y. Cho, *et al.* 2007. Defining the in vivo function of Siglec-F, a CD33-related Siglec expressed on mouse eosinophils. *Blood* **109:** 4280–4287.

62. Tateno, H., P.R. Crocker & J.C. Paulson. 2005. Mouse Siglec-F and human Siglec-8 are functionally convergent paralogs that are selectively expressed on eosinophils and recognize 6'-sulfo-sialyl Lewis X as a preferred glycan ligand. *Glycobiology* **15:** 1125–1135.

63. Dyer, K.D., M. Czapiga, B. Foster, *et al.* 2007. Eosinophils from lineage-ablated Delta dblGATA bone marrow progenitors: the dblGATA enhancer in the promoter of GATA-1 is not essential for differentiation ex vivo. *J. Immunol.* **179:** 1693–1699.

64. Ohnmacht, C., A. Pullner, N. van Rooijen & D. Voehringer. 2007. Analysis of eosinophil turnover in vivo reveals their active recruitment to and prolonged survival in the peritoneal cavity. *J. Immunol.* **179:** 4766–4774.

65. Bochner, B.S., R.A. Alvarez, P. Mehta, *et al.* 2005. Glycan array screening reveals a candidate ligand for Siglec-8. *J. Biol. Chem.* **280:** 4307–4312.

66. Cho, J.Y., D.J. Song, A. Pham, *et al.* 2010. Chronic OVA allergen challenged Siglec-F deficient mice have increased mucus, remodeling, and epithelial Siglec-F ligands which are up-regulated by IL-4 and IL-13. *Respir. Res.* **11:** 154.

67. Cohn, L., R.J. Homer, A. Marinov, *et al.* 1997. Induction of airway mucus production By T helper 2 (Th2) cells: a critical role for interleukin 4 in cell recruitment but not mucus production. *J. Exp. Med.* **186:** 1737–1747.

68. Gonzalo, J.A., C.M. Lloyd, L. Kremer, *et al.* 1996. Eosinophil recruitment to the lung in a murine model of allergic inflammation. The role of T cells, chemokines, and adhesion receptors. *J. Clin. Invest.* **98:** 2332–2345.

69. Sathish, J.G., J. Walters, J.C. Luo, *et al.* 2004. CD22 is a functional ligand for SH2 domain-containing protein-tyrosine phosphatase-1 in primary T cells. *J. Biol. Chem.* **279:** 47783–47791.

70. Yokoi, H., O.H. Choi, W. Hubbard, *et al.* 2008. Inhibition of FcepsilonRI-dependent mediator release and calcium flux from human mast cells by sialic acid-binding immunoglobulin-like lectin 8 engagement. *J. Allergy Clin. Immunol.* **121:** 499–505, e491.

71. Song, D.J., J.Y. Cho, S.Y. Lee, *et al.* 2009. Anti-Siglec-F antibody reduces allergen-induced eosinophilic inflammation and airway remodeling. *J. Immunol.* **183:** 5333–5341.

72. Song, D.J., J.Y. Cho, M. Miller, *et al.* 2009. Anti-Siglec-F antibody inhibits oral egg allergen induced intestinal eosinophilic inflammation in a mouse model. *Clin. Immunol.* **131:** 157–169.

73. Teeling, J.L. & V.H. Perry. 2009. Systemic infection and inflammation in acute CNS injury and chronic neurodegeneration: underlying mechanisms. *Neuroscience* **158:** 1062–1073.

74. Dheen, S.T., C. Kaur & E.A. Ling. 2007. Microglial activation and its implications in the brain diseases. *Curr. Med. Chem.* **14:** 1189–1197.

75. Ransohoff, R.M. & V.H. Perry. 2009. Microglial physiology: unique stimuli, specialized responses. *Annu. Rev. Immunol.* **27:** 119–145.

76. Hayakawa, T., T. Angata, A.L. Lewis, *et al.* 2005. A human-specific gene in microglia. *Science* **309:** 1693.

77. Wang, Y. & H. Neumann. 2010. Alleviation of neurotoxicity by microglial human Siglec-11. *J. Neurosci.* **30:** 3482–3488.

78. Lunnon, K., J.L. Teeling, A.L. Tutt, *et al.* 2011. Systemic inflammation modulates Fc receptor expression on microglia during chronic neurodegeneration. *J. Immunol.* **186:** 7215–7224.

Ann. N.Y. Acad. Sci. ISSN 0077-8923

ANNALS OF THE NEW YORK ACADEMY OF SCIENCES
Issue: *Glycobiology of the Immune Response*

Sulfated glycans control lymphocyte homing

Hiroto Kawashima[1] and Minoru Fukuda[2]

[1]Laboratory of Microbiology and Immunology, School of Pharmaceutical Sciences, University of Shizuoka, Shizuoka, Japan
[2]Glycobiology Unit, Sanford-Burnham Medical Research Institute, La Jolla, California

Address for correspondence: Hiroto Kawashima, Laboratory of Microbiology and Immunology, School of Pharmaceutical Sciences, University of Shizuoka, 52-1 Yada, Shizuoka 422-8526, Japan, kawashih@u-shizuoka-ken.ac.jp; or Minoru Fukuda, Glycobiology Unit, Sanford-Burnham Medical Research Institute, 10901 North Torrey Pines Road, La Jolla, CA 92037. minoru@sanfordburnham.org

Lymphocyte homing to the secondary lymphoid organs is pivotal for proper immune responses. Studies using sulfotransferase-deficient mice showed that 6-sulfo sialyl Lewis X (6-sulfo sLe^x), a major ligand for L-selectin that is expressed on the high endothelial venules (HEVs), plays critical roles in lymphocyte homing to the peripheral lymph nodes. More recent studies revealed that 6-sulfo sLe^x is essential for the homing of CD4$^+$CD25$^-$ conventional T cells to the nasal-associated lymphoid tissues (NALT) and is involved in nasal allergy. Further studies revealed that the homing of the CD4$^+$CD25$^+$ regulatory T cells to the NALT is dependent not only on the L-selectin-sulfated glycan interaction but also on P-selectin glycoprotein ligand-1 and CD44. These findings suggest that different carbohydrate-dependent homing mechanisms are utilized for different lymphocyte subsets. Recent studies indicated that the L-selectin–sulfated glycan interaction is also important for lymphocyte homing in chronic inflammation. In this review, the functions of the sulfated glycans in lymphocyte homing in physiological and pathological conditions are discussed.

Keywords: lymphocyte homing; L-selectin; high endothelial venule; sulfated glycan; sulfotransferase; regulatory T cell

Introduction

Lymphocytes migrate to the secondary lymphoid organs, such as the peripheral lymph nodes (PLNs) and Peyer's patches (PPs), where foreign antigens accumulate and where the immune responses primarily occur. The lymphocytes that migrate to PLNs and PPs through specialized blood vessels, called high endothelial venules (HEVs), emigrate through the efferent lymphatics to the lymph unless they encounter their cognate antigens. The lymphocytes that emigrate to the lymph return to the blood-stream through the thoracic duct through which the lymph fluid drains into the blood. The lymphocytes that return to the blood migrate again into the secondary lymphoid organs. This circulatory process is called lymphocyte recirculation or lymphocyte homing, which increases the chances that the lymphocytes will encounter antigens. Thus, lymphocyte homing is important for the immune system to recognize foreign antigens efficiently.

HEVs have a characteristic cuboidal morphology and a prominent Golgi complex where unique sulfated glycans are synthesized.[1] Lymphocyte migration is achieved by a series of interactions between lymphocytes and HEVs:[2] (i) lymphocyte rolling mediated by the interaction between L-selectin on the lymphocytes and a sulfated glycan, 6-sulfo sialyl Lewis X (Sialic acidα2-3Galβ1-4[Fucα1-3(sulfo-6)]GlcNAcβ1-R; 6-sulfo sLe^x) on the HEVs; (ii) activation of lymphocytes by chemokines presented on the surface of HEVs by heparan sulfate (HS);[3] (iii) firm attachment of lymphocytes mediated by integrins; and (iv) transmigration. Our recent findings suggested that different carbohydrate-dependent mechanisms can be utilized for the initial rolling interaction of a specific lymphocyte subset.[4] Recent studies also indicated that the sulfated glycans are often ectopically induced on HEV-like blood vessels at the sites of chronic inflammation and play important roles in pathological lymphocyte homing. In this review, we will discuss the roles

doi: 10.1111/j.1749-6632.2011.06356.x

of the sulfated glycans in lymphocyte homing under physiological and pathological conditions.

The homing receptor L-selectin and its ligands

The homing receptor L-selectin (LECAM-1, CD62L, LAM-1) recognizes carbohydrate ligands through the C-type lectin-like domain and mediates the initial tethering and rolling interaction of the lymphocytes on the surface of HEVs. L-selectin is localized to the tips of microvilli,[5] which is advantageous for the initial tethering to the HEV-borne ligands.

Various mucin-like glycoproteins, such as glycosylation-dependent cell adhesion molecule (GlyCAM)-1, CD34, podocalyxin-like protein, and Sgp200,[6] function as L-selectin ligands in the PLN HEVs. The binding of these mucin-like glycoproteins to L-selectin is dependent on their decoration with 6-sulfo sLe^x. The 6-sulfo sLe^x structure is present in the nonreducing termini of the core 2 and extended core 1 branch of the O-glycans. The biosynthetic pathway of these O-glycans is shown in Figure 1. The MECA-79 antibody, which is widely used to detect the HEV in lymph nodes or the HEV-like vessels at the sites of chronic inflammation, recognizes the O-glycans containing 6-sulfo GlcNAc in the extended core 1 structure.[7] Additionally, recent studies using mutant mice lacking the core 2 and extended core 1 O-glycans revealed that 6-sulfo sLe^x present on the N-glycans of the HEV glycoproteins is also involved in the binding to L-selectin.[8]

Pioneering studies by Dr. Rosen's group showed that sialic acid is essential for the specific interaction of lymphocytes with HEVs.[9] Studies using mice deficient for both fucosyltransferase IV and VII revealed that the fucosylation of the L-selectin ligands is also critical for their interaction with L-selectin.[10] Furthermore, in vitro biochemical studies indicated that the sulfation of the L-selectin ligands is also important for their interaction with L-selectin; GlyCAM-1 synthesized in the presence of sodium chlorate that inhibits the biosynthesis of 3'-phosphoadenosine 5'-phosphosulfate (PAPS), the high-energy donor of sulfate, did not interact with L-selectin.[11]

Sulfated glycans in lymphocyte homing to the PLNs

The addition of sulfate groups to the N- and O-glycans is catalyzed by sulfotransferases, which transfer a sulfate group from PAPS to a spe-

cific position on the acceptor oligosaccharide. The N-acetylglucosamine-6-O-sulfotransferases (Glc-NAc6STs) catalyze the 6-O-sulfation of N-acetylglucosamine (GlcNAc) on the acceptor glycans. So far, five GlcNAc6STs in humans and four in mice have been identified.[12,13] One of these, GlcNAc6ST-2 (also called HEC-GlcNAc6ST or L-selectin ligand sulfotransferase (LSST)), is known to be specifically expressed in the HEVs,[14,15] although it was recently found that this sulfotransferase is also expressed in the colon to catalyze the sulfation of the colonic mucins in mice.[16,17]

To determine the roles of the sulfation of the L-selectin ligands in lymphocyte homing, GlcNAc6ST-2-deficient (GlcNAc6ST-2 KO) mice were generated.[18,19] In the GlcNAc6ST-2 KO mice, the binding of the MECA-79 antibody to the HEVs was significantly diminished, although the binding was still observed in the abluminal lining of the HEVs. Thus, these studies indicated that GlcNAc6ST-2 is a major GlcNAc6ST, but not the sole GlcNAc6ST that is involved in the GlcNAc-6-O-sulfation of the mucin-like glycoproteins in HEVs.

Another member of the GlcNAc6ST family, GlcNAc6ST-1, is widely expressed in various tissues including the lymph node HEVs.[20] Our group[21] and others[22] crossbred the GlcNAc6ST-2 KO mice with the GlcNAc6ST-1–deficient (GlcNAc6ST-1 KO) mice to further determine the roles of the sulfation of the L-selectin ligands. In the GlcNAc6ST-1 and GlcNAc6ST-2 double-deficient (DKO) mice, the binding of the MECA-79 antibody to the PLN HEVs was completely abrogated, indicating that the GlcNAc-6-O-sulfation in the extended core 1 branch of the O-glycans in the HEVs was absent in these mutant mice. While GlcNAc6ST-1 and GlcNAc6ST-2 single-KO mice showed an approximately 20–50% reduction in lymphocyte homing, respectively, the DKO mice showed an approximately 75% reduction in lymphocyte homing. The contact hypersensitivity responses were also significantly diminished in the DKO mice due to a reduction in lymphocyte trafficking to the draining lymph nodes. These results demonstrate the essential role of GlcNAc6ST-1 and GlcNAc6ST-2 in lymphocyte homing and immune responses.

To determine the structural basis for these phenotypes, a detailed carbohydrate structural analysis of GlyCAM-1, a major L-selectin–binding glycoprotein in the HEVs, was performed.[21] The lymph

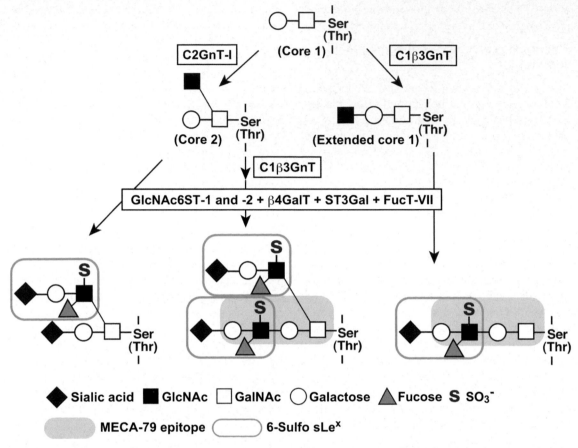

Figure 1. The biosynthesis of the O-glycans modified with 6-sulfo sLe[x]. The core 1 oligosaccharide can be converted to the core 2 O-glycan by the core 2 β1,6-N-acetylglucosaminyltransferase-I (C2GnT-I), sulfated by GlcNAc6ST-1 and GlcNAc6ST-2, galactosylated, sialylated, and fucosylated to form 6-sulfo sLe[x] in the core 2-branched O-glycans (left). Core 1 O-glycan can be extended by the core 1 β1,3-N-acetylglucosaminyltransferase (C1β3GnT), sulfated by GlcNAc6ST-1 and GlcNAc6ST-2, and galactosylated to form the minimum epitope for the MECA-79 antibody. The galactosylated oligosaccharide is further sialylated and fucosylated to form 6-sulfo sLe[x] in the extended core 1 O-glycans (right). C1β3GnT can act on the core 2-branched O-glycans to form the biantennary O-glycans containing both the core 2 branch and the core 1 extension. One or both of these branches can be modified to contain the 6-sulfo sLe[x] (*middle*). All of these O-glycans containing 6-sulfo sLe[x] function as L-selectin ligands. *β4GalT*, β1,4-galactosyltransferase. *ST3Gal*, α2,3-sialyltransferase. *FucT-VII*, fucosyltransferase VII. Modified from Hirakawa et al.[29]

nodes from wild-type (WT) and sulfotransferase-deficient mice were metabolically labeled with [³H]-galactose in organ culture. GlyCAM-1 was purified by affinity chromatography using anti-GlyCAM-1 antibody and then the O-glycans attached to GlyCAM-1 were released by β-elimination. After removing sialic acid by mild acid hydrolysis, the oligosaccharides were fractionated by gel filtration and anion-exchange column chromatography. The structure of each oligosaccharide was determined by exoglycosidase treatment and high performance liquid chromatography. As shown in Figure 2A, only

a very small amount of oligosaccharide containing GlcNAc-6-O-sulfate was detected in the DKO mice (see the 6th and 10th structures from the left). In contrast, some of the galactose-6-O-sulfated oligosaccharides significantly increased in the DKO mice (see the 5th, 7th, and 9th structures from the left). As summarized in Figure 2B, the 6-sulfo sialyl Lewis X structure was almost completely abrogated, whereas the unsulfated sialyl Lewis X (sLe[x]) was overexpressed in the DKO mice. These structural analyses suggest that the unsulfated sLe[x] structure that is known to interact with L-selectin[23] might

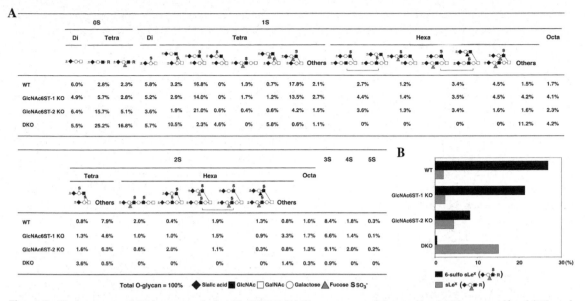

Figure 2. The structures of *O*-glycans attached to GlyCAM-1. (A) The structures of the unsulfated (0S), monosulfated (1S), and disulfated (2S) *O*-glycans on a disaccharide core (Di), tetrasaccharide core (tetra), or hexasaccharide core (hexa) structure that is attached to GlyCAM-1 from different mouse lines (WT, GlcNAc6ST-1 KO, GlcNAc6ST-2 KO and DKO mice). (B) The percentage of *O*-glycans attached to GlyCAM-1 containing 6-sulfo sLe[x] and sLe[x]. The total *O*-glycans attached to GlyCAM-1 = 100%. Modified from Kawashima *et al.*[21]

have supported the residual 25% lymphocyte trafficking to the PLNs observed in DKO mice. Indeed, a preinjection of the fucose- and sialic acid-specific lectins, *Aleuria aurantia* agglutinin (AAA) and *Maackia amurensis* agglutinin (MAA), respectively, significantly inhibited the lymphocyte homing to the PLNs in the DKO mice,[21] confirming that both fucose and sialic acid, constituents of sLe[x], are essential for the residual lymphocyte homing observed in these mutant mice. It was reported that a P-selectin glycoprotein ligand (PSGL)-1-like molecule, endoglycan, which is modified with the sLe[x] structure and sulfated tyrosines, interacts efficiently with L-selectin.[24,25] Therefore, it is possible that the unsulfated sLe[x] structure expressed in the DKO mice efficiently supports lymphocyte homing when it is present on specific glycoproteins, such as endoglycan.

Sulfated glycans in lymphocyte homing to the NALT and nasal allergy

The nasal-associated lymphoid tissue (NALT) is a mucosal lymphoid tissue in rodents that is considered to be the equivalent of the Waldeyer's ring in humans, which includes the tonsils and adenoids.[26] Recent studies indicated the presence of a NALT-

like structure in humans that is a morphologically distinct structure from the tonsils and adenoids and is disseminated in the nasal mucosa; this structure has the morphological features that are typical of the secondary lymphoid organs, such as HEVs and lymphoid follicles.[27] It is known that various antigens, including pollen and house dust, that are taken up from the nasal mucosal barrier often cause allergic rhinitis characterized by antigen-specific IgE production and nasal symptoms.[28]

The sulfated glycans recognized by the MECA-79 antibody and the recently generated antisulfated glycan monoclonal antibody (mAb) S2,[29] which is specifically reactive with sialylated 6-sulfo *N*-acetyllactosamine (LacNAc) and 6-sulfo sLe[x], are strongly expressed in the NALT HEVs of WT mice.[4,30] In consistent with this notion, studies by Dr. Ruddle's group clearly showed that GlcNAc6ST-2 is expressed in the NALT HEVs.[31–33] In contrast, strong staining with the antibody MECA-367, which reacts with mucosal addressin cell adhesion molecule-1 (MAdCAM-1),[34] was observed in the PPs, but not in the NALT of C57BL/6 WT mice.[4] Thus, although the NALT is a mucosa-associated lymphoid tissue (MALT) similar to the PPs, its expression pattern of the vascular addressins, which are the counter receptors for the lymphocyte homing receptors, is more similar to that of the PLNs. Consistently, our recent studies indicated that lymphocyte homing to the NALT was 90% lower in the DKO mice relative to the homing in the WT mice.[4] Furthermore, after an intranasal immunization with ovalbumin (OVA), the production of specific IgE antibodies and the number of sneezes was significantly lower in the DKO mice than in the WT mice. Consistently, the induction of the humoral immunity-related Th2 cytokine IL-4, which is a major mediator of IgE class switching in B cells,[35] was much lower in the DKO mice than in the WT mice, while the immune-suppressive cytokine IL-10 (Ref. 35) was more abundantly produced in the DKO mice.

PSGL-1 and CD44 in the homing of CD4⁺CD25⁺ T_reg cells to the NALT

Flow cytometric analyses showed that CD19⁺ B cells were more significantly diminished than CD3⁺ T cells in the NALT of the DKO mice.[4] This result was consistent with the previous results observed in the PLN in the DKO mice.[22] It is likely that this finding may be explained by B cells expressing lower levels of L-selectin relative to T cells and B cell homing being more susceptible to the reduction of L-selectin ligand activity in the HEVs than that of T cells.[22,36] Another notable difference was that the CD4⁺CD25⁻ conventional T (T_conv) cells in the NALT of the DKO mice were decreased, while the ratio of CD4⁺CD25⁺ regulatory T (T_reg) cells was approximately four times higher in the DKO mice.[4] A short-term homing assay indicated that the homing of the T_conv cells to the NALT was significantly decreased in the DKO mice, while the homing of the T_reg cells to the NALT in the DKO mice was similar to that observed in the WT mice.[4] These results collectively suggested that the homing of the T_reg cells to the NALT is less dependent on an L-selectin–sulfated glycan interaction than that of the T_conv cells. Indeed, the anti-L-selectin mAb MEL-14 (Ref. 37) significantly inhibited the homing of the T_conv cells to the NALT, but only partially inhibited that of the T_reg cells.

Similar to L-selectin, PSGL-1[38,39] and CD44[40] also function as carbohydrate-dependent rolling receptors on lymphocytes. We thus examined the effects of mAbs against these adhesion molecules and found that anti-PSGL-1, anti-P-selectin, and anti-

CD44 mAbs each partially inhibited the homing of the T_reg cells to the NALT.[4] Furthermore, the treatment of the T_reg cells with a mixture of the mAbs more significantly inhibited the homing of the T_reg cells to the NALT than treatment with the individual mAbs. Consistently, fucosyltransferase-VII, which is required for a post-translational modification of PSGL-1 to bind P-selectin,[41] was more abundant in the T_reg cells than in the T_conv cells.[4] In addition, the HEV cells isolated from the NALT by immunomagnetic selection expressed P-selectin mRNA. Furthermore, HA, a ligand for CD44,[42] was also expressed in the NALT HEVs, as determined by immunohistochemistry. Collectively, these results indicated that the homing of the T_reg cells to the NALT is less dependent on the L-selectin–sulfated glycan interaction but is more dependent on PSGL-1 and CD44 relative to the homing of the T_conv cells to the NALT.

It was reported that the homing of the T_reg cells to the PLNs requires L-selectin.[43] The same study also showed that 40% of the CD4⁺ T cells in the PLNs of L-selectin–deficient mice are T_reg cells, suggesting that adhesion molecules other than L-selectin might also be involved in the homing of the T_reg cells to the PLNs. However, as far as we know, no previous reports other than our paper[4] have shown the importance of the other carbohydrate-dependent adhesion pathways for the homing of T_reg cells. Because the appropriate localization is indispensable for the appropriate function of T_reg cells *in vivo*,[44] we believe that our recent work has clarified, at least in part, the address codes for T_reg cell migration and will be beneficial for manipulating the various functions of the T_reg cells, including the host response to tumors and the regulation of transplant rejection, autoimmunity, and allergy.[45]

Collectively, the above findings indicate that the sulfated glycans differentially regulate the homing of different lymphocyte subsets and support the model that T_reg cells accumulate in the NALT in the DKO mice because of their decreased dependence on the L-selectin-sulfated glycan adhesion pathway and their greater dependence on the PSGL-1/P-selectin and CD44-HA adhesion pathways; thus, they create an immune-suppressive environment in the NALT (Fig. 3). Therefore, blocking the sulfated glycan-mediated lymphocyte recruitment might serve as a novel therapeutic approach for modulating allergic rhinitis.

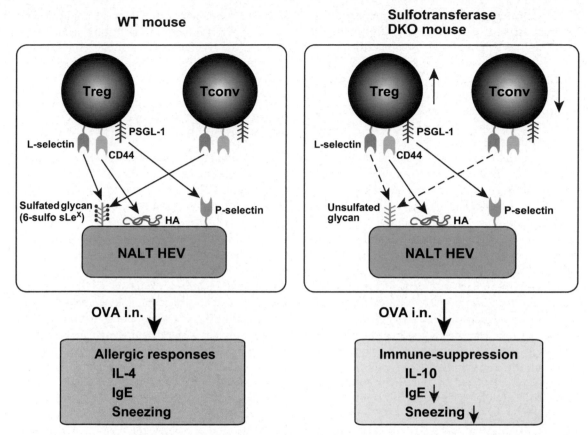

Figure 3. A proposed model for the molecular mechanisms of T_{reg} and T_{conv} cell homing to the NALT and the reduction of allergic responses in the DKO mice. In WT mice (left), the homing of the T_{reg} cells to the NALT is mediated by the L-selectin-6-sulfo sLe^x, CD44-HA, and PSGL-1/P-selectin interactions, whereas the homing of the T_{conv} cells to the NALT is mainly mediated by an L-selectin-6-sulfo sLe^x interaction (shown by the *arrows*). In the NALT of DKO mice (right), the HEV GlcNAc-6-O-sulfation of 6-sulfo sLe^x (shown by the *red* symbols) is eliminated, and the interaction between L-selectin and its ligand is weakened (*dotted arrows*). As a result, the T_{reg} cells are enriched in the NALT in the DKO mice, which creates an immuno-suppressive environment in the NALT. After an intranasal immunization (*OVA i.n.*), allergic responses, including IgE production and sneezing, are reduced in DKO mice, possibly through a shift in cell populations and the cytokine environment in the NALT. Modified from Ohmichi *et al.*[4]

Trafficking of activated lymphocytes to sites of inflammation

As described above, naive lymphocytes continuously recirculate until they encounter their cognate antigens in the secondary lymphoid organs. Once they encounter the antigens, they become activated and begin to express different adhesion molecules compared with those expressed in naive lymphocytes. Activated lymphocytes lose their expression of L-selectin and, in some cases, start to express cutaneous lymphocyte antigen (CLA),[46] which reacts with the mAb HECA-452.[47] Molecular characterization revealed that CLA is PSGL-1 modified with

the sLe^x structure.[48] Interleukin-12, which activates naive T cells and stimulates the generation of Th1 cells, induces the expression of fucosyltransferase VII, which is involved in sLe^x biosynthesis.[49]

The HECA-452–reactive activated lymphocytes specifically home to the inflammatory sites where the antigens enter, which is consistent with the expression of P- and E-selectin on the activated endothelial cells in the inflamed tissues. P-selectin expression on endothelial cells is due to the transport of P-selectin in intracellular Weibel–Palade bodies to the cell surface.[50] Because the *de novo* synthesis of P-selectin is not involved in the rapid expression of P-selectin, which takes place within 1 hour,

P-selectin can play a role in the acute phase of inflammation. This acute phase of P-selectin involvement is gradually replaced by E-selectin *de novo* synthesis and the surface expression of E-selectin in inflamed endothelial cells.[51] Because P-selectin–mediated cell adhesion is required for the rapid recruitment of leukocytes expressing P-selectin ligands, the interaction of P-selectin with its ligands is strong and is mediated by the high affinity ligands composed of sulfated tyrosine residues and the close neighboring sLe^x *O*-glycans on the NH_2-terminal domain of PSGL-1.[52] The collaboration of the sulfated tyrosine residues and sLe^x is critical for the P-selectin–mediated interaction.

A subpopulation of activated lymphocytes expresses the active forms of CD44. CD44 is a carbohydrate-binding protein that is expressed on naive lymphocytes in an inactive form that lacks ligand-binding activity; CD44 is converted to an active form with ligand-binding activity after lymphocyte activation. The precise mechanisms for CD44 activation are still unknown, but previous reports indicated that the ligand binding activity of CD44 is negatively regulated by sialic acid modification of CD44,[53] that CD44's association with the cytoskeleton is critical for its ligand-binding activity,[54] and that the modification of CD44 with sulfated *N*- and *O*-glycans enhances its ligand-binding activity.[55,56] In addition to the well-known CD44 ligand hyaluronic acid (HA), the chondroitin sulfate proteoglycans (CSPGs) that are modified with abundant sulfated glycans, such as versican[57] and aggrecan,[58] also function as ligands for CD44. It is also reported that a CSPG serglycin[59] functions as a CD44 ligand. *In vitro* rolling assays showed that CD44-HA[40] and CD44-CS[60] interactions mediate the lymphocyte rolling. Therefore, CD44 may function as a rolling receptor in a process mediated by the infiltration of L-selectin$^-$/active CD44$^+$ lymphocytes into the sites of inflammation by the interactions with a nonsulfated glycan HA and sulfated glycans CSs. Therefore, sulfated glycans may regulate CD44-dependent homing of activated lymphocytes in two ways: by their modification of CD44 and by their interaction with CD44.

Collectively, the above findings suggest that PSGL-1 and CD44 can be involved in the homing of activated lymphocytes to the sites of inflammation. Because the homing of T_{reg} cells to the NALT was also partially dependent on PSGL-1 and CD44,[4] the activated lymphocytes and T_{reg} cells might show a partially overlapping homing pattern, which may result in the regulation of immune responses and inflammation by T_{reg} cells under physiological and pathological situations.

Induction of sulfated glycans in HEV-like blood vessels at sites of chronic inflammation

The HEV-like blood vessels that express the sulfated glycans are induced at various sites of chronic inflammation. For example, the MECA-79–reactive HEV-like blood vessels are induced in human gastric mucosa infected with *Helicobacter pylori*,[61] in ulcerative colitis[62] and in rheumatoid arthritis.[63,64] Furthermore, MECA-79 antigen and GlcNAc6ST-2 are expressed in ectopically induced HEV-like blood vessels in the pancreas of diabetic NOD mice and RIP-BLC mice in which chemokine CXCL13 is expressed in the islets.[65] It was shown that the MECA-79 antibody has a significant therapeutic effect in a sheep model of asthma.[66] The intravenous injection of MECA-79 significantly inhibited both the late-phase airway response and also airway hyperresponsiveness in allergic sheep and contributed to the accumulation of leukocytes in bronchoalveolar lavage fluid. However, the physiological role of the MECA-79–reactive sulfated glycans that are expressed in the other chronic inflammation is still unknown. In addition, recent studies showed that the HEV-like structure of the MALT lymphoma lacks the MECA-79 antigen and that 6-sulfo sLe^x on the core 2-branched *O*-glycans is apparently the sole L-selectin ligand.[67] Therefore, it is likely that MECA-79-reactive sulfated glycans are not the only L-selectin binding sulfated glycans ectopically expressed in the HEV-like blood vessels and that the mAbs that react with 6-sulfo sLe^x on the core 2-branched *O*-glycans might be useful in assessing the function of the sulfated glycans in some diseases.

Recently, antisulfated glycan mAbs that strongly blocked L-selectin–mediated lymphocyte homing were generated using sulfotransferase DKO mice by our group[29] and others.[68] Because these mAbs bound not only mouse but also human tissues,[29,68] these mAbs will be useful for assessing the ectopic expression of sulfated glycans in pathophysiological situations. As mentioned above, MECA-79 binds only sulfated extended core 1 *O*-glycans[7] and not sulfated core 2 O-glycans or N-glycans. In addition, MECA-79 does not recognize terminal sialic acids.[7] In contrast, the newly generated antisulfated glycan mAbs S2[29] and CL40[68] bound sialylated 6-sulfo LacNAc on both N-glycans and various O-glycans including the core 2-branched O-glycans, and thus can recognize glycan epitopes that are more closely related to the L-selectin-binding sites compared to MECA-79. Therefore, it is expected that these newly generated mAbs will be useful in assessing the roles of the sulfated glycans in various HEV-like blood vessels in chronic inflammation.

Conclusion

Our studies using the sulfotransferase DKO mice have provided a link between carbohydrate structural changes and the alteration of lymphocyte homing under physiological conditions. Using the same DKO mice, we recently found that the sulfated glycans are important for the homing of the T_{conv} cells and for allergic rhinitis. Furthermore, our studies revealed that the homing of the CD4$^+$CD25$^+$ T_{reg} cells to the NALT utilizes CD44 and PSGL-1 in addition to 6-sulfo sLe^x, but the conventional CD4$^+$CD25$^-$ T_{conv} cells lack those additional adhesions. These findings indicate that different carbohydrate-dependent homing mechanisms may be utilized for the homing of different lymphocyte subsets. A recent study also showed the roles of the sulfated glycans in chronic inflammation. Further studies using newly generated antisulfated glycan mAbs should further reveal the function of the sulfated glycans in regulating lymphocyte homing in health and disease.

Acknowledgments

We would like to thank Drs. Jotaro Hirakawa and Yasuyuki Imai, and Ms. Yukari Ohmichi (University of Shizuoka) for their collaboration. This work was supported in part by a Grant-in-Aid for Scientific Research (B) and by a Grant-in-Aid for Challenging Exploratory Research from the Ministry of Education, Culture, Sports, Science and Technology, Japan (21390023 and 22659018, respectively, to H.K.); the Shizuoka Research Institute (SRI) academic research grant (to H.K.); and the NIH grant PO1CA71932 (to M.F.).

Conflicts of interest

The authors declare no conflicts of interest.

References

1. Girard, J.P. & T.A. Springer. 1995. High endothelial venules (HEVs): specialized endothelium for lymphocyte migration. *Immunol. Today* **16:** 449–457.
2. Springer, T.A. 1994. Traffic signals for lymphocyte recirculation and leukocyte emigration: the multistep paradigm. *Cell* **76:** 301–314.
3. Bao, X., E.A. Moseman, H. Saito, *et al.* 2010. Endothelial heparan sulfate controls chemokine presentation in recruitment of lymphocytes and dendritic cells to lymph nodes. *Immunity* **33:** 817–829.
4. Ohmichi, Y., J. Hirakawa, Y. Imai, *et al.* 2011. Essential role of peripheral node addressin in lymphocyte homing to nasal-associated lymphoid tissues and allergic immune responses. *J. Exp. Med.* **208:** 1015–1025.
5. von Andrian, U.H., S.R. Hasslen, R.D. Nelson, *et al.* 1995. A central role for microvillous receptor presentation in leukocyte adhesion under flow. *Cell* **82:** 989–999.
6. Rosen, S.D. 2004. Ligands for L-selectin: homing, inflammation, and beyond. *Annu. Rev. Immunol.* **22:** 129–156.
7. Yeh, J.C., N. Hiraoka, B. Petryniak, *et al.* 2001. Novel sulfated lymphocyte homing receptors and their control by a Core1 extension β1,3-N-acetylglucosaminyltransferase. *Cell* **105:** 957–969.
8. Mitoma, J., X. Bao, B. Petryanik, *et al.* 2007. Critical functions of N-glycans in L-selectin-mediated lymphocyte homing and recruitment. *Nat. Immunol.* **8:** 409–418.
9. Rosen, S.D., M.S. Singer, T.A. Yednock & L.M. Stoolman. 1985. Involvement of sialic acid on endothelial cells in organ-specific lymphocyte recirculation. *Science* **228:** 1005–1007.
10. Homeister, J.W., A.D. Thall, B. Petryniak, *et al.* 2001. The α(1,3)fucosyltransferases FucT-IV and FucT-VII exert collaborative control over selectin-dependent leukocyte recruitment and lymphocyte homing. *Immunity* **15:** 115–126.
11. Imai, Y., L.A. Lasky & S.D. Rosen. 1993. Sulphation requirement for GlyCAM-1, an endothelial ligand for L-selectin. *Nature* **361:** 555–557.
12. Fukuda, M., N. Hiraoka, T.O. Akama & M.N. Fukuda. 2001. Carbohydrate-modifying sulfotransferases: structure, function, and pathophysiology. *J. Biol. Chem.* **276:** 47747–47750.
13. Hemmerich, S. & S.D. Rosen. 2000. Carbohydrate sulfotransferases in lymphocyte homing. *Glycobiology* **10:** 849–856.
14. Bistrup, A., S. Bhakta, J.K. Lee, *et al.* 1999. Sulfotransferases of two specificities function in the reconstitution of high endothelial cell ligands for L-selectin. *J. Cell Biol.* **145:** 899–910.
15. Hiraoka, N., B. Petryniak, J. Nakayama, *et al.* 1999. A novel, high endothelial venule-specific sulfotransferase expresses 6-sulfo sialyl Lewis^x, an L-selectin ligand displayed by CD34. *Immunity* **11:** 79–89.
16. Kawashima, H., J. Hirakawa, Y. Tobisawa, *et al.* 2009. Conditional gene targeting in mouse high endothelial venules. *J. Immunol.* **182:** 5461–5468.
17. Tobisawa, Y., Y. Imai, M. Fukuda & H. Kawashima. 2010. Sulfation of colonic mucins by N-acetylglucosamine 6-O-sulfotransferase-2 and its protective function in experimental colitis in mice. *J. Biol. Chem.* **285:** 6750–6760.

18. Hemmerich, S., A. Bistrup, M.S. Singer, *et al.* 2001. Sulfation of L-selectin ligands by an HEV-restricted sulfotransferase regulates lymphocyte homing to lymph nodes. *Immunity* **15**: 237–247.

19. Hiraoka, N., H. Kawashima, B. Petryniak, *et al.* 2004. Core 2 branching β1,6-*N*-acetylglucosaminyltransferase and high endothelial venule-restricted sulfotransferase collaboratively control lymphocyte homing. *J. Biol. Chem.* **279**: 3058–3067.

20. Uchimura, K., K. Kadomatsu, F.M. El-Fasakhany, *et al.* 2004. *N*-acetylglucosamine 6-*O*-sulfotransferase-1 regulates expression of L-selectin ligands and lymphocyte homing. *J. Biol. Chem.* **279**: 35001–35008.

21. Kawashima, H., B. Petryniak, N. Hiraoka, *et al.* 2005. *N*-acetylglucosamine-6-*O*-sulfotransferases 1 and 2 cooperatively control lymphocyte homing through L-selectin ligand biosynthesis in high endothelial venules. *Nat. Immunol.* **6**: 1096–1104.

22. Uchimura, K., J.M. Gauguet, M.S. Singer, *et al.* 2005. A major class of L-selectin ligands is eliminated in mice deficient in two sulfotransferases expressed in high endothelial venules. *Nat. Immunol.* **6**: 1105–1113.

23. Foxall, C., S.R. Watson, D. Dowbenko, *et al.* 1992. The three members of the selectin receptor family recognize a common carbohydrate epitope, the sialyl Lewis[x] oligosaccharide. *J. Cell Biol.* **117**: 895–902.

24. Fieger, C.B., C.M. Sassetti & S.D. Rosen. 2003. Endoglycan, a member of the CD34 family, functions as an L-selectin ligand through modification with tyrosine sulfation and sialyl Lewis[x]. *J. Biol. Chem.* **278**: 27390–27398.

25. Leppanen, A., V. Parviainen, E. Ahola-Iivarinen, *et al.* 2010. Human L-selectin preferentially binds synthetic glycosulfopeptides modeled after endoglycan and containing tyrosine sulfate residues and sialyl Lewis[x] in core 2 *O*-glycans. *Glycobiology* **20**: 1170–1185.

26. Hellings, P., M. Jorissen & J.L. Ceuppens. 2000. The Waldeyer's ring. *Acta Otorhinolaryngol. Belg.* **54**: 237–241.

27. Debertin, A.S., T. Tschernig, H. Tonjes, *et al.* 2003. Nasal-associated lymphoid tissue (NALT): frequency and localization in young children. *Clin. Exp. Immunol.* **134**: 503–507.

28. Gelfand, E.W. 2004. Inflammatory mediators in allergic rhinitis. *J. Allergy Clin. Immunol.* **114**: S135–138.

29. Hirakawa, J., K. Tsuboi, K. Sato, *et al.* 2010. Novel anti-carbohydrate antibodies reveal the cooperative function of sulfated *N*- and *O*-glycans in lymphocyte homing. *J. Biol. Chem.* **285**: 40864–40878.

30. Csencsits, K.L., M.A. Jutila & D.W. Pascual. 1999. Nasal-associated lymphoid tissue: phenotypic and functional evidence for the primary role of peripheral node addressin in naive lymphocyte adhesion to high endothelial venules in a mucosal site. *J. Immunol.* **163**: 1382–1389.

31. Drayton, D.L., G. Bonizzi, X. Ying, *et al.* 2004. IκB kinase complex α kinase activity controls chemokine and high endothelial venule gene expression in lymph nodes and nasal-associated lymphoid tissue. *J. Immunol.* **173**: 6161–6168.

32. Ying, X., K. Chan, P. Shenoy, *et al.* 2005. Lymphotoxin plays a crucial role in the development and function of nasal-associated lymphoid tissue through regulation of

chemokines and peripheral node addressin. *Am. J. Pathol.* **166**: 135–146.

33. Liao, S., K. Bentley, M. Lebrun, *et al.* 2007. Transgenic LacZ under control of Hec-6st regulatory sequences recapitulates endogenous gene expression on high endothelial venules. *Proc. Natl. Acad. Sci. USA* **104**: 4577–4582.

34. Nakache, M., E.L. Berg, P.R. Streeter & E.C. Butcher. 1989. The mucosal vascular addressin is a tissue-specific endothelial cell adhesion molecule for circulating lymphocytes. *Nature* **337**: 179–181.

35. Zhu, J. & W.E. Paul. 2008. CD4 T cells: fates, functions, and faults. *Blood* **112**: 1557–1569.

36. Gauguet, J.M., S.D. Rosen, J.D. Marth & U.H. von Andrian. 2004. Core 2 branching β1,6-*N*-acetylglucosaminyltransferase and high endothelial cell *N*-acetylglucosamine-6-sulfotransferase exert differential control over B- and T-lymphocyte homing to peripheral lymph nodes. *Blood* **104**: 4104–4112.

37. Gallatin, W.M., I.L. Weissman & E.C. Butcher. 1983. A cell-surface molecule involved in organ-specific homing of lymphocytes. *Nature* **304**: 30–34.

38. Moore, K.L., K.D. Patel, R.E. Bruehl, *et al.* 1995. P-selectin glycoprotein ligand-1 mediates rolling of human neutrophils on P-selectin. *J. Cell Biol.* **128**: 661–671.

39. Alon, R., D.A. Hammer & T.A. Springer. 1995. Lifetime of the P-selectin-carbohydrate bond and its response to tensile force in hydrodynamic flow. *Nature* **374**: 539–542.

40. DeGrendele, H.C., P. Estess, L.J. Picker & M.H. Siegelman. 1996. CD44 and its ligand hyaluronate mediate rolling under physiologic flow: a novel lymphocyte-endothelial cell primary adhesion pathway. *J. Exp. Med.* **183**: 1119–1130.

41. Snapp, K.R., A.J. Wagers, R. Craig, *et al.* 1997. P-selectin glycoprotein ligand-1 is essential for adhesion to P-selectin but not E-selectin in stably transfected hematopoietic cell lines. *Blood* **89**: 896–901.

42. Aruffo, A., I. Stamenkovic, M. Melnick, *et al.* 1990. CD44 is the principal cell surface receptor for hyaluronate. *Cell* **61**: 1303–1313.

43. Venturi, G.M., R.M. Conway, D.A. Steeber & T.F. Tedder. 2007. CD25[+]CD4[+] regulatory T cell migration requires L-selectin expression: L-selectin transcriptional regulation balances constitutive receptor turnover. *J. Immunol.* **178**: 291–300.

44. Huehn, J. & A. Hamann. 2005. Homing to suppress: address codes for T$_{reg}$ migration. *Trends Immunol.* **26**: 632–636.

45. Sakaguchi, S. 2005. Naturally arising Foxp3-expressing CD25[+]CD4[+] regulatory T cells in immunological tolerance to self and non-self. *Nat. Immunol.* **6**: 345–352.

46. Picker, L.J., L.W. Terstappen, L.S. Rott, *et al.* 1990. Differential expression of homing-associated adhesion molecules by T cell subsets in man. *J. Immunol.* **145**: 3247–3255.

47. Duijvestijn, A.M., E. Horst, S.T. Pals, *et al.* 1988. High endothelial differentiation in human lymphoid and inflammatory tissues defined by monoclonal antibody HECA-452. *Am. J. Pathol.* **130**: 147–155.

48. Fuhlbrigge, R.C., J.D. Kieffer, D. Armerding & T.S. Kupper. 1997. Cutaneous lymphocyte antigen is a specialized form of PSGL-1 expressed on skin-homing T cells. *Nature* **389**: 978–981.

49. Wagers, A.J., C.M. Waters, L.M. Stoolman & G.S. Kansas. 1998. Interleukin 12 and interleukin 4 control T cell adhesion to endothelial selectins through opposite effects on α1, 3-fucosyltransferase VII gene expression. *J. Exp. Med.* **188:** 2225–2231.

50. Bonfanti, R., B.C. Furie, B. Furie & D.D. Wagner. 1989. PADGEM (GMP140) is a component of Weibel-Palade bodies of human endothelial cells. *Blood* **73:** 1109–1112.

51. Bevilacqua, M.P., S. Stengelin, M.A. Gimbrone Jr. & B. Seed. 1989. Endothelial leukocyte adhesion molecule 1: an inducible receptor for neutrophils related to complement regulatory proteins and lectins. *Science* **243:** 1160–1165.

52. Leppanen, A., P. Mehta, Y.B. Ouyang, *et al.* 1999. A novel glycosulfopeptide binds to P-selectin and inhibits leukocyte adhesion to P-selectin. *J. Biol. Chem.* **274:** 24838–24848.

53. Skelton, T.P., C. Zeng, A. Nocks & I. Stamenkovic. 1998. Glycosylation provides both stimulatory and inhibitory effects on cell surface and soluble CD44 binding to hyaluronan. *J. Cell Biol.* **140:** 431–446.

54. Perschl, A., J. Lesley, N. English, *et al.* 1995. Role of CD44 cytoplasmic domain in hyaluronan binding. *Eur. J. Immunol.* **25:** 495–501.

55. Maiti, A., G. Maki & P. Johnson. 1998. TNF-α induction of CD44-mediated leukocyte adhesion by sulfation. *Science* **282:** 941–943.

56. Delcommenne, M., R. Kannagi & P. Johnson. 2002. TNF-α increases the carbohydrate sulfation of CD44: induction of 6-sulfo *N*-acetyl lactosamine on *N*- and *O*-linked glycans. *Glycobiology* **12:** 613–622.

57. Kawashima, H., M. Hirose, J. Hirose, *et al.* 2000. Binding of a large chondroitin sulfate/dermatan sulfate proteoglycan, versican, to L-selectin, P-selectin, and CD44. *J. Biol. Chem.* **275:** 35448–35456.

58. Fujimoto, T., H. Kawashima, T. Tanaka, *et al.* 2001. CD44 binds a chondroitin sulfate proteoglycan, aggrecan. *Int. Immunol.* **13:** 359–366.

59. Toyama-Sorimachi, N., H. Sorimachi, Y. Tobita, *et al.* 1995. A novel ligand for CD44 is serglycin, a hematopoietic cell lineage-specific proteoglycan. Possible involvement in lymphoid cell adherence and activation. *J. Biol. Chem.* **270:** 7437–7444.

60. Murai, T., N. Sougawa, H. Kawashima, *et al.* 2004. CD44-chondroitin sulfate interactions mediate leukocyte rolling under physiological flow conditions. *Immunol. Lett.* **93:** 163–170.

61. Kobayashi, M., J. Mitoma, N. Nakamura, *et al.* 2004. Induction of peripheral lymph node addressin in human gastric mucosa infected by *Helicobacter pylori. Proc. Natl. Acad. Sci. USA* **101:** 17807–17812.

62. Suzawa, K., M. Kobayashi, Y. Sakai, *et al.* 2007. Preferential induction of peripheral lymph node addressin on high endothelial venule-like vessels in the active phase of ulcerative colitis. *Am. J. Gastroenterol.* **102:** 1499–1509.

63. Michie, S.A., P.R. Streeter, P.A. Bolt, *et al.* 1993. The human peripheral lymph node vascular addressin. An inducible endothelial antigen involved in lymphocyte homing. *Am. J. Pathol.* **143:** 1688–1698.

64. Pablos, J.L., B. Santiago, D. Tsay, *et al.* 2005. A HEV-restricted sulfotransferase is expressed in rheumatoid arthritis synovium and is induced by lymphotoxin-α/β and TNF-α in cultured endothelial cells. *BMC Immunol.* **6:** 6.

65. Bistrup, A., D. Tsay, P. Shenoy, *et al.* 2004. Detection of a sulfotransferase (HEC-GlcNAc6ST) in high endothelial venules of lymph nodes and in high endothelial venule-like vessels within ectopic lymphoid aggregates: relationship to the MECA-79 epitope. *Am. J. Pathol.* **164:** 1635–1644.

66. Rosen, S.D., D. Tsay, M.S. Singer, *et al.* 2005. Therapeutic targeting of endothelial ligands for L-selectin (PNAd) in a sheep model of asthma. *Am. J. Pathol.* **166:** 935–944.

67. Kobayashi, M., J. Mitoma, H. Hoshino, *et al.* 2011. Prominent expression of sialyl Lewis[x]-capped core 2-branched *O*-glycans on high endothelial venule-like vessels in gastric MALT lymphoma. *J. Pathol.* **224:** 67–77.

68. Arata-Kawai, H., M.S. Singer, A. Bistrup, *et al.* 2011. Functional contributions of *N*- and *O*-glycans to L-selectin ligands in murine and human lymphoid organs. *Am. J. Pathol.* **178:** 423–433.

Ann. N.Y. Acad. Sci. ISSN 0077-8923

Acute phase glycoproteins: bystanders or participants in carcinogenesis?

Eugene Dempsey[1] and Pauline M. Rudd[1,2]

[1]Dublin–Oxford Glycobiology Laboratory, UCD Conway Institute of Biomolecular and Biomedical Research, University College Dublin, Dublin, Ireland. [2]Dublin–Oxford Glycobiology Laboratory, National Institute for Bioprocessing Research and Training, University College Dublin, Dublin, Ireland

Address for correspondence: Pauline M. Rudd, Dublin–Oxford Glycobiology Laboratory, UCD Conway Institute of Biomolecular and Biomedical Research, University College Dublin, Belfield, Dublin 4, Ireland. pauline.rudd@nibrt.ie

Acute phase proteins (APPs) are a group of serum proteins that undergo dramatic changes in concentration during times of inflammation. Many APPs are heavily glycosylated, and their sugar content and complexity change in the presence of cancer-induced chronic inflammation. These changes in glycosylation are currently being exploited in the search for novel biomarkers of cancer. Like other posttranslational modifications, such as phosphorylation, changes in glycosylation can profoundly alter the function of a protein. We hypothesize that besides being a rich source of potential biomarkers APPs may also play an active role in tumorigenesis. The glycan content of the APPs haptoglobin and kininogen, for example, is altered in many types of cancer. These APPs can interact with a number of receptors on macrophages in the tumor microenvironment, potentially modulating macrophage activity and thereby contributing to tumor cell survival, growth, and metastasis.

Keywords: cancer; acute phase proteins; glycosylation; macrophage; haptoglobin; kininogen

Acute phase proteins and inflammation

The acute phase response is a systemic reaction to the presence of infection, tissue damage, or cancer.[1] A major component of this response is the dramatic change in the liver's production of a number of proteins referred to as acute phase proteins (APPs). These are evolutionarily conserved plasma proteins which have a wide range of functions and are defined as a group by the fact that their levels change by >25% after an inflammatory stimulus.[2] APPs have a vast array of functions including participating in pathogen opsonization, promoting angiogenesis, and modulating pro- and anti-inflammatory responses.[3–5] Several inflammatory cytokines can induce changes in APP production including tumor necrosis factor α (TNF-α), interleukin-1β (IL-1β), and IL-6.[6] IL-6 and related family members have been deemed the most potent stimulators of the acute phase response.[7] Despite APPs being directly linked to the presence of inflammation, their effects on cells of the immune system are still poorly characterized.

Glycosylation of APPs

Many APPs are heavily glycosylated, and our group and others have shown that their glycan profile changes in the presence of tumors.[8–28] Table 1 outlines some common cancers and their association with altered glycosylation of certain APPs. During this screening for biomarkers of cancer, the most consistent N-linked glycosylation changes in APPs were increases in both sialylation and fucosylation.

Sialylation is the addition of sialic acid groups as terminal modifications to both *N*- and *O*-glycans by members of the sialyltransferase family.[29] The addition of sialic acid groups plays a diverse role in glycoprotein function, including conformation stabilization, protease resistance, and altering binding affinities through charge.[30,31] The ability of sialylation to confer protease resistance on proteins can improve the half-life of glycoproteins.[32]

Fucosylation has been termed one of the most important types of glycosylation changes to occur during cancer and inflammation.[33] The fucosyltransferase (FUT) family of enzymes is responsible for

doi: 10.1111/j.1749-6632.2011.06420.x

Table 1. Glycosylation changes in APPs associated with common cancers

Cancer	APPs	Glycosylation changes	References
Breast	Haptoglobin	Increased SLe[x]	Abd Hamid[9]
	AGP	Increased SLe[x]	
Ovarian	Haptoglobin	Increased SLe[x]	Saldova[115]
	AGP	Increased SLe[x]	
	α-1 anti-chymotrypsin	Increased SLe[x]	
	Haptoglobin	Increased fucosylation	Thompson[20]
	Haptoglobin	Increased branching	Turner[21]
	α-1 anti-trypsin	Decreased branching, increased fucosylation	
Prostate	Haptoglobin	Increased branching, increased fucosylation	Fujimura[13]
	Haptoglobin	Increased monosialylation of tri-antennary structures	Yoon[12]
	Kininogen 1	Increased fucosylation and sialylation	Zhao[28]
	Haptoglobin	Increased fucosylation and sialylation	
	C1 inhibitor	Decreased sialylation	
Pancreatic	Haptoglobin	Increased core fucosylation	Sarrats[8]
	AGP	Increased SLe[x], increased core fucosylation	
	Haptoglobin	Increased fucosylation	Okuyama[17]
	Haptoglobin	Increased fucosylation	Matsumoto[24]
	Haptoglobin	Increased fucosylation	Nakano[25]
	Haptoglobin	Increased core and anntennary fucosylation	Lin[23]
Gastric	Haptoglobin	Increased SLe[x]	Bones[10]
	AGP	Increased SLe[x]	
	Kininogen 1	Increased SLe[x]	
Hepatic	α-1 anti-trypsin	Increased core fucosylation	Comunale[11]
	Haptoglobin	Increase sialylation and fucosylation	Ang[15]
	Kininogen 1	Increased fucosylation	Wang[16]
	Haptoglobin	Increased fucosylation, Increased branching	Zhang[18]
	Complement C3	Increased fucosylation	Lui[19]
Colon	Kininogen 1	Increased fucosylation and sialylation	Qui[14]
	Complement C3	Increased fucosylation and sialylation	
	Haptoglobin	Increased fucosylation	Park[22]
Lung	Haptoglobin	Increased fucosylation	Tsai[42]
	Haptoglobin	Increased fucosylation and sialylation	Hoagland[26]
	Haptoglobin	Increased SLe[x]	Arnold[27]

AGP, alpha-1-acid glycoprotein.

the addition of fucose to glycoproteins. Specifically, FUT8 can add $\alpha(1,6)$ fucose to the innermost N-acetylglucosamine residue of the N-glycan, giving rise to core fucosylated structures. The presence of a core fucose has been shown to be important in a number of signaling pathways, including TRAIL-mediated apoptosis of tumor cells.[34]

It is difficult to speculate on the exact role that increases in fucose or sialic acid content will have on the function of APPs; however, evidence from the functional roles these sugar moieties have for IgG molecule activity[35] integrin signaling,[36] and EGFR signaling[37–39]—among many other signaling pathways—suggests that such glycan modification imparts altered functionality on APPs. That a similar trend of glycan changes is seen in a diverse array of cancers suggests that modifications to APPs are in some way beneficial to the developing tumor.

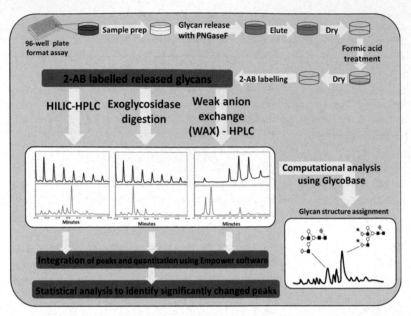

Figure 1. Schematic description of the high-throughput method based on the HILIC–HPLC technology. In brief, glycans are cleaved from the protein backbone by enzymatic or chemical methods and subsequently the addition of fluorescent label is accomplished by 2-aminobenzamide (2-AB). As 2-AB–labeled glycans are separated by HILIC–HPLC, accurate and quantitative measurement of the relative amounts of individual glycans is achieved by the 1:1 stoichiometric and nonselective labeling. HILIC data can also be complemented by weak anionic exchange chromatography (WAX), which provides information regarding negatively charged glycans. The HILIC methodology uses a 2-AB labeled dextran ladder (2-AB–glucose homopolymer) as an external standard, to which the retention time of each species is related by fitting a fifth-order polynomial distribution curve. The retention time is then transformed into glucose units (GU), which can be used as reference standard values because the calibration minimizes system-to-system and/or day-to-day variations, making the data comparable and robust. For each monosaccharide present in the structure, an incremental value can be obtained that is not only reproducible but also predictive. Glycan structures can then be characterized after enzymatic digestion of pools of glycans by diverse exoglycosidases (in arrays or singly), which specifically cleave glycosidic bonds of individual monosaccharide units from the terminal residue. Typically, a glycosylation profile can be obtained from 25 ng of glycoprotein. This technology is also suitable for release and preparation of N-glycans for other analytical techniques.

High throughput characterization of N-glycan structures on secreted glycoproteins

Until relatively recently large scale analysis of glycans was a difficult task to perform. While still requiring expertise, technological advances now mean glycan analysis is beginning to achieve parity with both genomic and proteomic analysis.[40] For complete characterization of glycans, an orthogonal approach must be taken using a combination of different separation technologies, such as high-performance liquid chromatography (HPLC) or capillary electrophoresis and mass spectrometry (MS).[41] A plethora of powerful MS technologies have been developed in recent years, contributing enormously to the advancement of glycomics.[42–44] However, as a technology, its adaptation to the high-throughput analysis of large sample cohorts, essential for clinical applications, is still at an early stage.

A robust, fully automatable technology platform was presented in 2008[45] and later demonstrated to be a key tool for the analysis of glycans in high-throughput fashion (Fig. 1). This platform includes computer software for the detailed analysis of low femtomole quantities of N-linked sugars released from glycoproteins. It is based on sample immobilization in 96-well plates, glycan release and fluorescent labeling, quantitative HPLC analysis utilizing HILIC (hydrophilic interaction liquid chromatography) and structural assignment of peaks from HPLC profiles via a database (GlycoBase)[46] of more than 350 N-glycan structures. This strategy was applied in the identification and screening of disease biomarkers[47,48] as well as large scale GWAS studies.[40]

APPs and macrophages in the tumor microenvironment

Two hallmarks of cancer are immune evasion by cancer cells and the presence of tumor-promoting inflammation.[49] Although a number of different immune cells—including various T cell subsets, neutrophils, mast cells, and eosinophils—are found associated with the tumor microenvironment,[50] arguably it is the influx of macrophages that plays the most significant role in tumor metastasis. In mouse models in which macrophages are deficient, tumors grow at a slower rate and do not metastasize.[51] To further highlight the significance of macrophage-induced metastasis, approximately 90% of cancer deaths are related to metastatic disease.[52] Tumor associated macrophages (TAMs) have been shown to promote immune evasion, cell proliferation, and increased migratory potential of cancer cells through the release of cytokines (IL-10 and transforming growth factor (TGF)-β), growth factors (e.g., vascular endothelial growth factor (VEGF)), and chemokines (C–C motif ligand 2 (CCL2)).[53,54] Macrophages can differentiate as either classically activated (M1) or alternatively activated (M2) depending on the cytokines present during their activation. In certain cancers, TAMs can represent >50% of the tumor volume,[54,55] and a significant proportion of these are thought to be of an M2-like phenotype.[56–58] The natural function of M2 macrophages is to promote tissue repair, angiogenesis, and extracellular remodeling; however, within the context of the tumor microenvironment these activities promote tumor formation.[54] The APPs haptoglobin (Hp) and kininogen can interact with a number of receptors, such as CD163 (discussed later), present on TAMs; such receptors may have the potential to augment the immune response by influencing the secretion of chemokines, cytokines, or factors that influence angiogenesis, thereby creating an environment beneficial to tumor cell survival (Fig. 2).

Hp and the tumor microenvironment

The α (light) and β (heavy) chains of Hp are produced by cleavage of a large polypeptide chain encoded by two separate but highly linked genetic loci on chromosome 16q22.[59,60] Humans have a polymorphism in the α chain loci that produces different alleles, Hp1 and Hp2.[61,62] This polymorphism results in three different common phenotypes: Hp1–1

has an α_1 chain linked to a β chain, Hp2–1 has an α_2 chain linked to a β chain, and Hp2–2 is comprised of multiple $\alpha_2\beta$ dimers joined by disulphide bonds to form complexes between 170–900 KDa.[63] The various forms of Hp have been reported to have different abilities to bind hemoglobin and different properties regarding inflammatory and angiogenesis functions.[64,65] Hp has four N-glycosylation sites, all of which are present on the β chain.[25]

The best-characterized physiological function of Hp is its ability to bind free hemoglobin (Hb), thereby preventing it from causing oxidative damage to tissues. However, Hp knockout mice have stunted lymphoid organ development and a reduction in the number of both T and B lymphocytes,[66] highlighting that the functions of Hp are likely to be more complex than is currently understood. Hp has been reported to bind to several important receptors on leukocytes, including CD163,[67] CD22,[68] CCR2,[69] and CD18/CD11b.[70]

CD163 and Hp. CD163, a scavenger receptor expressed on monocytes and macrophages, is responsible for the internalization of Hp–Hb complexes.[71] These Hp–Hb complexes can then be degraded, thereby providing a source of iron to the macrophage. Iron metabolism is becoming recognized as a major contributing factor to carcinogenesis.[72] In addition to having high levels of CD163, M2 macrophages display high ferroportin levels (iron release) and low ferritin levels (iron retention).[52,73,74] This combination of iron-regulating proteins gives M2 macrophages a high capacity to release iron; and in the context of TAMs, this iron-regulating proteins can lead to release of iron directly into the tumor microenvironment (Fig. 3). Iron has been shown to influence angiogenesis, wound healing, cell proliferation, and extracellular matrix remodeling.[75] Recalcati *et al.* recently demonstrated that the preconditioned medium from M2 macrophages can promote tumor growth, and that this activity was abrogated by an iron chelator.[76]

It has been demonstrated that Hp can be detected in breast tumor interstitial fluid,[77] indicating its availability for interaction with TAMs. With a high concentration of M2 macrophages resident within tumors, it is possible that Hp could play a significant role in the delivery of iron to a growing tumor. In addition, there have also been reports of

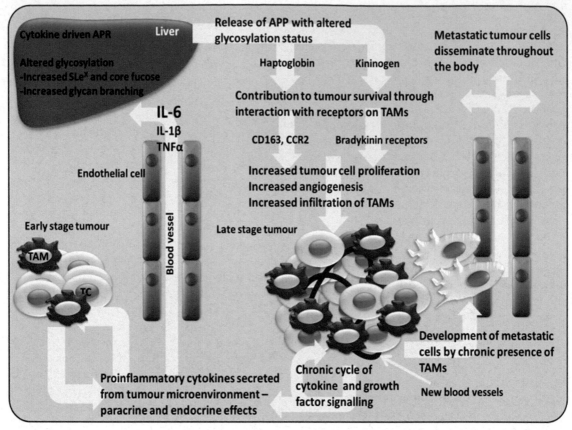

Figure 2. IL-6 secreted by tumor cell and TAMs can stimulate the APP production. The APPs haptoglobin and kininogen have the potential to influence the secretion of proangiogenic by interacting with receptor on tumor-associated macrophages. This creates a chronic cycle that favors tumor growth, angiogenesis, and, eventually, tumor metastasis.

breast cancer cells that acquire the ability to express CD163,[78,79] which may provide them an alternative method for acquiring iron from Hp–Hb complexes.

When cross-linked CD163 can stimulate intracellular signaling in macrophages, leading to enhanced expression of IL-6 and macrophage colony-stimulating factor (M-CSF).[80] Chronic secretion of IL-6 by TAMs provides a feed-forward loop resulting in continual production of APPs by the liver (Fig. 2). It has also been demonstrated that CD163 activation by Hp–Hb complexes results in the release of IL-10.[81] Increased IL-10 secretion by M2 macrophages is an important mechanism by which they dampen the adaptive immune response: by suppressing T cell activation. Glycosylation has been reported to play an important role in ligand–receptor interactions.[38,39,82] Thus, the glycosylation status of Hp may play a crucial role in the intracellular signaling of CD163 by determining the ex-

tent to which the activation signal is transmitted to macrophages.

CCR2 and Hp. Chemokine receptor 2 (CCR2), which is highly expressed on M2 macrophages,[83] can also bind Hp. CCR2 and its ligand chemokine CCL2, a potent chemoattractant for the migration of activated monocytes, are commonly associated with macrophage infiltration in a number of tumor types, including prostate,[84] colon,[85] and breast,[86] among others. Maffei et al. have shown that the binding of Hp to CCR2 leads to CCR2 internalization and increased monocyte migration as a result of extracellular signal regulated kinase (ERK) activation.[69] The glycosylation status of Hp may also play an important role from its ability to interact with CCR2, thus augmenting a signaling pathway that is central to the recruitment of macrophages into the tumor microenvironment.

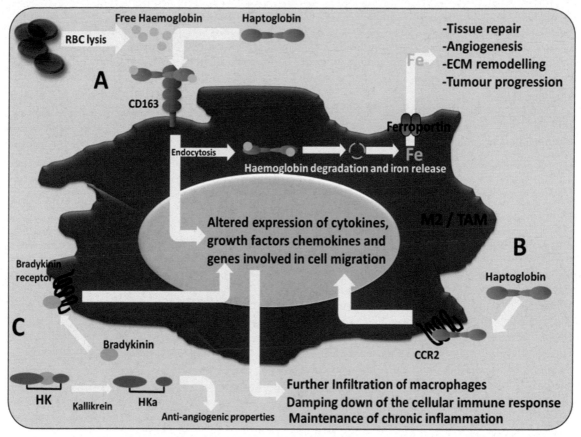

Figure 3. The potential influence of haptoglobin and kininogen on TAMs. (A) The interaction of haptoglobin with CD163 may provide a mechanism by which TAMs can deliver a source of iron to growing tumor cells. Intracellular signaling by CD163 also stimulates the release of IL-10 from macrophages, a cytokine crucial for suppressing the adaptive immune response thus allowing tumor cells to evade detection. (B) Haptoglobin's interaction with CCR2 can induce recruitment of additional macrophages into the tumor microenvironment. (C) Stimulation of the bradykinin receptors by BK can stimulate macrophages to release a number of cytokines (e.g., IL-1 and TNF-α) and chemotactic factors (e.g., IL-8).

Kininogen and the tumor microenvironment

The kininogen-1 gene produces both high molecular weight kininogen (HK) and low molecular weight kininogen (LK) through alternative splicing, with both proteins having identical heavy chains and either a 52 KDa or a 4 KDa light chain, respectively.[87] In rat models of inflammatory bowel disease, where the reading frame for HK is knocked-out, there is less inflammation and fewer intestinal lesions,[88,89] indicating HK has the potential to regulate inflammation.

HK can be cleaved by serine proteases of the kallikrein family,[90] resulting in the release of bioactive peptide called bradykinin (BK). A nine amino acid peptide, BK signals through BK receptor 2 (BDKRB2); removal of an arginine from BK produces des-ARG-BK, which signals through BK receptor 1 (BDKRB1). BK has been demonstrated to have both proinflammatory and proangiogenic effects,[91,92] to induce cell migration,[93,94] and to play a role in a number of diseases, including cancer.[95,96] In addition, BK regulates the production of TNF-α and IL-1,[97] as well as IL-8 and CCL3,[98] by macrophages, thereby stimulating a proinflammatory environment. IL-8 secretion by M2 macrophages has been shown to be a major contributing factor in their ability to stimulate angiogenesis.[99]

The cleavage of BK from HK leaves the heavy and light chains joined by a disulphide bond, a protein referred to as HKa.[100] Furthermore, the removal of the BK causes a major conformational change that results in HKa having a significantly different structure from that of HK,[101] conferring unique properties to HKa. HKa displays an anticell-adhesion effect (compared to HK)[102] and antiangiogenic properties by selectively causing apoptosis in proliferating endothelial cells.[103,104]

The reported opposing effects of BK and HKa on inflammation are likely to be a mechanism of maintaining balance at inflammatory sites. However, this creates a scenario where altered glycosylation of HK could have an impact on its processing by kallikrein enzymes during release of the BK peptide and HKa.

Further perspectives

Many functions of the immune system are heavily influenced by the glycan content of key proteins; clear examples of this are the interactions between PSGL-1 and selectins that facilitate leukocyte extravasation[105] and glycosylation of IgG molecules that modulate antibody function.[106,107] Because of the ubiquitous presence of glycans on key molecules controlling immune cell functions, for the coming decade we envision that joint efforts between glycobiologists and immunologists will yield significant understanding of key aspects of the immune system. As technological advances have allowed glycobiology to join the "omic's" club,[40,108,109] a future challenge will be to move beyond structural glycomics to develop a better understanding of functional glycomics. In this short review, we have touched on just one group of proteins—acute phase glycoproteins—for which a better understanding of how glycosylation impacts their function could provide new insights into the roles they play in tumor promotion. The detailed analysis of the glycosylation status of APPs is yielding many novel biomarkers for various cancers. And through the example of Hp, it is possible to identify a number of pathways that it may influence in the tumor microenvironment.

One mechanism through which glycosylation may influence APP function is by improving APP stability. Increased sialylation has been demonstrated to significantly improve the half-life of soluble proteins,[110–112] thereby augmenting their pharmokinetics. Sialylation and fucosylation can influence many receptor–ligand interactions;[34,113,114] and as Hp can bind with at least four known receptors, changes in its glycosylation status are likely to bear some influence on these interactions. A shift in the strength of signaling through one receptor over another could determine the overall response of M2 macrophages in the tumor microenvironment. Ultimately, how glycosylation changes alter the function of Hp, HK, and other APPs is still largely unknown. Nevertheless, fast-moving technological advances for glycan analysis, and a greater awareness by scientists in different disciplines of the impact glycosylation has on protein function, will no doubt lead to the generation of a wealth of information in this area during the coming years.

Acknowledgments

Eugene Dempsey was supported by Cancer Research Ireland (Grant CRI07RUD). The authors would like to thank Dr. Karina Mariño for her help in preparing this manuscript.

Conflicts of interest

The authors declare no conflicts of interest.

References

1. Gruys, E., M.J. Toussaint, T.A. Niewold & S.J. Koopmans. 2005. Acute phase reaction and acute phase proteins. *J. Zhejiang Univ. Sci. B.* **6:** 1045–1056.

2. Gabay, C. & I. Kushner. 1999. Acute-phase proteins and other systemic responses to inflammation. *N. Engl. J. Med.* **340:** 448–454.

3. Schultz, D.R. & P.I. Arnold. 1990. Properties of four acute phase proteins: C-reactive protein, serum amyloid A protein, alpha 1-acid glycoprotein, and fibrinogen. *Semin. Arthritis. Rheum.* **20:** 129–147.

4. Krzystek-Korpacka, M., M. Matusiewicz, D. Diakowska, *et al.* 2008. Acute-phase response proteins are related to cachexia and accelerated angiogenesis in gastroesophageal cancers. *Clin. Chem. Lab. Med.* **46:** 359–364.

5. Mullan, R.H., B. Bresnihan, L. Golden-Mason, *et al.* 2006. Acute-phase serum amyloid A stimulation of angiogenesis, leukocyte recruitment, and matrix degradation in rheumatoid arthritis through an NF-kappaB-dependent signal transduction pathway. *Arthritis Rheum.* **54:** 105–114.

6. Prowse, K.R. & H. Baumann. 1989. Interleukin-1 and interleukin-6 stimulate acute-phase protein production in primary mouse hepatocytes. *J. Leukoc. Biol.* **45:** 55–61.

7. Fattori, E., M. Cappelletti, P. Costa, *et al.* 1994. Defective inflammatory response in interleukin 6-deficient mice. *J. Exp. Med.* **180:** 1243–1250.

8. Sarrats, A., R. Saldova, E. Pla, *et al.* 2010. Glycosylation of liver acute-phase proteins in pancreatic cancer and chronic pancreatitis. *Proteomics Clin. Appl.* **4:** 432–448.

9. Abd Hamid, U.M., L. Royle, R. Saldova, *et al.* 2008. A strategy to reveal potential glycan markers from serum glycoproteins associated with breast cancer progression. *Glycobiology.* **18:** 1105–1118.

10. Bones, J., J.C. Byrne, N. O'Donoghue, *et al.* 2011. Glycomic and glycoproteomic analysis of serum from patients with stomach cancer reveals potential markers arising from host

defense response mechanisms. *J. Proteome Res.* **10**: 1246–1265.

11. Comunale, M.A., L. Rodemich-Betesh, J. Hafner, *et al.* 2010. Linkage specific fucosylation of alpha-1-antitrypsin in liver cirrhosis and cancer patients: implications for a biomarker of hepatocellular carcinoma. *PLoS One.* **5**: e12419.

12. Yoon, S.J., S.Y. Park, P.C. Pang, *et al.* 2010. N-glycosylation status of beta-haptoglobin in sera of patients with prostate cancer vs. benign prostate diseases. *Int. J. Oncol.* **36**: 193–203.

13. Fujimura, T., Y. Shinohara, B. Tissot, *et al.* 2008. Glycosylation status of haptoglobin in sera of patients with prostate cancer vs. benign prostate disease or normal subjects. *Int. J. Cancer.* **122**: 39–49.

14. Qiu, Y., T.H. Patwa, L. Xu, *et al.* 2008. Plasma glycoprotein profiling for colorectal cancer biomarker identification by lectin glycoarray and lectin blot. *J. Proteome Res.* **7**: 1693–1703.

15. Ang, I.L., T.C.W. Poon, P.B.S. Lai, *et al.* 2006. Study of serum haptoglobin and its glycoforms in the diagnosis of hepatocellular carcinoma: a glycoproteomic approach. *J. Proteome Res.* **5**: 2691–2700.

16. Wang, M., R.E. Long, M.A. Comunale, *et al.* 2009. Novel fucosylated biomarkers for the early detection of hepatocellular carcinoma. *Cancer Epidemiol. Biomarkers Prev.* **18**: 1914–1921.

17. Okuyama, N., Y. Ide, M. Nakano, *et al.* 2006. Fucosylated haptoglobin is a novel marker for pancreatic cancer: a detailed analysis of the oligosaccharide structure and a possible mechanism for fucosylation. *Int. J. Cancer.* **118**: 2803–2808.

18. Zhang, S., H. Shu, K. Luo, *et al.* 2011. N-linked glycan changes of serum haptoglobin beta chain in liver disease patients. *Mol. Biosyst.* **7**: 1621–1628.

19. Liu, Y., J. He, C. Li, *et al.* 2010. Identification and confirmation of biomarkers using an integrated platform for quantitative analysis of glycoproteins and their glycosylations. *J. Proteome Res.* **9**: 798–805.

20. Thompson, S., E. Dargan & G.A. Turner. 1992. Increased fucosylation and other carbohydrate changes in haptoglobin in ovarian cancer. *Cancer Lett.* **66**: 43–48.

21. Turner, G.A., M.T. Goodarzi & S. Thompson. 1995. Glycosylation of alpha-1-proteinase inhibitor and haptoglobin in ovarian cancer: evidence for two different mechanisms. *Glycoconj J.* **12**: 211–218.

22. Park, S.Y., S.J. Yoon, Y.T. Jeong, *et al.* 2010. N-glycosylation status of beta-haptoglobin in sera of patients with colon cancer, chronic inflammatory diseases and normal subjects. *Int. J. Cancer.* **126**: 142–155.

23. Lin, Z., D.M. Simeone, M.A. Anderson, *et al.* 2011. Mass spectrometric assay for analysis of haptoglobin fucosylation in pancreatic cancer. *J. Proteome Res.* **10**: 2602–2611.

24. Matsumoto, H., S. Shinzaki, M. Narisada, *et al.* 2010. Clinical application of a lectin-antibody ELISA to measure fucosylated haptoglobin in sera of patients with pancreatic cancer. *Clin. Chem. Lab. Med.* **48**: 505–512.

25. Nakano, M., T. Nakagawa, T. Ito, *et al.* 2008. Site-specific analysis of N-glycans on haptoglobin in sera of patients with pancreatic cancer: a novel approach for the development of tumor markers. *Int. J. Cancer.* **122**: 2301–2309.

26. Hoagland, L.F., M.J. Campa, E.B. Gottlin, *et al.* 2007. Haptoglobin and posttranslational glycan-modified derivatives as serum biomarkers for the diagnosis of nonsmall cell lung cancer. *Cancer.* **110**: 2260–2268.

27. Arnold, J.N., R. Saldova, M.C. Galligan, *et al.* 2011. Novel glycan biomarkers for the detection of lung cancer. *J. Proteome Res.* **10**: 1755–1764.

28. Zhao, J., T.H. Patwa, W. Qiu, *et al.* 2007. Glycoprotein microarrays with multi-lectin detection: unique lectin binding patterns as a tool for classifying normal, chronic pancreatitis and pancreatic cancer sera. *J. Proteome Res.* **6**: 1864–1874.

29. Harduin-Lepers, A., V. Vallejo-Ruiz, M.A. Krzewinski-Recchi, *et al.* 2001. The human sialyltransferase family. *Biochimie.* **83**: 727–737.

30. Millar, J.S. 2001. The sialylation of plasma lipoproteins. *Atherosclerosis.* **154**: 1–13.

31. Schauer, R. 2009. Sialic acids as regulators of molecular and cellular interactions. *Curr. Opin. Struct. Biol.* **19**: 507–514.

32. Morell, A.G., G. Gregoriadis, I.H. Scheinberg, *et al.* 1971. The role of sialic acid in determining the survival of glycoproteins in the circulation. *J. Biol. Chem.* **246**: 1461–1467.

33. Miyoshi, E., K. Moriwaki & T. Nakagawa. 2008. Biological function of fucosylation in cancer biology. *J. Biochem.* **143**: 725–729.

34. Moriwaki, K., M. Narisada, T. Imai, *et al.* 2010. The effect of epigenetic regulation of fucosylation on TRAIL-induced apoptosis. *Glycoconj J.* **27**: 649–659.

35. Anthony, R.M., F. Nimmerjahn, D.J. Ashline, *et al.* 2008. Recapitulation of IVIG anti-inflammatory activity with a recombinant IgG Fc. *Science.* **320**: 373–376.

36. Seales, E.C., G.A. Jurado, B.A. Brunson, *et al.* 2005. Hypersialylation of beta1 integrins, observed in colon adenocarcinoma, may contribute to cancer progression by upregulating cell motility. *Cancer Res.* **65**: 4645–4652.

37. Liu, Y.C., H.Y. Yen, C.Y. Chen, *et al.* 2011. Sialylation and fucosylation of epidermal growth factor receptor suppress its dimerization and activation in lung cancer cells. *Proc. Natl. Acad. Sci. U. S. A.* **108**: 11332–11337.

38. Matsumoto, K., H. Yokote, T. Arao, *et al.* 2008. N-Glycan fucosylation of epidermal growth factor receptor modulates receptor activity and sensitivity to epidermal growth factor receptor tyrosine kinase inhibitor. *Cancer Sci.* **99**: 1611–1617.

39. Wang, X., J. Gu, H. Ihara, *et al.* 2006. Core fucosylation regulates epidermal growth factor receptor-mediated intracellular signaling. *J. Biol. Chem.* **281**: 2572–2577.

40. Lauc, G., A. Essafi, J.E. Huffman, *et al.* 2010. Genomics meets glycomics-the first GWAS study of human N-Glycome identifies HNF1alpha as a master regulator of plasma protein fucosylation. *PLoS Genet.* **6**: e1001256.

41. Marino, K., J. Bones, J.J. Kattla & P.M. Rudd. 2010. A systematic approach to protein glycosylation analysis: a path through the maze. *Nat. Chem. Biol.* **6**: 713–723.

42. Tsai, H.-Y., K. Boonyapranai, S. Sriyam, *et al.* 2011. Glycoproteomics analysis to identify a glycoform on haptoglobin associated with lung cancer. *Proteomics.* **11**: 2162–2170.

43. Ivancic, M.M., H.S. Gadgil, H.B. Halsall & M.J. Treuheit. 2010. LC/MS analysis of complex multiglycosylated human Î±1-acid glycoprotein as a model for developing

identification and quantitation methods for intact glycopeptide analysis. *Anal. Biochem.* **400:** 25–32.

44. Imre, T., T. Kremmer, K. Héberger, *et al.* 2008. Mass spectrometric and linear discriminant analysis of N-glycans of human serum alpha-1-acid glycoprotein in cancer patients and healthy individuals. *J. Proteomics.* **71:** 186–197.

45. Royle, L., M.P. Campbell, C.M. Radcliffe, *et al.* 2008. HPLC-based analysis of serum N-glycans on a 96-well plate platform with dedicated database software. *Anal. Biochem.* **376:** 1–12.

46. Campbell, M.P., L. Royle, C.M. Radcliffe, *et al.* 2008. GlycoBase and autoGU: tools for HPLC-based glycan analysis. *Bioinformatics.* **24:** 1214–1216.

47. Saldova, R., J.M. Reuben, U.M. Abd Hamid, *et al.* 2011. Levels of specific serum N-glycans identify breast cancer patients with higher circulating tumor cell counts. *Ann. Oncol.* **22:** 1113–1119.

48. Stanta, J.L., R. Saldova, W.B. Struwe, *et al.* 2010. Identification of N-glycosylation changes in the CSF and serum in patients with schizophrenia. *J. Proteome Res.* **9:** 4476–4489.

49. Hanahan, D. & R.A. Weinberg. 2011. Hallmarks of cancer: the next generation. *Cell.* **144:** 646–674.

50. Mantovani, A., P. Allavena, A. Sica & F. Balkwill. 2008. Cancer-related inflammation. *Nature.* **454:** 436–444.

51. Lin, E.Y. & J.W. Pollard. 2004. Macrophages: modulators of breast cancer progression. *Novartis Found. Symp.* **256:** 158–168; discussion 168–172, 259–269.

52. Richardson, D.R., D.S. Kalinowski, S. Lau, *et al.* 2009. Cancer cell iron metabolism and the development of potent iron chelators as anti-tumor agents. *Biochim. Biophys. Acta.* **1790:** 702–717.

53. Joyce, J.A. & J.W. Pollard. 2009. Microenvironmental regulation of metastasis. *Nat. Rev. Cancer.* **9:** 239–252.

54. Solinas, G., G. Germano, A. Mantovani, *et al.* 2009. Tumor-associated macrophages (TAM) as major players of the cancer-related inflammation. *J. Leukoc. Biol.* **86:** 1065–1073.

55. Lewis, C.E. & J.W. Pollard. 2006. Distinct role of macrophages in different tumor microenvironments. *Cancer Res.* **66:** 605–612.

56. Kurahara, H., H. Shinchi, Y. Mataki, *et al.* 2011. Significance of M2-polarized tumor-associated macrophage in pancreatic cancer. *J. Surg. Res.* **167:** e211–219.

57. Solinas, G., S. Schiarea, M. Liguori, *et al.* 2010. Tumor-conditioned macrophages secrete migration-stimulating factor: a new marker for M2-polarization, influencing tumor cell motility. *J. Immunol.* **185:** 642–652.

58. Allavena, P., A. Sica, C. Garlanda & A. Mantovani. 2008. The Yin-Yang of tumor-associated macrophages in neoplastic progression and immune surveillance. *Immunol. Rev.* **222:** 155–161.

59. Robson, E.B., P.E. Polani, S.J. Dart, *et al.* 1969. Probable assignment of the alpha locus of haptoglobin to chromome 16 in man. *Nature.* **223:** 1163–1165.

60. Raugei, G., G. Bensi, V. Colantuoni, *et al.* 1983. Sequence of human haptoglobin cDNA: evidence that the alpha and beta subunits are coded by the same mRNA. *Nucleic Acids Res.* **11:** 5811–5819.

61. Levy, A.P., R. Asleh, S. Blum, *et al.* 2010. Haptoglobin: basic and clinical aspects. *Antioxid. Redox. Signal.* **12:** 293–304.

62. Marquez, L., C. Shen, I. Cleynen, *et al.* 2011. Effects of haptoglobin polymorphisms and deficiency on susceptibility to inflammatory bowel disease and on severity of murine colitis. *Gut.* [Epub ahead of print.] doi:10.1136/gut.2011.240978.

63. Sadrzadeh, S.M. & J. Bozorgmehr. 2004. Haptoglobin phenotypes in health and disorders. *Am. J. Clin. Pathol.* **121**(Suppl): S97–S104.

64. Van Vlierberghe, H., M. Langlois & J. Delanghe. 2004. Haptoglobin polymorphisms and iron homeostasis in health and in disease. *Clin. Chim. Acta.* **345:** 35–42.

65. Guetta, J., M. Strauss, N.S. Levy, *et al.* 2007. Haptoglobin genotype modulates the balance of Th1/Th2 cytokines produced by macrophages exposed to free hemoglobin. *Atherosclerosis.* **191:** 48–53.

66. Huntoon, K.M., Y. Wang, C.A. Eppolito, *et al.* 2008. The acute phase protein haptoglobin regulates host immunity. *J. Leukoc. Biol.* **84:** 170–181.

67. Kristiansen, M., J.H. Graversen, C. Jacobsen, *et al.* 2001. Identification of the haemoglobin scavenger receptor. *Nature.* **409:** 198–201.

68. Hanasaki, K., L.D. Powell & A. Varki. 1995. Binding of human plasma sialoglycoproteins by the B cell-specific lectin CD22. Selective recognition of immunoglobulin M and haptoglobin. *J. Biol. Chem.* **270:** 7543–7550.

69. Maffei, M., M. Funicello, T. Vottari, *et al.* 2009. The obesity and inflammatory marker haptoglobin attracts monocytes via interaction with chemokine (C-C motif) receptor 2 (CCR2). *BMC Biol.* **7:** 87.

70. El Ghmati, S.M., E.M. Van Hoeyveld, J.G. Van Strijp, *et al.* 1996. Identification of haptoglobin as an alternative ligand for CD11b/CD18. *J. Immunol.* **156:** 2542–2552.

71. Kaempfer, T., E. Duerst, P. Gehrig, *et al.* 2011. Extracellular hemoglobin polarizes the macrophage proteome toward Hb-clearance, enhanced antioxidant capacity and suppressed HLA class 2 expression. *J. Proteome. Res.* **10:** 2397–2408.

72. Toyokuni, S. 2009. Role of iron in carcinogenesis: cancer as a ferrotoxic disease. *Cancer Sci.* **100:** 9–16.

73. Torti, S.V. & F.M. Torti. 2011. Ironing out cancer. *Cancer Res.* **71:** 1511–1514.

74. Wang, J. & K. Pantopoulos. 2011. Regulation of cellular iron metabolism. *Biochem. J.* **434:** 365–381.

75. Cairo, G., S. Recalcati, A. Mantovani & M. Locati. 2011. Iron trafficking and metabolism in macrophages: contribution to the polarized phenotype. *Trends. Immunol.* **32:** 241–247.

76. Recalcati, S., M. Locati, A. Marini, *et al.* 2010. Differential regulation of iron homeostasis during human macrophage polarized activation. *Eur. J. Immunol.* **40:** 824–835.

77. Celis, J.E., P. Gromov, T. Cabezon, *et al.* 2004. Proteomic characterization of the interstitial fluid perfusing the breast tumor microenvironment: a novel resource for biomarker and therapeutic target discovery. *Mol. Cell Proteomics.* **3:** 327–344.

78. Shabo, I. & J. Svanvik. 2011. Expression of macrophage antigens by tumor cells. *Adv. Exp. Med. Biol.* **714:** 141–150.

79. Shabo, I., O. Stal, H. Olsson, *et al.* 2008. Breast cancer expression of CD163, a macrophage scavenger receptor, is related

to early distant recurrence and reduced patient survival. *Int. J. Cancer*. **123:** 780–786.

80. Van den Heuvel, M.M., C.P. Tensen, J.H. van As, *et al.* 1999. Regulation of CD163 on human macrophages: cross-linking of CD163 induces signaling and activation. *J. Leukoc. Biol.* **66:** 858–866.

81. Philippidis, P., J.C. Mason, B.J. Evans, *et al.* 2004. Hemoglobin scavenger receptor CD163 mediates interleukin-10 release and heme oxygenase-1 synthesis: antiinflammatory monocyte-macrophage responses in vitro, in resolving skin blisters in vivo, and after cardiopulmonary bypass surgery. *Circ. Res.* **94:** 119–126.

82. Margraf-Schonfeld, S., C. Bohm & C. Watzl. 2011. Glycosylation affects ligand binding and function of the activating natural killer cell receptor 2B4 (CD244) protein. *J. Biol. Chem.* **286:** 24142–24149.

83. Mantovani, A., S. Sozzani, M. Locati, *et al.* 2002. Macrophage polarization: tumor-associated macrophages as a paradigm for polarized M2 mononuclear phagocytes. *Trends Immunol.* **23:** 549–555.

84. Zhang, J., Y. Lu & K.J. Pienta. 2010. Multiple roles of chemokine (C-C motif) ligand 2 in promoting prostate cancer growth. *J. Natl. Cancer Inst.* **102:** 522–528.

85. Popivanova, B.K., F.I. Kostadinova, K. Furuichi, *et al.* 2009. Blockade of a chemokine, CCL2, reduces chronic colitis-associated carcinogenesis in mice. *Cancer Res.* **69:** 7884–7892.

86. Qian, B.Z., J. Li, H. Zhang, *et al.* 2011. CCL2 recruits inflammatory monocytes to facilitate breast-tumor metastasis. *Nature*. **475:** 222–225.

87. Kitamura, N., H. Kitagawa, D. Fukushima, *et al.* 1985. Structural organization of the human kininogen gene and a model for its evolution. *J. Biol. Chem.* **260:** 8610–8617.

88. Isordia-Salas, I., R.A. Pixley, H. Parekh, *et al.* 2003. The mutation Ser511Asn leads to N-glycosylation and increases the cleavage of high molecular weight kininogen in rats genetically susceptible to inflammation. *Blood*. **102:** 2835–2842.

89. Isordia-Salas, I., R.A. Pixley, F. Li, *et al.* 2002. Chronic intestinal inflammation and angiogenesis in genetically susceptible rats is modulated by kininogen deficiency. *Int. Immunopharmacol.* **2:** 1895–1905.

90. Bryant, J.W. & Z. Shariat-Madar. 2009. Human plasma kallikrein-kinin system: physiological and biochemical parameters. *Cardiovasc. Hematol. Agents Med. Chem.* **7:** 234–250.

91. Hayashi, I., H. Amano, S. Yoshida, *et al.* 2002. Suppressed angiogenesis in kininogen-deficiencies. *Lab. Invest.* **82:** 871–880.

92. Guo, Y.L. & R.W. Colman. 2005. Two faces of high-molecular-weight kininogen (HK) in angiogenesis: bradykinin turns it on and cleaved HK (HKa) turns it off. *J. Thromb. Haemost.* **3:** 670–676.

93. Erices, R., J. Corthorn, F. Lisboa & G. Valdes. 2011. Bradykinin promotes migration and invasion of human immortalized trophoblasts. *Reprod. Biol. Endocrinol.* **9:** 97.

94. Montana, V. & H. Sontheimer. 2011. Bradykinin promotes the chemotactic invasion of primary brain tumors. *J. Neurosci.* **31:** 4858–4867.

95. Ikeda, Y., I. Hayashi, E. Kamoshita, *et al.* 2004. Host stromal bradykinin B2 receptor signaling facilitates tumor-associated angiogenesis and tumor growth. *Cancer Res.* **64:** 5178–5185.

96. Stewart, J.M., L. Gera, D.C. Chan, *et al.* 2005. Combination cancer chemotherapy with one compound: pluripotent bradykinin antagonists. *Peptides*. **26:** 1288–1291.

97. Tiffany, C.W. & R.M. Burch. 1989. Bradykinin stimulates tumor necrosis factor and interleukin-1 release from macrophages. *FEBS Lett.* **247:** 189–192.

98. Sato, E., S. Koyama, H. Nomura, *et al.* 1996. Bradykinin stimulates alveolar macrophages to release neutrophil, monocyte, and eosinophil chemotactic activity. *J. Immunol.* **157:** 3122–3129.

99. Medina, R.J., L. O'Neill C, M. O'Doherty T, *et al.* 2011. Myeloid angiogenic cells act as alternative M2 macrophages and modulate angiogenesis through interleukin-8. *Mol. Med.* **17:** 1045–1055.

100. Sun, D. & K.R. McCrae. 2006. Endothelial-cell apoptosis induced by cleaved high-molecular-weight kininogen (HKa) is matrix dependent and requires the generation of reactive oxygen species. *Blood*. **107:** 4714–4720.

101. Weisel, J.W., C. Nagaswami, J.L. Woodhead, *et al.* 1994. The shape of high molecular weight kininogen. Organization into structural domains, changes with activation, and interactions with prekallikrein, as determined by electron microscopy. *J. Biol. Chem.* **269:** 10100–10106.

102. Guo, Y.L., S. Wang, D.J. Cao & R.W. Colman. 2003. Apoptotic effect of cleaved high molecular weight kininogen is regulated by extracellular matrix proteins. *J. Cell. Biochem.* **89:** 622–632.

103. Cao, D.J., Y.L. Guo & R.W. Colman. 2004. Urokinase-type plasminogen activator receptor is involved in mediating the apoptotic effect of cleaved high molecular weight kininogen in human endothelial cells. *Circ Res.* **94:** 1227–1234.

104. Zhang, J.C., K. Claffey, R. Sakthivel, *et al.* 2000. Two-chain high molecular weight kininogen induces endothelial cell apoptosis and inhibits angiogenesis: partial activity within domain 5. *FASEB J.* **14:** 2589–2600.

105. Carlow, D.A., K. Gossens, S. Naus, *et al.* 2009. PSGL-1 function in immunity and steady state homeostasis. *Immunol. Rev.* **230:** 75–96.

106. Nimmerjahn, F., R.M. Anthony & J.V. Ravetch. 2007. Agalactosylated IgG antibodies depend on cellular Fc receptors for in vivo activity. *Proc. Natl. Acad. Sci. U. S. A.* **104:** 8433–8437.

107. Anthony, R.M. & J.V. Ravetch. 2010. A novel role for the IgG Fc glycan: the anti-inflammatory activity of sialylated IgG Fcs. *J. Clin. Immunol.* **30(Suppl 1):** S9–14.

108. Pucic, M., A. Knezevic, J. Vidic, *et al.* 2011. High throughput isolation and glycosylation analysis of IgG-variability and heritability of the IgG glycome in three isolated human populations. *Mol. Cell Proteomics.* **10:** M111 010090.

109. Igl, W., O. Polasek, O. Gornik, *et al.* 2011. Glycomics meets lipidomics—associations of N-glycans with classical lipids, glycerophospholipids, and sphingolipids in three European populations. *Mol. Biosyst.* **7:** 1852–1862.

110. Webster, R., J. Taberner, A. Edgington, *et al.* 1999. Role of sialylation in determining the pharmacokinetics of neutrophil inhibitory factor (NIF) in the Fischer 344 rat. *Xenobiotica.* **29:** 1141–1155.

111. Richards, A.A., M.L. Colgrave, J. Zhang, *et al.* 2010. Sialic acid modification of adiponectin is not required for multi-merization or secretion but determines half-life in circulation. *Mol. Endocrinol.* **24:** 229–239.

112. Flintegaard, T.V., P. Thygesen, H. Rahbek-Nielsen, *et al.* 2010. N-glycosylation increases the circulatory half-life of human growth hormone. *Endocrinology.* **151:** 5326–5336.

113. Lin, H., D. Wang, T. Wu, *et al.* 2011. Blocking core fuco-sylation of TGF-beta1 receptors downregulates their functions and attenuates the epithelial-mesenchymal transition of renal tubular cells. *Am. J. Physiol. Renal. Physiol.* **300:** F1017–1025.

114. Zhao, Y., S. Itoh, X. Wang, *et al.* 2006. Deletion of core fucosylation on alpha3beta1 integrin down-regulates its functions. *J. Biol. Chem.* **281:** 38343–38350.

115. Saldova, R., L. Royle, C.M. Radcliffe, *et al.* 2007. Ovarian cancer is associated with changes in glycosylation in both acute-phase proteins and IgG. *Glycobiology* **17:** 1344–1356.

Ann. N.Y. Acad. Sci. ISSN 0077-8923

ANNALS OF THE NEW YORK ACADEMY OF SCIENCES

Issue: *Glycobiology of the Immune Response*

Glycans, galectins, and HIV-1 infection

Sachiko Sato,[1] Michel Ouellet,[2] Christian St-Pierre,[1] and Michel J. Tremblay[2]

[1]Glycobiology and Bioimaging Laboratory Research Centre for Infectious Diseases, Laval University, Quebec, Canada.
[2]Laboratory of Human ImmunoRetrovirology, Research Centre for Infectious Diseases, Laval University, Quebec, Canada.

Address for correspondence: Sachiko Sato, Faculty of Medicine, Glycobiology and Bioimaging Laboratory, Research Centre for Infectious Diseases, Laval University, 2705 boul. Laurier, Quebec, Qc, Canada G1V 4G2. Sachiko.Sato@crchul.ulaval.ca
Michel J. Tremblay, Faculty of Medicine, Laboratory of Human ImmunoRetrovirology, Research Centre for Infectious Diseases, Laval University, 2705 boul. Laurier, Quebec, Qc, Canada G1V 4G2. michel.j.tremblay@crchul.ulaval.ca

During sexual transmission, HIV-1 must overcome physiological barriers to establish a founder cell population. Viral adhesion represents a bottleneck for HIV-1 propagation that the virus widens by exploiting some specific host factors. Recognition of oligomannosyl glycans of gp120 by C-type lectins is one such example. Recent works suggest that complex glycans of gp120 are recognized by another host lectin, galectin-1. This interaction results in rapid association of HIV-1 to susceptible cells and facilitates infection. The peculiar presentation of complex glycans on gp120 seems to impart specificity for galectin-1, as another member of the same family, galectin-3, is unable to bind gp120 or enhance HIV-1 infection. Other studies have shown that galectin-9 could also increase HIV-1 infectivity but via an indirect mechanism. Thus, current research suggests that galectins play various roles in HIV-1 pathogenesis. Drug discovery approaches targeting host lectins at early steps could benefit the current arsenal of antiretrovirals.

Keywords: glycobiology; HIV-1; lectin; galectin

AIDS and HIV-1 infection

In the summer of 1981, the world first became aware of an emerging disease that would later be named acquired immune deficiency syndrome (AIDS). Since then, AIDS has killed more than 25 million people.[1,2] Human immunodeficiency virus type-1 (HIV-1), a lentivirus that belongs to the retrovirus family, is the causing agent of AIDS. The number of newly infected individuals has steadily, yet slowly, been declining since its peak of 3.2 million reached in 1997, down to 2.6 million in 2009, possibly due to restless public awareness and prevention campaigns.[1,2] However, 33.3 million people are still living with HIV-1 today.[1,2] Since this retrovirus mainly infects cells of the immune system—namely CD4[+] T cells and macrophages—and cripples the adaptive immune system, HIV-1 infections have an enormous impact on the emergence and spread of other infectious diseases, especially in developing countries.[3] First, in many of those pandemic areas access to medication is difficult and HIV-infected individuals rapidly develop AIDS; they thus become especially sensitive to infection by other pathogens—opportunistic or not—present in the environment. The ensuing immunocompromised status leads to high loads of coinfecting pathogens, a phenomenon that increases the chances of transmission to immunocompetent hosts. Immunosuppression of millions of people in a population can also reduce the effectiveness of immunization campaigns. In the worst possible scenario, one can hypothesize that bacteria, fungi, and protozoa in the environment, as well as zoonotic pathogens, could have a reservoir of ~30 million immunocompromised individuals in which to evolve and adapt to become human pathogens. It is thus paramount, not only for the most afflicted developing countries but also for industrialized countries, to control the HIV-1 pandemic in order to reduce the risk of widespread emergence of other pathogenic agents.[3]

HIV-1 replication cycle

The HIV-1 transmission process[4] is initiated by the binding of the viral envelope glycoprotein complex

doi: 10.1111/j.1749-6632.2012.06475.x

HIV-1 replicative cycle

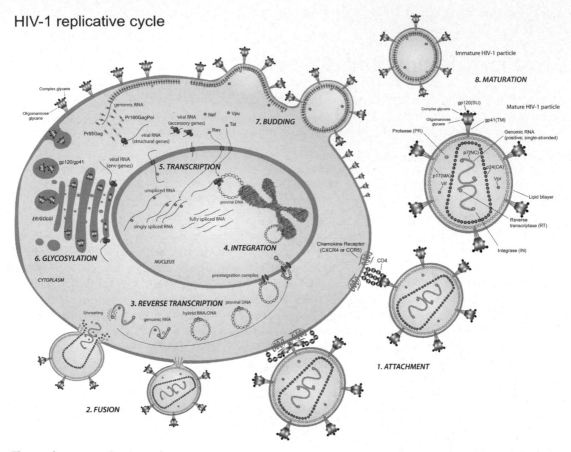

Figure 1. HIV-1 replication cycle.

(Env; gp120, and gp41) to CD4, which is located on the surface of susceptible cells (Fig. 1-1, Attachment). Viral entry requires at least two host surface proteins, CD4 and a coreceptor of the chemokine receptor family, mainly CCR5 and/or CXCR4. Upon binding to CD4, trimeric gp120 undergoes a conformational change that leads to further interactions with its coreceptor. The transmembrane component of the envelope protein, gp41, can then initiate fusion of the viral membrane with the plasma membrane of the host cell. After fusion and delivery of the viral capsid in the cytoplasm, decapsidation leads to the release of viral enzymes, proteins, and genomic RNA inside the cell (Fig. 1-2, Fusion). Reverse transcription of the viral genomic single-stranded positive RNA is then initiated to yield a double-stranded proviral DNA to be imported in the nucleus and integrated into host chromosome (Fig. 1-3 and 1-4, Reverse Transcription and Integration). Active transcription from the integrated proviral DNA occurs

in the presence of NF-κB and viral Tat (Fig. 1-5, Transcription). Splicing of viral mRNA yields early accessory proteins like Tat, Rev, and Nef, which help in transcription, splicing, and modification of the cellular machinery, respectively. Accumulation of Rev protects the viral mRNA from splicing, thus yielding increasingly longer mRNAs able to code for structural and envelope proteins, and finally viral genomic RNAs ready to be encapsidated. Full length viral RNAs bind specifically to viral structural proteins Pr55^{Gag} and Pr160$^{Gag-Pol}$ via Ψ sequences to be encapsidated in the nascent viral particle. The synthesis of the Env precursor protein gp160 occurs in the secretory pathway after docking of ribosomes translating viral RNAs encoding Env to the endoplasmic reticulum (ER). Extensive glycosylation of more than 25 N-glycosylation sites (Asn-X-Ser/Thr) of the Env precursor occurs cotranslationally in the ER and N-linked glycans are further processed in the lumen of the ER and the Golgi complex[5,6] (Fig. 1-6,

Glycosylation). The Env precursor is also cleaved in these compartments by the furin host protease to generate a trimeric gp120/gp41 complex that is then transported to the plasma membrane. Assembly of Pr55^{Gag} and Pr160$^{Gag–Pol}$ associated to viral genomic RNAs occurs at the inner surface of the plasma membrane and induces a characteristic curvature of the membrane that leads to budding of the virion (Fig. 1-7, Budding). During this budding process, HIV-1 also acquires membrane glycoproteins of cellular origin into the viral surface.[7–9] Since HIV-1 carries very few Env trimers or spikes (as few as 14 ± 7) on its surface,[10] the thick glycocalyx present on the virus surface is mainly composed of host membrane components. Host-derived glycans attached to membrane glycoproteins may thus also contribute to conceal antigenic surfaces.[6] Once released, the viral particle achieves maturation through the action of the viral protease that cleaves the polyproteins Pr55^{Gag} and Pr160$^{Gag–Pol}$ into structural and enzymatic components that assemble to form a mature HIV-1 particle (Fig. 1-8, Maturation).

Glycosylation of gp120 of HIV-1

Despite high genetic variability among different isolates and clades of HIV-1, the N-glycosylation sites of gp120 are spatially conserved.[11] The glycans of the Env complex are synthesized by the host glycosylation machinery in the ER and Golgi complex. Like other host glycoproteins, two types of N-linked glycans are found on gp120, namely oligomannose-type glycans, which are rich in mannose residues (Man$_{5\sim9}$GlcNAc$_2$), and complex-type glycans, which carry 2–6 β-galactoside residues (lactosamine residue [GalGlcNAc; LacNAc]) on their trimannose core structure, Man$_3$GlcNAc$_2$ (Fig. 2A).[12–18] Glycosylation of gp120 has at least two unique features that distinguish it from that of host membrane proteins.[6] Those features are observed in both recombinant and viral gp120.[12–18] Since N-linked glycan profiles of host glycoproteins found in HIV-1–infected cells are suggested to be quite similar to those of noninfected cells,[17] it is unlikely that such unique patterns of gp120 glycosylation are entirely the result of viral modulation of the host glycosylation pathways. It has instead been proposed that the distinctive glycosylation patterns of gp120 compared to host glycoproteins may be due to the unusually dense arrangement of N-linked gly-

cans of gp120, which limits the accessibility of the glycan processing enzymes.[6,18]

First, gp120 contains high levels of oligomannose-type glycans, which are considered as incomplete processing forms of glycans that are rarely found in the extracellular space.[12–17] Indeed, 56–79% of the N-linked glycans of gp120 on HIV-1 are oligomannose-type glycans.[17] Such levels of oligomannose-type glycans on HIV-1 gp120 have attracted a lot of attention and their biological significance has been extensively studied either as pathogen-associated molecular patterns (PAMPs) or as an antigen. Interestingly, recent studies also confirmed that intact virus-associated gp120 also contains steady levels of complex-type glycans (21–44% of N-glycans).[16,17,19] In this case, however, the biological significance of complex-type glycans in HIV-1 pathology remains unexplored.

Second, oligomannose- and complex-type glycans are spatially distributed on the surface of gp120 and form distinct homogenous patches (Fig. S1). Oligomannose-type glycans are clustered in an area distal to the CD4 binding site and often associate with the immunologically silent face of gp120.[11,20] Complex-type glycans patches are found proximal to the CD4 binding site (Fig. S1).[20] It appears that high levels of glycan–glycan interactions occur between neighboring glycans, thus forming tight clusters. These clusters can mask the peptide backbone structure from antibodies.[6,21] Since glycans attached to host glycoproteins normally exhibit considerable conformational flexibility because of their highly hydrophilic nature, such clustered arrangement of glycans on gp120 is considered unique and peculiar. Using the 2G12 neutralizing antibody that binds to the oligomannose-type glycans of gp120, it has been established that the high concentration and clustering of oligomannose-type glycans on the surface of gp120 leads to its recognition as nonself glycans, such as those found on many bacterial surfaces.[11,22] Our recent study on the interaction between galectins and gp120 also suggests the presence of unusual clusters of complex-type glycans on gp120 (see later for more details).[16] Densely packed glycan patches on gp120 may thus also exhibit non-self properties. Importantly, such unique glycan presentation depends on the tertiary structure of gp120 and requires it to be intact and properly folded.[16,23,24]

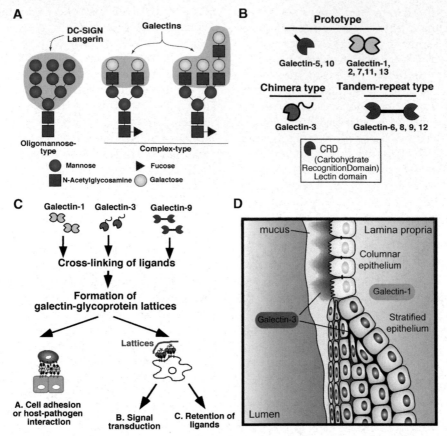

Figure 2. N-linked glycans of gp120 and description of galectins. (A) N-linked glycans of gp120 and the sacharride moieties that could be recognized by lectins are highlighted. (B) Galectin family. (C) Cross-linking of the ligands by galectins. (D) Expression of galectins in epithelia and gut-associated lymphoid tissue.

C-type lectins and HIV-1 infection

Since glycans found on host extracellular glycoproteins are mostly complex-type glycans, glycoproteins carrying high levels of oligomannose-type glycans can be recognized as PAMPs in some cases. HIV-1 is indeed recognized by dendritic cells (DCs) and Langerhans cells (LCs) via specific C-type lectins, such as DC-specific ICAM3-grabbing nonintegrin (DC-SIGN), DC immunoreceptor (DCIR), and Langerin.[25–27] DC-SIGN is highly expressed in monocyte-derived DCs and dermal and mucosal DCs but not in LCs.[28] DC-SIGN binds to oligomannose-type glycans of gp120 and facilitates both direct infection of DCs (*cis*-infection) as well as DC-mediated HIV-1 transfer toward CD4+ T cells (*trans*-infection).[25,29] DCIR is expressed in antigen-presenting cells, including myeloid and plasmacy-toid DCs, as well as macrophages, but very few in LCs.[30] This lectin has recently been suggested to be involved in both *cis-* and *trans*-infection by HIV-1 using monocyte-derived DCs.[27] It remains elusive whether the oligomannose-type glycans of gp120 (Fig. 2A) are the primary HIV-1 ligands for DCIR or not. LCs express another type of C-type lectin, Langerin, which also binds the oligomannose-type glycans of gp120 (Fig. 2A). However, in contrast to DC-SIGN and DCIR, interaction of HIV-1 with Langerin leads to the rapid degradation of the virus following uptake.[26] Thus, the clustered oligomannose-type glycans on gp120 have multi-faceted roles in HIV-1 pathogenesis. Our recent work on clustered complex-type glycans of HIV-1 gp120 suggests that these complex-type patches can also be exploited by the virus and that they may favor the transmission or replication of HIV-1.[16]

Host membrane proteins that assist HIV-1 infection

In addition to C-type lectins, two classes of host proteins, integrins and syndecans, are known to facilitate HIV-1 infection. Two integrin family members, $\alpha_4\beta_7$ and $\alpha_L\beta_2$ are listed as assisting HIV-1 entry into susceptible cells. CD4$^+$ T cells express $\alpha_L\beta_2$ (LFA-1), which promotes HIV-1 attachment by binding to ICAM-1, an adhesion molecule acquired by HIV-1 during the budding process.[7,9,31] Activation of CD4$^+$ T cells leads to a conformational switch of LFA-1 toward a high affinity state, which further improves the LFA-1–ICAM-1–mediated enhancement of HIV-1 infection. Additionally, the $\alpha_4\beta_7$ gut-homing receptor is highly expressed in memory CD4$^+$ T cells, and this integrin can interact directly with gp120. This interaction has also been shown to activate LFA-1 and to facilitate virus entry in gut-homing CD4$^+$ T cells.[32,33] Another family that promotes HIV-1 infection is syndecans.[34] Syndecans belong to a class of proteoglycans that carry covalently linked, heparan sulfate glycosaminoglycans, to which gp120 directly binds.[34,35] Interestingly, HIV-1 retains its infectivity for as long as a week when bound to cell-surface syndecans and can be readily transmitted upon contact with CD4$^+$ T cells.[34–36] In addition to these host membrane-bound proteins, the host soluble lectin galectin-1 also contributes to HIV-1 binding to CD4$^+$ susceptible cells and promotes HIV-1 infection.[16,19,37–39] The possible role played by galectin-1 in the pathogenesis of HIV-1 infection is described later.

Selective transmission of R5-tropic isolates of HIV-1

HIV-1 uses either CCR5 or CXCR4 as its coreceptor for entry. HIV-1 variants that use the CCR5 are called R5 variants, while those that use CXCR4 are referred to as X4 variants. Although both R5 and X4 HIV-1 variants are present in the body fluids (semen, blood, cervicovaginal secretion) of transmitting individuals, current studies suggest that in the majority of cases the transmitted/founder virus found in the GALT of recently infected individuals is an R5 variant. These data suggest highly selective mechanisms favoring R5 over X4 during transmission events and/or initial replication in the GALT.[40–42] This selection appears to occur in multiple layers. First, HIV-1 infection occurs initially in

GALT-associated CD4$^+$ T cells, which express CCR5. Second, because gp120 from X4 variants has more exposed cationic charge than R5 variants, X4 HIV-1 may bind more strongly to polyanionic mucin and be subject to a preferential clearance.[42–44]

In addition to the selectivity in the transmission of HIV-1 R5 variants, several studies suggest that the gp120 molecules of transmitted/founder R5 HIV-1 contain reduced N-linked glycosylation sites in their V1/V2 regions.[42,45–47] There thus seems to be another selection process among R5 variants. These hypoglycosylated variants do not display any direct replicative advantage using a HeLa cell-based reporter cell line expressing CD4 and CCR5[47] but instead display a unique sensitivity to antibody neutralization.[45] Thus, there may be additional selective pressure for the HIV-1 R5 variant that carries a reduced number of glycosylation sites in spite of its increased susceptibility to neutralizing antibodies. One of such possibilities is related to the recognition of oligomannose-type glycans by host C-type lectins. Although HIV-1 replication is enhanced through *trans-* and *cis*-infection mediated by DC-SIGN and DCIR in DCs, recent studies suggest that early HIV-1 infection exclusively occurs in CD4$^+$ T cells rather than macrophages or DCs, which reside in the subepithelium.[42,48,49] In addition, a recent elegant work by the group of Sanders also demonstrated that virions carrying gp120 with higher numbers of oligomannose-type glycans are more efficiently endocytosed through DC-SIGN and more proficiently processed for antigen presentation than HIV-1 containing gp120 with heterogeneous glycans (complex- and oligomannose-type).[50] In addition, *trans*-infection of those HIV-1 with oligomannose-type rich gp120 is relatively inefficient.[50] Finally, Langerin-expressing LCs residing in the genital mucosal epithelia are most probably the first DC subset to encounter HIV-1 and could contribute to the rapid destruction of HIV-1 to provide a barrier to transmission.[26] Thus, it is possible that at the initial stage of the infection process, HIV-1 with reduced oligomannose-type glycans has some advantage for the establishment of an infection by escaping its uptake by antigen-presenting cells in the genital epithelia. An alternative hypothesis was recently proposed by the group lead by Fauci.[33] His group argues that due to fewer glycosylation sites, gp120 molecules located on transmitted/founder virus could bind more efficiently to $\alpha_4\beta_7$ integrin, resulting in the activation of LFA-1 and enhancement of infection via stronger ICAM-1/LFA-1 interactions.[33,51] Thus, despite the fact that the presence of highly constrained oligomannose-type glycans on gp120 renders HIV-1 significantly resistant to neutralizing immunity,[6] HIV-1 does not fully utilize its glycan shield during a sexual transmission event. Instead, transmitted/founder virus selectively lowers this oligomannose-type glycan shield in order to both evade the innate mucosal defense system and facilitate its interaction with its target cells, activated CD4$^+$ T cells. Importantly, those HIV-1 particles that penetrate across genital epithelia have to interact rapidly with susceptible cells or with cells expressing alternative receptors (such as syndecans), since cell-free HIV-1 virion becomes inactive in a relatively short period of time.

Biological processes involved in the transmission and initial replication of HIV-1

The frequency of HIV-1 transmission following an unprotected sexual intercourse is quite low (0.05–0.1%) when compared to other sexually transmitted viruses, such as the hepatitis B virus.[52–57] Few virus particles penetrate across genital epithelial layers to reach the mucosa-associated lymphoid tissues (MALT) and especially the gut-associated lymphoid tissues (GALT) that are rich in HIV-1–susceptible CD4$^+$ T cells.[49,58–61] Indeed, recent works have established that more than 75% of HIV-1 infection cases are initiated by a single founder virus.[33,42,49,62] Further, unlike other enveloped viruses like influenza, HIV-1 virions carry as low as 14 ± 7 Env molecules that are required for viral entry via CD4.[10] Viral attachment to CD4 thus displays an intrinsically weak avidity, especially *in vivo* where it occurs under suboptimal conditions.[7–9,34,63] Notably, during this initial replication period, HIV-1 susceptible cells in the GALT express low levels of CD4 and CCR5 and are less competent than fully activated T cells.[64–66] In addition, GALT-associated LCs could capture cell-free HIV-1 and degrade it.[26] Despite this relatively unfavorable environment, when HIV-1 transmission occurs, HIV-1 infect CD4$^+$ T cells to create a replicative focus that will expand to establish a self-propagating infection within a day post infection.[61,64–67] More than 90% of memory CD4$^+$ T cells, which reside in the GALT, are depleted by local viral replication within one month of infection.[58–61] Thus, contrary to the traditional view of HIV-1 in-fection characterized by the slow decline of CD4$^+$ T cells, recent findings of this rapid and extensive removal of the major portion of the immunological memory initiated by only a few viral particles reemphasizes the need to control viral replication as early as possible and ideally to prevent transmission altogether. Since HIV-1 establishes its infection within a day under those restrictive conditions, it is possible that the virus exploits additional host factors to achieve rapid viral attachment to susceptible cells, but such possibility has remained more or less unexplored until recently. As mentioned later, galectin-1, a soluble host β-galactoside–binding lectin, strongly increases the association of HIV-1 viral particles to susceptible cells and this activity is observed within minutes.[16,19,37,39] Another member of the same family, galectin-3 could not display the same activity while a recent work suggests that galectin-9 also promotes HIV-1 infection.[68]

Galectins

Galectins form one host soluble lectin family defined by an ability to bind β-galactoside–containing glycans and a conserved amino-acid sequence in their carbohydrate recognition domain (CRD; Fig. 2B).[69] Extracellular functions of galectins rely on the multivalency of glycan binding.[69–71] Galectin-1 is dimer, and galectin-3 forms oligomer through its N-terminal domain. Galectin-9 contains two CRDs connected by a linker.[72–74] Therefore, galectin-1, 3, and 9 can crosslink specific ligands. When those galectins bind to ligands from different entities, they can mediate cell/cell or cell/pathogen interactions (Fig. 2C).[74–76] Studies from different laboratories including ours suggest that galectin-1, but not galectin-3, promotes HIV-1 binding to susceptible CD4-expressing human cells and directly enhances virus infection.[16,19,37–39,68] Since the roles of galectins in host–pathogen interactions and immune responses are extensively reviewed in this issue and elsewhere,[72,77–79] only the aspects of galectins that are related to HIV-1 infection will be introduced and discussed here.

Galectin expression

Galectin-1 is expressed in the thymus and by lymphoid parenchymal epithelial cells, endothelial cells, trophoblasts, activated T and B cells, macrophages, follicular DCs, and CD4$^+$CD25$^+$ regulatory T cells.[80–86] Among subtypes of CD4$^+$ T cells, galectin-1 is highly expressed and secreted by Th1 cells.[87] Significant accumulation of galectin-1 is found in the lamina muscularis mucosae, just beneath the epithelium and the lamina propria, where HIV-1 susceptible CD4[+] T cells can be found[88–90] (Fig. 2D). Lymphoid tissues, such as tonsils and MALT, contain concentrations of galectin-1 that can reach 20 μM.[19,86] HIV-1 Tat was reported to induce galectin-3 expression.[91] Our preliminary data[a] suggest that galectin-1 expression is higher in HIV-1-infected CD4[+] T cells compared to noninfected bystander cells or untreated cells.[92] Galectin-3 is expressed in endothelial cells, macrophages, and microglia, and their activation enhances its expression level.[73,93–97] Galectin-3 is also highly expressed and secreted at the apical side (luminal side) of the mucosal epithelium (Fig. 2D)[88,94] while galectin-9 is expressed by T cells, eosinophils, endothelical cells, DCs, and macrophages, as well as by the epithelium of the gastrointestinal tract.[98,99]

Galectin-1 promotes HIV-1 infection

We previously established that galectin-1, but not galectin-3, accelerates (as high as 40-fold) the binding kinetics of HIV-1 to susceptible cells, and facilitates robust HIV-1 replication.[16,19,37,39] These activities of galectin-1 in HIV-1 infection are completely inhibited by lactose, a β-galactoside–containing saccharide, but not mannose, confirming that galectin-1 recognizes β-galactoside residues expressed on HIV-1.[16,19,37,39] Importantly, even in this condition, HIV-1 entry strictly relies on the interaction between gp120 and CD4/coreceptor, suggesting that galectin-1 only assists the initial virus binding event without affecting the rest of the HIV-1 entry process. An incubation of less than two minutes is sufficient to detect an increase in HIV-1 binding to susceptible cells by galectin-1 (Ref. 19). Further, in the presence of galectin-1, more than 30% of the initial virus input is associated with peripheral blood monocuclear cells (PBMCs) after 30 min, while less than 1% of the virus is bound in the absence of galectin-1. This activity can be observed at 4 °C, suggesting that increased binding kinetics is due to direct cross-linking of HIV-1 to cells, rather than galectin-1–induced signal transduction.[19] As shown in Table 1, galectin-1 can promote the infection of HIV-1 X4, X4R5, and R5 variants in various susceptible cells.[16,19,37,39] Our recent work suggests that galectin-1 directly binds to HIV-1 virus particles.[16] This interaction is sensitive to lactose but not mannose,[16,39] confirming that galectin-1 utilizes its β-galactoside–binding activity to recognize HIV-1. In contrast to our studies, the group led by Mahal reports that galectin-1 binds to mannose residues on HIV-1, based on the results obtained by a lectin microarray system where galectin-1 is densely plotted on the array.[100] This report contradicts the previously established glycan-binding specificity of galectin-1.[73,101–105] In addition, galectin-1 is sensitive to oxidation and is liable to become oxidized in dry conditions such as those required for array spotting. Indeed, caution on the quality of galectin-1 in assays has been recently raised by several groups.[39,106] Thus, further investigation under strict quality control is necessary to verify whether galectin-1 may aberrantly change its specificity from β-galactoside to mannose upon immobilization on microarray plates (see discussion in Ref. 39). Nonetheless, densely clustered immobilized galectin-1 would simply not be expected in a physiological environment, especially in the context of HIV-1 infection. In a physiological setting, galectin-1 preferentially binds to β-galactoside residues, such as those presented on the surface of HIV-1.

As listed in Table 2, galectin-1 can directly interact with several HIV-1 variants produced in both HEK 293T cells or primary PBMCs. Importantly, pseudovirus that lacks gp120 cannot be recognized by galectin-1, and purified gp120 interacts with galectin-1 but not galectin-3. Thus, gp120 is the specific viral ligand for galectin-1 (see later for detailed glycan-binding specificity). It is expected that differences in cell surface protein glycosylation are present in different susceptible cells and virus-producing cells. Despite this fact, galectin-1 can cross-link HIV-1 to all susceptible cells tested and promote their infection, suggesting that susceptible cells and HIV-1 both carry N-linked glycans that have affinity for galectin-1. Indeed, recent structural studies of gp120 glycans purified from HIV-1 produced in 293T cells as well as PBMCs indicate that HIV-1 gp120 carries significant levels (as high as 44%) of N-linked complex-type glycans.[17] Thus, while further studies are warranted, galectin-1 can be listed as one

---

[a]Michel Ouellet, Michael Imbeault, and Michel J. Tremblay, unpublished observations.

Table 1. Cells in which galectin-1 promotes HIV-1 infection

Infected cells	Virus strain	Virus producer cells	Reference
LuSIV cells	NL4-3 (X4)	293T	17
1G5 cells	NL4-3 (X4)	PBMCs	17
PBMCs	NL4-3 (X4)	293T	17
CD4⁺ T cells	93US151 (R5×4)	PBMCs	17
CD4⁺ T cells	NL4-3 (X4)	293T	16
CD4⁺ T cells	NL4-3Bal (R5)	293T	16
Macrophages	NL4-3Bal (R5)	Macrophages	35
Macrophages	Virus peudotyped with JR-FL env (R5)	293T	35
Tonsillar tissues (cultured *ex vivo*)	NL4-3 (X4)	293T	17

potential host factor that could directly contribute to HIV-1 transmission or to its early expansion in the GALT.

Complex-type glycan clusters of gp120 are found on a domain close to the CD4 binding site (Fig. S1). It has been recently proposed that one of the host ligands of galectin-1 is the CD4 glycoprotein.[16] Indeed, a significant increase in gp120 binding to CD4 is observed upon adding galectin-1 to a column of immobilized CD4 through which soluble recombinant gp120 is run. This suggests that galectin-1 can directly cross-link gp120 to CD4. Furthermore, using CD4-expressing and CD4-deficient cell lines, we have shown that galectin-1 cannot promote HIV-1 binding to cells that do not express CD4. Together, these results suggest that CD4 is one of the host ligands of galectin-1, and galectin-1 facilitates HIV-1 infection through direct cross-linking of gp120 and CD4 (Fig. 3).

Recently, the group lead by Baum reported that galectin-9 also potentiates HIV-1 infection.[68] In the case of galectin-9, however, this assistance requires a longer exposure than galectin-1. Interestingly, they found that galectin-9 binds to host surface protein disulfide isomerase (PDI), leading to an increased retention of PDI on the cell surface. Since surface PDI promotes HIV-1 infection in T cells through alteration of the redox state surrounding the cell, galectin-9–induced changes in the membrane dynamics of PDI creates a favorable environment for HIV-1 entry. Thus, at least two galectins found in the genital tract or GALT can facilitate HIV-1 infection.

Glycan-binding specificity of galectins

In order to better delineate the distinctive activity of galectin-1 toward HIV-1 binding to CD4⁺ susceptible cells, we would first like to briefly introduce the general glycans-binding specificity of galectins. The minimal binding unit recognized by galectins is a β-galactoside (galactose residue linked to a glycan through β linkage), such as *N*-acetyllactosamine (LacNAc).[73,101–103,107] While β-galactoside-containing glycans are often found in complex-type glycans attached to proteins (Fig. 2A), each galectin binds to a relatively limited set of ligands. This is likely due to a distinctive presentation of their CRDs, as well as to the structural differences found in the CRD of each galectin.[104] Modifications of galactose residues found on β-galactosides drastically alter their affinity for galectins.[104] For example, an additional β-linked galactose residue on β-galactosides increases their affinity for galectin-3, but abolishes their affinity for galectin-1.[108,109] In contrast, α2–6 (but not α2–3) sialic acid modification dramatically reduces the affinity for galectin-1, galectin-3, and galectin-9[104] Due to this difference in affinity, galectins can bind to Th1 and Th17 but not Th2 polarized cells (in both mice and humans) since glycans of Th1 and Th17 carry α2–3 sialyl modifications while those of Th2 cells are α2–6.[110] This is one good example of seemingly minor peripheral differences in glycan structures that regulate important immune functions and alter the biological activity of lectins. Interestingly, HIV-1 infection reduces sialylation of surface glycoproteins, thereby likely increasing galectin-1 binding.[111] We previously reported that bacterial and influenza-derived sialidases (neuraminidases) can significantly increase HIV-1 infection.[112,113] Individuals who are chronically infected with HIV-1 are susceptible to various infections with pathogens that carry sialidase.[112,114–116] Thus, the level of

Table 2. Interaction between galectins and HIV-1

	Producer cells	Galectin binding
HIV-1 virus (NL4-3, X4)	293T	galectin-1 > galectin-3
HIV-1 virus (89.6, R5×4)	PBMCs	galectin- 1> galectin-3
HIV-1 virus (NL4-3Bal, R5)	PBMCs	galectin-1 >> galectin-3
Envelope-deficient HIV-1 virus (NL4-3, X4)	293T	Galectin-1 binding is low and galectin-3 binding remains similar to the virus with Env
Detergent lysates (Triton X100-soluble) of HIV-1 virus (NL4-3, X4)	293T	galectin-1 > galectin-3
Recombinant gp120 (96ZM651, X4)	293	galectin-1 > galectin-3
Recombinant gp120 (NL4-3Bal, R5)	293	galectin-1 > galectin-3
DTT-treated recombinant gp120 (96ZM651, X4)	293	galectin-1 = galectin-3
DTT-treated recombinant gp120 (NL4-3Bal, R5)	293	galectin-1 = galectin-3
N-linked glycans of recombinant gp120 (96ZM651, X4)	293	galectin-1 = galectin-3
N-linked glycans of recombinant gp120 (NL4-3Bal, R5)	293	galectin-1 = galectin-3

sialylation of both HIV-1 virions and host susceptible cells in lymph nodes might be significantly reduced and thus more susceptible to be recognized by galectins.

Differential recognition of CD4, the main HIV-1 receptor, by galectin-1 and galectin-3

Although both galectin-1 and galectin-3 bind to LacNAc residues, significant difference in their binding preferences have been reported. We previously reported that galectin-3 binds to human amniotic fluid fibronectin but fails to bind to plasma fibronectin.[107] Glycan profile analysis shows that amniotic fluid fibronectin carry tri- and tetra-antennary complex-type glycans (i.e., they carry three or four LacNAc per glycan) while serum fibronectin instead contains bi-antennary complex-type glycans (i.e., they carry only two LacNAc per glycan).[107,117,118] Therefore, those data suggest that galectin-3 cannot steadily bind to bi-antennary complex-type glycans attached on a protein despite the presence of LacNAc. Recently, by using glycan microassays, Cummings and his colleagues reported a more sustained binding of galectin-1 to bi-antennary complex-type glycans compared to galectin-3 at low lectin concentrations.[105] While this preferential binding is less clear at higher lectin concentrations,[105] our unpublished results using affinity chromatography analysis also confirm such preference that galectin-1 but not galectin-3 is better retained by a column of immobilized asialotrans-

ferrin that carries only bi-antennary complex-type glycans.[b] In contrast, both galectin-1 and galectin-3 are retained similarly by a column of immobilized asialofetuin that carries tri-antennary complex-type glycans.[c]

More relevant to HIV-1 infection, recombinant CD4 contains only two bi-antennary complex-type glycans,[14] like plasma fibronectin and transferin. Indeed, we recently reported that galectin-1, but not galectin-3, strongly binds to recombinant CD4, the main host receptor for HIV-1.[16] Further, galectin-1 increases HIV-1 binding to CD4-expressing cells.[d,16] Thus, the selective recognition of CD4 by galectin-1 but not galectin-3 is likely due to the difference in their binding preferences for bi-antennary complex-type glycans presented on the surface of CD4. In contrast to those results based on biological relevant assays, binding affinity of galectin-1 for bi-antennary glycans is consistently lower than galectin-3,[104,105] underlining some difficulty in the usage of affinity constants or dissociation rates obtained by kinetics assays to access or predict the interaction between natural galectin ligands and galectins.

---

[b]Christian St-Pierre and Sachiko Sato, unpublished observations.
[c]Christian St-Pierre and Sachiko Sato, unpublished observations.
[d]Michel Ouellet, Michel J. Tremblay, and Sachiko Sato, unpublished observations.

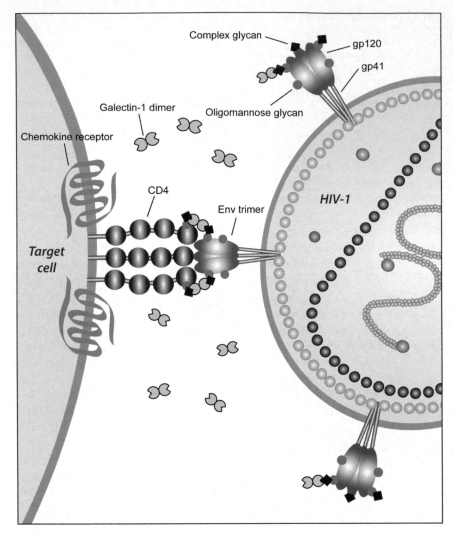

Figure 3. Galectin-1–mediated HIV-1 adhesion to CD4⁺ T cells.

Differential gp120 recognition by galectin-1 and galectin-3

Being lactosamine-binding proteins, it would be expected that all galectins exhibit higher affinity for glycans that contain a higher number of lactosamine residues (like polylactosamine or tri-/tetra-antennary complex type). Indeed, this polylactosamine effect is significant for galectin-3 and galectin-9, but not for galectin-1. Thus, galectin-1 exhibits constant K_d for lactosamine-containing glycans regardless of the number of lactosamine residues. In contrast, the affinity of galectin-3 for glycans increases with good corre-

lation upon increasing their number of lactosamine residues.[101,102,104,107] Importantly, since galectin-3 recognizes both internal and external lactosamine residues, this lectin can bind to glycan as long as the glycan contains a couple of internal lactosamine residues.[105,107] However, it remains elusive which lactosamine residues, internal or external, are more critical to form persistent interaction of the glycan with galectin-3. In contrast, it is likely that galectin-1 preferentially binds to peripheral Gal residues presented by lactosamine.[110]

Complex-type glycans of HIV-1 gp120 contain bi-, tri-, and tetra-antennary chains.[12–17] Depending on clinical HIV-1 quasispecies or variants,

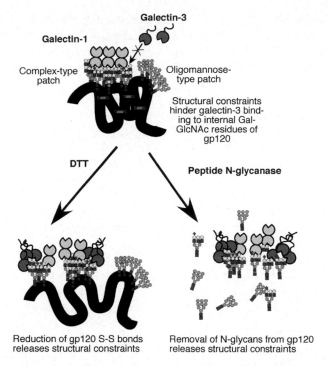

Galectin-3 preferentially binds to internal Gal-GalNAc residues of complex-type glycans while galectin-1 preferentially binds to external Gal-GlcNAc residues.

Figure 4. Glycan density-dependent gp120 recognition by galectin-1.

virus-producing cells, and gp120-producing cells, some differences in the composition of complex-type glycans have been reported.[12–17] Our recent work using recombinant gp120 derived from R5 and X4 variants reveals that galectin-1 firmly binds to each of these gp120 variants, while galectin-3 fails to form a tight interaction with any of them. Since this interaction is disrupted by lactose but not mannose, the binding of galectin-1 to gp120 depends on the presence of β-galactoside residues. Curiously, compositions (percentage of total N-linked glycans) of complex-type glycans of gp120 X4 and R5 carrying bi-antennary and tri-/tetra-antennary, both of which exhibit high affinity for galectin-1 and/or galectin-3, are 16:10 and 9:18, respectively.[16] Thus, the obtained binding data are not reconciled with the established glycan specificities based on biochemical kinetic analysis. Indeed this observed preference of galectin-1 for gp120 compared to that of galectin-3, is not likely due to a *de facto* preference of a galectin toward specific glycans displayed by gp120. Importantly, when complex-type glycans released from gp120 are subject to galectin binding

analysis, both galectin-1 and galectin-3 bind similarly to those protein-free glycans (Table 2),[16] fitting well with their previously established binding specificity. Similarly, when disulfide bonds of recombinant gp120 are partially denatured by Dithiothreitol, the above distinct galectin-1 specific recognition of gp120 is also lost and both galectin-1 and galectin-3 equally interact with the partially denatured gp120 (Table 2). Together, it seems to be the spatial organization of complex-type glycans on gp120 that significantly contributes to the repulsion of galectin-3 and the firm binding to galectin-1 (Fig. 4). Such peculiar presentation of complex-type glycans is likely to be supported by the local tertiary structure of the peptide backbone. As such, studies of antibodies suggest that N-linked glycans are constrained within tight clusters on gp120, where unusual sugar–sugar interactions are formed between neighboring glycans as mentioned above.[6,11,20,21,23] Thus, like oligomannose-type glycans, complex-type glycans appear to be also tightly constrained, forming a cluster where only peripheral or external β-galactoside residues are available for lectin binding. In this context, binding by galectin-1 is clearly favored (Fig. 4). The failure of galectin-3 to strongly bind to gp120 also seems to suggest that internal lactosamine residues of gp120 complex-type glycans are sequestered within the glycan patch or via glycan–protein interactions. This interpretation is also strengthened by results showing that the presence of galectin-3 neither interferes nor promotes galectin-1–enhanced HIV-1 infectivity.[38] Galectin-3 was previously considered to interact with both external and internal β-galactoside residues similarly. Unexpectedly, the failure of galectin-3 to bind to intact gp120 suggests that, at least in the case of gp120, the presence of external β-galactoside residues is not sufficient *per se* to ensure a strong binding by galectin-3. Additional studies are warranted to understand if such binding requirement of galectin-3 to have access to internal β-galactosides is a common phenomenon or not.

As mentioned above, clusters of common host glycans could display nonself properties to the innate immune system.[11,22,119] In the case of HIV-1 gp120, clustering of complex-type glycan prevents it from being bound by galectin-3, which is abundantly expressed at the apical surface of the mucosal membranes, including in the genital epithelia, the entry site of HIV-1.[88,120] Since galectin-3 is believed to be an active participant of the mucosal clearance system,[121] such exclusion is exploited to avoid binding of gp120 by galectin-3, which would lead to a biological dead-end for the virus.

Conclusion and future directions

Unlike the traditional view of HIV-1 infection characterized by a slow decline of $CD4^+$ T cells from circulation, recent advance indicates that extensive infection and removal of local $CD4^+$ T cells in GALT occurs within the first month of infection. Since only a few virus particles successfully penetrate the genital epithelia, it thus becomes critical to understand the basic mechanisms underlying the rapid establishment of HIV-1 in the GALT. As HIV-1 is known to carry relatively few Env spikes that are essential for its binding and entry into host cells, it appears important to study how it exploits host factors that may enhance its association with its initial target cells and/or successfully evade the innate immune system. This could open unexplored and interesting

avenues of therapeutic intervention targeting transmission and/or the early stage of infection where HIV-1 is more vulnerable to host natural defenses. Recent studies suggest that in addition to C-type lectins of DCs, galectins expressed in the GALT may facilitate HIV-1 transmission and early infection. Galectin-1 greatly increases the association of HIV-1 to susceptible cells through cross-linking of gp120 and CD4 (Fig. 3). The roles of galectin-1 as an HIV-1 assisting molecule should be further investigated to gain deeper insights in the context of viral transmission or early replication stages using different *in vitro* and *ex vivo* systems. In addition, the relevance of these findings in the pathobiology of the virus would need to be further verified *in vivo*, with a model of rhesus macaques infected by the simian immunodeficiency virus, a close relative of HIV-1. Recent advances in synthetic glycochemistry now makes it possible to develop highly specific galectin-1 antagonists that do not interfere with other galectins such as galectin-3.[38,122–125] The use of specific inhibitors or antagonists of galectin-1 thus becomes an attractive therapeutic strategy to further reduce the transmission rate or to limit early replication in the GALT following sexual transmission. However, before such chemical intervention can be accepted and used, it is crucial to better understand the more subtle aspects of galectin-1–mediated facilitation of HIV-1 infection in conjunction with the many other immunological activities that galectin-1 may have in the GALT (see extensive reviews by Rabinovich and others in Refs. 72, 77–79).

Acknowledgments

We acknowledge numerous contributions from various laboratories whose references were not cited in the present review due to space limitations. We would like to acknowledge Dr. T. Endo and our collaborators in the HIV-1 project for fruitful discussion. This study was supported by an operating grant to S.S. and M.J.T. from the Canadian Institutes of Health Research (CIHR) Operating Grant Program (MOP-89743). C.S.-P. is the recipient of a CIHR Doctoral Fellowship, while S.S. holds a Scholarship Award (senior level) from the Fonds de la Recherche en Santé du Québec. M.J.T. is the recipient of the Canada Research Chair in Human Immuno—Retrovirology (Tier 1 level).

Supporting information

Additional supporting information may be found in the online version of this article:

Figure S1. The glycan clustering on the surface of gp120 (trimerized form). The glycans are clustered roughly in three domains, oligomannose-type glycan (pink), which is away from CD4 binding sites (∗) and complex-type glyans (blue) are relatively close to CD4 binding pockets. Remaining surface of the peptide are marked in red. (Modified from Fig. 3 of Ref. 43 and Fig. 2 of Ref. 6).

Please note: Wiley-Blackwell is not responsible for the content or functionality of any supporting materials supplied by the authors. Any queries (other than missing material) should be directed to the corresponding author for the article.

Conflicts of interest

The authors declare no conflicts of interest.

References

1. UNAIDS, J.U.N.P.o.H.A. 2009. *2009 AIDS Epidemic Update*. Joint United Nations Programme on HIV/AIDS. Geneva, Switzerland.
2. UNAIDS, J.U.N.P.o.H.A. 2010. Global report: UNAIDS report on the global AIDS epidemic 2010. Available at: http://www.unaids.org/en/media/unaids/contentassets/documents/unaidspublication/2010/20101123˙global report˙en.pdf.˳ Accessed 23 Nov 2010.
3. Weiss, R.A. 2001. Gulliver's travels in HIVland. *Nature* **410**: 963–967.
4. Freed, E.O. & Martin M.A. 2007. HIVs and their replication. In *Fields Virology*, 5th ed. D.M. Knipc & P.M. Howley, Eds.: 2107—2185. Lippincott Williams & Wilkins. Philadelphia.
5. Zhu, X., C. Borchers, R.J. Bienstock & K.B. Tomer. 2000. Mass spectrometric characterization of the glycosylation pattern of HIV-gp120 expressed in CHO cells. *Biochemistry* **39**: 11194–11204.
6. Scanlan, C.N., J. Offer, N. Zitzmann & R.A. Dwek. 2007. Exploiting the defensive sugars of HIV-1 for drug and vaccine design. *Nature* **446**: 1038–1045.
7. Tremblay, M.J., J.F. Fortin & R. Cantin. 1998. The acquisition of host-encoded proteins by nascent HIV-1. *Immunol. Today* **19**: 346–351.
8. Ugolini, S., I. Mondor & Q.J. Sattentau. 1999. HIV-1 attachment: another look. *Trends Microbiol.* **7**: 144–149.
9. Cantin, R., S. Methot & M.J. Tremblay. 2005. Plunder and stowaways: incorporation of cellular proteins by enveloped viruses. *J. Virol.* **79**: 6577–6587.
10. Zhu, P., J. Liu, J. Bess, Jr., *et al.* 2006. Distribution and three-dimensional structure of AIDS virus envelope spikes. *Nature* **441**: 847–852.
11. Pantophlet, R. & D.R. Burton. 2006. GP120: target for neutralizing HIV-1 antibodies. *Annu. Rev. Immunol.* **24**: 739–769.
12. Leonard, C.K., M.W. Spellman, L. Riddle, *et al.* 1990. Assignment of intrachain disulfide bonds and characterization of potential glycosylation sites of the type 1 recombinant human immunodeficiency virus envelope glycoprotein (gp120) expressed in Chinese hamster ovary cells. *J. Biol. Chem.* **265**: 10373–10382.
13. Li, H., C.F. Xu, S. Blais, *et al.* 2009. Proximal glycans outside of the epitopes regulate the presentation of HIV-1 envelope gp120 helper epitopes. *J. Immunol.* **182**: 6369–6378.
14. Yuen, C.T., S.A. Carr & T. Feizi. 1990. The spectrum of N-linked oligosaccharide structures detected by enzymic microsequencing on a recombinant soluble CD4 glycoprotein from Chinese hamster ovary cells. *Eur. J. Biochem.* **192**: 523–528.
15. Pang, P.C., H.R. Morris, S.M. Haslam, *et al.* 2007. Glycomics analysis of HIV-1 gp120 glycoforms. *Glycobiology* **17**: 1274.
16. St-Pierre, C., H. Manya, M. Ouellet, *et al.* 2011. Host-soluble galectin-1 promotes HIV-1 replication through a direct interaction with glycans of viral gp120 and host CD4. *J. Virol.* **85**: 11742–11751.
17. Bonomelli, C., K.J. Doores, D.C. Dunlop, *et al.* 2011. The glycan shield of HIV is predominantly oligomannose independently of production system or viral clade. *PLoS One* **6**: e23521.
18. Doores, K.J., C. Bonomelli, D.J. Harvey, *et al.* 2010. Envelope glycans of immunodeficiency virions are almost entirely oligomannose antigens. *Proc. Natl. Acad. Sci. U. S. A.* **107**: 13800–13805.
19. Ouellet, M., S. Mercier, I. Pelletier, *et al.* 2005. Galectin-1 acts as a soluble host factor that promotes HIV-1 infectivity through stabilization of virus attachment to host cells. *J. Immunol.* **174**: 4120–4126.
20. Wyatt, R., P.D. Kwong, E. Desjardins, *et al.* 1998. The antigenic structure of the HIV gp120 envelope glycoprotein. *Nature* **393**: 705–711.
21. Chen, B., E.M. Vogan, H. Gong, *et al.* 2005. Determining the structure of an unliganded and fully glycosylated SIV gp120 envelope glycoprotein. *Structure* **13**: 197–211.
22. Calarese, D., C. Scanlan, H.-K. Lee, *et al.* 2006. Toward a carbohydrate-based HIV-1 vaccine. In *Carbohydrate Drug Design*, 161–185. American Chemical Society. Washington.
23. Astronomo, R.D. & D.R. Burton. 2010. Carbohydrate vaccines: developing sweet solutions to sticky situations? *Nat. Rev. Drug. Discov.* **9**: 308–324.
24. Agrawal-Gamse, C., R.J. Luallen, B. Liu, *et al.* 2011. Yeast-elicited cross-reactive antibodies to HIV Env glycans efficiently neutralize virions expressing exclusively high-mannose N-linked glycans. *J. Virol.* **85**: 470–480.
25. Geijtenbeek, T.B., D.S. Kwon, R. Torensma, *et al.* 2000. DC-SIGN, a dendritic cell-specific HIV-1-binding protein that enhances trans-infection of T cells. *Cell* **100**: 587–597.
26. de Witte, L., A. Nabatov, M. Pion, *et al.* 2007. Langerin is a natural barrier to HIV-1 transmission by Langerhans cells. *Nat. Med.* **13**: 367–371.

27. Lambert, A.A., C. Gilbert, M. Richard, *et al.* 2008. The C-type lectin surface receptor DCIR acts as a new attachment factor for HIV-1 in dendritic cells and contributes to trans- and cis-infection pathways. *Blood* **112:** 1299–1307.

28. Piguet, V. & R.M. Steinman. 2007. The interaction of HIV with dendritic cells: outcomes and pathways. *Trends Immunol.* **28:** 503–510.

29. Wu, L. & V.N. KewalRamani. 2006. Dendritic-cell interactions with HIV: infection and viral dissemination. *Nat. Rev. Immunol.* **6:** 859–868.

30. Bates, E.E., N. Fournier, E. Garcia, *et al.* 1999. APCs express DCIR, a novel C-type lectin surface receptor containing an immunoreceptor tyrosine-based inhibitory motif. *J. Immunol.* **163:** 1973–1983.

31. Hildreth, J.E. & R.J. Orentas. 1989. Involvement of a leuko-cyte adhesion receptor (LFA-1) in HIV-induced syncytium formation. *Science* **244:** 1075–1078.

32. Arthos, J., C. Cicala, E. Martinelli, *et al.* 2008. HIV-1 envelope protein binds to and signals through integrin alpha4beta7, the gut mucosal homing receptor for peripheral T cells. *Nat. Immunol.* **9:** 301–309.

33. Cicala, C., J. Arthos & A.S. Fauci. 2011. HIV-1 envelope, integrins and co-receptor use in mucosal transmission of HIV. *J. Transl. Med.* **9**(Suppl. 1): S2.

34. Bobardt, M.D., A.C. Saphire, H.C. Hung, *et al.* 2003. Syndecan captures, protects, and transmits HIV to T lympho-cytes. *Immunity* **18:** 27–39.

35. Gallay, P. 2004. Syndecans and HIV-1 pathogenesis. *Microbes. Infect.* **6:** 617–622.

36. de Witte, L., M. Bobardt, U. Chatterji, *et al.* 2007. Syndecan-3 is a dendritic cell-specific attachment receptor for HIV-1. *Proc. Natl. Acad. Sci. U S A.* **104:** 19464–19469.

37. Mercier, S., C. St-Pierre, I. Pelletier, *et al.* 2008. Galectin-1 promotes HIV-1 infectivity in macrophages through stabilization of viral adsorption. *Virology* **371:** 121–129.

38. St-Pierre, C., M. Ouellet, D. Giguere, *et al.* 2012. Galectin-1 specific inhibitors as a new class of compounds to treat HIV-1 infection. *Antimicrob. Agents Chemother.* **56:** 154–162.

39. St-Pierre, C., M. Ouellet, M.J. Tremblay & S. Sato. 2010. Galectin-1 and HIV-1 infection. *Methods Enzymol.* **480:** 267–294.

40. Keele, B.F. & C.A. Derdeyn. 2009. Genetic and antigenic features of the transmitted virus. *Curr. Opin. HIV AIDS* **4:** 352–357.

41. Salazar-Gonzalez, J.F., M.G. Salazar, B.F. Keele, *et al.* 2009. Genetic identity, biological phenotype, and evolutionary pathways of transmitted/founder viruses in acute and early HIV-1 infection. *J. Exp. Med.* **206:** 1273–1289.

42. Grivel, J.C., R.J. Shattock & L.B. Margolis. 2011. Selective transmission of R5 HIV-1 variants: where is the gatekeeper? *J. Transl. Med.* **9** (Suppl. 1): S6.

43. Kwong, P.D., R. Wyatt, Q.J. Sattentau, J. Sodroski *et al.* 2000. Oligomeric modeling and electrostatic analysis of the gp120 envelope glycoprotein of human immunodeficiency virus. *J. Virol.* **74:** 1961–1972.

44. Moulard, M., H. Lortat-Jacob, I. Mondor, *et al.* 2000. Selective interactions of polyanions with basic surfaces on human immunodeficiency virus type 1 gp120. *J. Virol.* **74:** 1948–1960.

45. Derdeyn, C.A., J.M. Decker, F. Bibollet-Ruche, *et al.* 2004. Envelope-constrained neutralization-sensitive HIV-1 after heterosexual transmission. *Science* **303:** 2019–2022.

46. Liu, Y., M.E. Curlin, K. Diem, *et al.* 2008. Env length and N-linked glycosylation following transmission of human immunodeficiency virus Type 1 subtype B viruses. *Virology* **374:** 229–233.

47. Alexander, M., R. Lynch, J. Mulenga, *et al.* 2010. Donor and recipient envs from heterosexual human immunodeficiency virus subtype C transmission pairs require high receptor levels for entry. *J. Virol.* **84:** 4100–4104.

48. Zhang, Z., T. Schuler, M. Zupancic, *e al.* 1999. Sexual transmission and propagation of SIV and HIV in resting and activated CD4+ T cells. *Science* **286:** 1353–1357.

49. Haase, A.T. 2010. Targeting early infection to prevent HIV-1 mucosal transmission. *Nature* **464:** 217–223.

50. van Montfort, T., D. Eggink, M. Boot, *et al.* 2011. HIV-1 N-glycan composition governs a balance between dendritic cell-mediated viral transmission and antigen presentation. *J. Immunol.* **187:** 4676–4685.

51. Cicala, C., E. Martinelli, J. P. McNally, *et al.* 2009. The integrin alpha4beta7 forms a complex with cell-surface CD4 and defines a T-cell subset that is highly susceptible to infection by HIV-1. *Proc. Natl. Acad. Sci. U. S. A.* **106:** 20877–20882.

52. European Study Group on Heterosexual Transmission of HIV. 1992. Comparison of female to male and male to female transmission of HIV in 563 stable couples. *BMJ* **304:** 809–813.

53. Leynaert, B., A.M. Downs & I. de Vincenzi. 1998. Heterosexual transmission of human immunodeficiency virus: variability of infectivity throughout the course of infection. European Study Group on Heterosexual Transmission of HIV. *Am J Epidemiol* **148:** 88–96.

54. Varghese, B., J.E. Maher, T.A. Peterman, *et al.* 2002. Reducing the risk of sexual HIV transmission: quantifying the per-act risk for HIV on the basis of choice of partner, sex act, and condom use. *Sex. Transm. Dis.* **29:** 38–43.

55. Boily, M.C., R.F. Baggaley, L. Wang, *et al.* 2009. Heterosexual risk of HIV-1 infection per sexual act: systematic review and meta-analysis of observational studies. *Lancet Infect. Dis.* **9:** 118–129.

56. Shepard, C.W., E.P. Simard, L. Finelli, *et al.* 2006. Hepatitis B virus infection: epidemiology and vaccination. *Epidemiol. Rev.* **28:** 112–125.

57. Piot, P., C. Goilav & E. Kegels. 1990. Hepatitis B: transmission by sexual contact and needle sharing. *Vaccine* **8**(Suppl.): S37–40; discussion S41–33.

58. Mattapallil, J.J., D.C. Douek, B. Hill, *et al.* 2005. Massive infection and loss of memory CD4+ T cells in multiple tissues during acute SIV infection. *Nature* **434:** 1093–1097.

59. J.M. Brenchley, T.W. Schacker, L.E. Ruff, *et al.* 2004. CD4+ T cell depletion during all stages of HIV disease occurs predominantly in the gastrointestinal tract. *J. Exp. Med.* **200:** 749–759.

60. Veazey, R.S., M. DeMaria, L.V. Chalifoux, *et al.* 1998. Gastrointestinal tract as a major site of CD4+ T cell depletion and viral replication in SIV infection. *Science* **280:** 427–431.

61. Hel, Z., J.R. McGhee & J. Mestecky. 2006. HIV infection: first battle decides the war. *Trends Immunol.* **27:** 274–281.

62. Keele, B.F., E.E. Giorgi, J.F. Salazar-Gonzalez, *et al.* 2008. Identification and characterization of transmitted and early founder virus envelopes in primary HIV-1 infection. *Proc. Natl. Acad. Sci. U. S. A.* **105**: 7552–7557.

63. Mondor, I., S. Ugolini & Q.J. Sattentau. 1998. Human immunodeficiency virus type 1 attachment to HeLa CD4 cells is CD4 independent and gp120 dependent and requires cell surface heparans. *J. Virol.* **72**: 3623–3634.

64. Karlsson Hedestam, G.B., R.A. Fouchier, S. Phogat, *et al.* 2008. The challenges of eliciting neutralizing antibodies to HIV-1 and to influenza virus. *Nat. Rev. Microbiol.* **6**: 143–155.

65. Haase, A.T. 2005. Perils at mucosal front lines for HIV and SIV and their hosts. *Nat. Rev. Immunol.* **5**: 783–792.

66. McMichael, A. J., P. Borrow, G.D. Tomaras, *et al.* 2010. The immune response during acute HIV-1 infection: clues for vaccine development. *Nat. Rev. Immunol.* **10**: 11–23.

67. Shattock, R.J. & J.P. Moore. 2003. Inhibiting sexual transmission of HIV-1 infection. *Nat. Rev. Microbiol.* **1**: 25–34.

68. Bi, S., P.W. Hong, B. Lee & L.G. Baum. 2011. Galectin-9 binding to cell surface protein disulfide isomerase regulates the redox environment to enhance T-cell migration and HIV entry. *Proc. Natl. Acad. Sci. U. S. A.* **108**: 10650–10655.

69. Hirabayashi, J. & K. Kasai. 1993. The family of metazoan metal-independent beta-galactoside-binding lectins: structure, function and molecular evolution. *Glycobiology* **3**: 297–304.

70. Klysov, A.A., Z.J. Witczak & D. Platt. 2008. *Galectins.* John Wiley & Sons, Inc. Hoboken, NJ.

71. Leffler, H. 2004. Special issue on galectins. *Glycoconj. J.* **19**: 433–638.

72. Rabinovich, G.A. & M.A. Toscano. 2009. Turning 'sweet' on immunity: galectin-glycan interactions in immune tolerance and inflammation. *Nat. Rev. Immunol.* **9**: 338–352.

73. Sato, S. & J. Nieminen. 2004. Seeing strangers or announcing "danger": galectin-3 in two models of innate immunity. *Glycoconj. J.* **19**: 583–591.

74. Nieminen, J., A. Kuno, J. Hirabayashi & S. Sato. 2007. Visualization of galectin-3 oligomerization on the surface of neutrophils and endothelial cells using fluorescence resonance energy transfer. *J. Biol. Chem.* **282**: 1374–1383.

75. Nieminen, J., C. St-Pierre & S. Sato. 2005. Galectin-3 interacts with naive and primed neutrophils, inducing innate immune responses. *J. Leukoc. Biol.* **78**: 1127–1135.

76. Rabinovich, G.A., M.A. Toscano, S.S. Jackson & G.R. Vasta. 2007. Functions of cell surface galectin-glycoprotein lattices. *Curr. Opin. Struct. Biol.* **17**: 513–520.

77. Liu F.T. & G.A. Rabinovich. 2010. Galectins: regulators of acute and chronic inflammation. *Ann. N. Y. Acad. Sci.* **1183**: 158–182.

78. Sato, S., C. St-Pierre, P. Bhaumik & J. Nieminen. 2009. Galectins in innate immunity: dual functions of host soluble beta-galactoside-binding lectins as damage-associated molecular patterns (DAMPs) and as receptors for pathogen-associated molecular patterns (PAMPs). *Immunol. Rev.* **230**: 172–187.

79. Vasta, G.R. 2009. Roles of galectins in infection. *Nat. Rev. Microbiol.* **7**: 424–438.

80. Rabinovich, G., L. Castagna, C. Landa, *et al.* 1996. Regulated expression of a 16-kd galectin-like protein in activated rat macrophages. *J. Leukoc. Biol.* **59**: 363–370.

81. Baum, L.G., M. Pang, N.L. Perillo, *et al.* 1995. Human thymic epithelial cells express an endogenous lectin, galectin-1, which binds to core 2 O-glycans on thymocytes and T lymphoblastoid cells. *J. Exp. Med.* **181**: 877–887.

82. Zuniga, E., G.A. Rabinovich, M.M. Iglesias & A. Gruppi. 2001. Regulated expression of galectin-1 during B-cell activation and implications for T-cell apoptosis. *J. Leukoc. Biol.* **70**: 73–79.

83. Blaser, C., M. Kaufmann, C. Muller, *et al.* 1998. Beta-galactoside-binding protein secreted by activated T cells inhibits antigen-induced proliferation of T cells. *Eur. J. Immunol.* **28**: 2311–2319.

84. Dettin, L., N. Rubinstein, A. Aoki, *et al.* 2003. Regulated expression and ultrastructural localization of galectin-1, a proapoptotic beta-galactoside-binding lectin, during spermatogenesis in rat testis. *Biol. Reprod* **68**: 51–59.

85. Jeschke, U., T. Reimer, C. Bergemann, *et al.* 2004. Binding of galectin-1 (gal-1) on trophoblast cells and inhibition of hormone production of trophoblast tumor cells in vitro by gal-1. *Histochem. Cell. Biol.* **121**: 501–508.

86. Stillman, B.N., D.K. Hsu, M. Pang, *et al.* 2006. Galectin-3 and galectin-1 bind distinct cell surface glycoprotein receptors to induce T cell death. *J. Immunol.* **176**: 778–789.

87. Toscano, M.A., L. Campagna, L.L. Molinero, *et al.* 2011. Nuclear factor (NF)-kappaB controls expression of the immunoregulatory glycan-binding protein galectin-1. *Mol. Immunol.* **48**: 1940–1949.

88. Nio-Kobayashi, J., H., Takahashi-Iwanaga & T. Iwanaga. 2009. Immunohistochemical localization of six galectin subtypes in the mouse digestive tract. *J. Histochem. Cytochem.* **57**: 41–50.

89. Singh, B.N., G.R. Hayes, J.J. Lucas, *et al.* 2009. Structural details and composition of Trichomonas vaginalis lipophosphoglycan in relevance to the epithelial immune function. *Glycoconj. J.* **26**: 3–17.

90. Fichorova, R.N. 2009. Impact of T. vaginalis infection on innate immune responses and reproductive outcome. *J. Reprod. Immunol.* **83**: 185–189.

91. Fogel, S., M. Guittaut, A. Legrand, *et al.* 1999. The tat protein of HIV-1 induces galectin-3 expression. *Glycobiology* **9**: 383–387.

92. Imbeault, M., R. Lodge, M. Ouellet & M.J. Tremblay. 2009. Efficient magnetic bead-based separation of HIV-1-infected cells using an improved reporter virus system reveals that p53 up-regulation occurs exclusively in the virus-expressing cell population. *Virology* **393**: 160–167.

93. Sato, S. & R.C. Hughes. 1994. Regulation of secretion and surface expression of Mac-2, a galactoside-binding protein of macrophages. *J. Biol. Chem.* **269**: 4424–4430.

94. Sundblad, V., D.O. Croci & G.A. Rabinovich. 2011. Regulated expression of galectin-3, a multifunctional glycan-binding protein, in haematopoietic and non-haematopoietic tissues. *Histol. Histopathol* **26**: 247–265.

95. Lalancette-Hébert, M., G. Gowing, A. Simard, *et al.* 2007. Selective ablation of proliferating microglial cells exacerbates ischemic injury in the brain. *J. Neurosci.* **27:** 2596–2605.

96. Liu, F.T. 2005. Regulatory roles of galectins in the immune response. *Int. Arch. Allergy. Immunol.* **136,** 385–400.

97. Sato, S. 2002. Galectin as a molecule of danger signal, which could evoke immune response to infection. *Trends Glycosci. Glycotechnol.* **14:** 285–301.

98. Wada, J., K. Ota, A. Kumar, *et al.* 1997. Developmental regulation, expression, and apoptotic potential of galectin-9, a beta-galactoside binding lectin. *J. Clin. Invest.* **99:** 2452–2461.

99. Matsumoto, R., H. Matsumoto, M. Seki, *et al.* 1998. Human ecalectin, a variant of human galectin-9, is a novel eosinophil chemoattractant produced by T lymphocytes. *J. Biol. Chem.* **273:** 16976–16984.

100. Krishnamoorthy, L., J.W. Bess, Jr., A.B. Preston, *et al.* 2009. HIV-1 and microvesicles from T cells share a common glycome, arguing for a common origin. *Nat. Chem. Biol.* **5:** 244–250.

101. Sparrow, C.P., H. Leffler & S.H. Barondes. 1987. Multiple soluble beta-galactoside-binding lectins from human lung. *J. Biol. Chem.* **262:** 7383–7390.

102. Barondes, S.H., D.N. Cooper, M.A. Gitt & H. Leffler. 1994. Galectins. Structure and function of a large family of animal lectins. *J. Biol. Chem.* **269:** 20807–20810.

103. Rabinovich, G.A., L.G. Baum, N. Tinari, *et al.* 2002. Galectins and their ligands: amplifiers, silencers or tuners of the inflammatory response? *Trends Immunol.* **23:** 313–320.

104. Hirabayashi, J., T. Hashidate, Y. Arata, *et al.* 2002. Oligosaccharide specificity of galectins: a search by frontal affinity chromatography. *Biochim. Biophys. Acta.* **1572:** 232–254.

105. Stowell, S.R., C.M. Arthur, P. Mehta, *et al.* 2008. Galectin-1, -2, and -3 exhibit differential recognition of sialylated glycans and blood group antigens. *J. Biol. Chem.* **283:** 10109–10123.

106. Stowell, S.R., S. Karmakar, C.J. Stowell, *et al.* 2007. Human galectin-1, -2, and -4 induce surface exposure of phosphatidylserine in activated human neutrophils but not in activated T cells. *Blood* **109:** 219–227.

107. Sato, S. & R.C. Hughes. 1992. Binding specificity of a baby hamster kidney lectin for H type I and II chains, polylactosamine glycans, and appropriately glycosylated forms of laminin and fibronectin. *J. Biol. Chem.* **267:** 6983–6990.

108. Pelletier, I. & S. Sato. 2002. Specific recognition and cleavage of galectin-3 by *Leishmania major* through species-specific polygalactose epitope. *J. Biol. Chem.* **277:** 17663–17670.

109. Pelletier, I., T. Hashidate, T. Urashima, *et al.* 2003. Specific recognition of Leishmania major poly-beta-galactosyl epitopes by galectin-9: possible implication of galectin-9 in interaction between L. major and host cells. *J. Biol. Chem.* **278:** 22223–22230.

110. Toscano, M.A., G.A. Bianco, J.M. Ilarregui, *et al.* 2007. Differential glycosylation of T(H)1, T(H)2 and T(H)-17

111. Lanteri, M., V. Giordanengo, N. Hiraoka, *et al.* 2003. Altered T cell surface glycosylation in HIV-1 infection results in increased susceptibility to galectin-1-induced cell death. *Glycobiology* **13:** 909–918.

112. Sun, J., B. Barbeau, S. Sato, *et al.* 2002. Syncytium formation and HIV-1 replication are both accentuated by purified influenza and virus-associated neuraminidase. *J. Biol. Chem.* **277:** 9825–9833.

113. Sun, J., B. Barbeau, S. Sato & M.J. Tremblay. 2001. Neuraminidase from a bacterial source enhances both HIV-1-mediated syncytium formation and the virus binding/entry process. *Virology* **284:** 26–36.

114. Roggentin, P., B. Rothe, J.B. Kaper, *et al.* 1989. Conserved sequences in bacterial and viral sialidases. *Glycoconj. J.* **6:** 349–353.

115. Roggentin, P., R. Schauer, L.L. Hoyer & E.R. Vimr. 1993. The sialidase superfamily and its spread by horizontal gene transfer. *Mol. Microbiol.* **9:** 915–921.

116. Soong, G., A. Muir, M.I. Gomez, *et al.* 2006. Bacterial neuraminidase facilitates mucosal infection by participating in biofilm production. *J. Clin. Invest.* **116:** 2297–2305.

117. Takasaki, S., K. Yamashita, K. Suzuki & A. Kobata. 1980. Structural studies of the sugar chains of cold-insoluble globulin isolated from human plasma. *J. Biochem.* **88:** 1587–1594.

118. Takamoto, M., T. Endo, M. Isemura, *et al.* 1989. Structures of asparagine-linked oligosaccharides of human placental fibronectin. *J. Biochem.* **105:** 742–750.

119. Varki, A., M.E. Etzler, R.D. Cummings & J.D. Esko. 2009. Discovery and classification of glycan-binding proteins. In *Essentials of Glycobiology*, 2nd ed. A. Varki, R.D. Cummings, J. Esko, H.H. Freeze, P. Stanley, C.R. G. Bertozzi, W. Hart, & M.E. Etzler, Eds.: 375–386. Cold Spring Harbor Laboratory Press. New York.

120. Nio-Kobayashi, J. & T. Iwanaga. 2010. Immunohistochemical localization of galectins in mouse genital tract. *Kaibougaku Zasshi (Acta Anatomica Nipponica)* **85:** 58–59.

121. Argueso, P., A. Guzman-Aranguez, F. Mantelli, *et al.* 2009. Association of cell surface mucins with galectin-3 contributes to the ocular surface epithelial barrier. *J. Biol. Chem.* **284:** 23037–23045.

122. Tejler, J., E. Tullberg, T. Frejd, *et al.* 2006. Synthesis of multivalent lactose derivatives by 1,3-dipolar cycloadditions: selective galectin-1 inhibition. *Carbohydr. Res.* **341:** 1353–1362.

123. Giguere, D., R. Patnam, M.A. Bellefleur, *et al.* 2006. Carbohydrate triazoles and isoxazoles as inhibitors of galectins-1 and -3. *Chem. Commun. (Camb).* **22:**2379–2381.

124. Giguere, D., S. Sato, St-Pierre, C., *et al.* 2006. Aryl O- and S-galactosides and lactosides as specific inhibitors of human galectins-1 and -3: role of electrostatic potential at O-3. *Bioorg. Med. Chem. Lett.* **16:** 1668–1672.

125. Giguere, D., M.A. Bonin, P. Cloutier, *et al.* 2008. Synthesis of stable and selective inhibitors of human galectins-1 and -3. *Bioorg. Med. Chem.* **16:** 7811–7823.

Ann. N.Y. Acad. Sci. ISSN 0077-8923

ANNALS OF THE NEW YORK ACADEMY OF SCIENCES
Issue: *Glycobiology of the Immune Response*

An evolutionary perspective on C-type lectins in infection and immunity

Linda M. van den Berg, Sonja I. Gringhuis, and Teunis B.H. Geijtenbeek

Center for Experimental and Molecular Medicine, Academic Medical Center, University of Amsterdam, the Netherlands

Address for correspondence: Teunis B.H. Geijtenbeek, Center for Experimental and Molecular Medicine, Academic Medical Center (AMC), Meibergdreef 9, 1105 AZ Amsterdam, the Netherlands. t.b.geijtenbeek@amc.uva.nl

Host–pathogen interactions have coevolved for many years. On the one hand, the human immune system consists of innate and adaptive immune cells that function to defeat pathogens, and on the other hand, pathogens have coevolved to use the system for their own propagation. C-type lectins are conserved receptors recognizing carbohydrate structures on viruses, bacteria, parasites, and fungi. C-type lectins such as DC-SIGN, langerin, and dectin-1 are expressed by dendritic cell subsets and macrophages. Pathogen recognition by C-type lectins triggers signaling pathways that lead to the expression of specific cytokines which subsequently instruct adaptive T helper immune responses. T helper cell differentiation is crucial for initiating proper adaptive immune responses; some pathogens, however, use pattern recognition receptors like C-type lectins to subvert immune responses for survival. This review provides an update on the role of C-type lectins in HIV-1, mycobacterial, and *Candida* infections, and the coevolution of hosts and pathogens.

Keywords: C-type lectin; dectin-1; DC-SIGN; langerin; Langerhans cell; dendritic cell

Introduction

During evolution the mammalian immune system evolved to defeat pathogens at utmost efficiency. Therefore the human immune system contains a diversity of specialized innate and adaptive immune cells that recognize evolutionary conserved pathogen-associated molecular patterns (PAMPs) by means of pattern recognition receptors (PRRs). PRRs recognize conserved patterns originating from pathogens such as bacterial cell wall structures, viral RNA/DNA, viral envelope structures, and fungal structures. Dendritic cells (DCs), Langerhans cells (LCs), and macrophages are innate immune cells located throughout the body that act as sentinels of the immune system.[1] DCs and LCs express a variety of PRRs, including toll-like receptors (TLRs), NOD-like receptors (NLRs), and C-type lectin receptors (CLRs), through which they capture and internalize invading pathogens and, subsequently, present pathogen-derived antigens on MHC class I and MHC class II molecules to CD8+ and CD4+ T cells, respectively.[2]

Depending on the different PRRs triggered by the PAMPs of the pathogen, DCs and LCs initiate signals that induce naive CD4+ T cell differentiation into distinct T helper cells.[3] T helper cells are named and classified according to their cytokine profile and the type of infection that is combated. T helper 1 cells (Th1) produce interferon-γ (IFN-γ), which activates macrophages and cytotoxic T cells to fight intracellular pathogens. Th2 cells secrete IL-4, IL-5, and IL-13 to activate B cells and humoral immune responses against extracellular pathogens such as helminths and bacteria.[4] The IL-17–secreting Th-17 cells mobilize phagocytes and are required for anti-fungal and antibacterial immunity.[5] Thus, pathogen recognition by PRRs on DCs and LCs results in cytokine production that is crucial for T helper cell differentiation and pathogen eradication.

The C-type lectin receptor (CLR) family comprises a large group of PRRs present on DCs and LCs that shape the immune response. Here we focus on the molecular signaling and the subsequent immunological responses induced by the

doi: 10.1111/j.1749-6632.2011.06392.x

Ca^{2+}-dependent CLRs langerin and DC-SIGN, and the Ca^{2+}-independent CLR dectin-1. Although the host's immune system is optimized for pathogen eradication, pathogens have coevolved, and the more successful pathogens have developed ways to utilize the immune system for their own survival and even propagation. This review gives an update on antiviral, antifungal, and antibacterial immune responses elicited by CLRs and the advantages and disadvantages for host and pathogen.

Carbohydrate recognition by DC-SIGN, langerin, and dectin-1

Originally the term *C-type lectin* referred to proteins with a carbohydrate recognition domain (CRD), or C-type lectin-like domain (CTLD), which bind carbohydrate structures ("lectin") in a Ca^{2+}-dependent manner ("C-type").[6] The mammalian CLR family is divided into 17 subgroups (Fig. 1) based on their phylogenetic relationships and domain struc-

tures.[7–9] However, comparison of CTLD homology revealed that not all CTLDs bind carbohydrate structures or are Ca^{2+} dependent.[10] Since Ca^{2+}-dependent carbohydrate binding is conserved from sponges to human (Fig. 1), it is likely to be the ancestral function.[6] Most C-type lectin family members are adhesion receptors. However, types II, IV, and V CLRs, present on immune cells (Fig. 1), also function as PRRs and induce signaling and immune responses.[6] Within the CTLD, the highly conserved Glu–Pro–Asn (EPN) and Gln–Pro–Asp (QPD) motifs are essential for recognizing mannose- and galactose-containing ligands.[11]

CTLDs of langerin versus DC-SIGN

Langerin[12] and DC-SIGN[13,14] bind ligands in a Ca^{2+}-dependent way, contain EPN motifs, and belong to the type-II CLRs—the asialoglycoprotein receptor family with one CRD (Fig. 1).[15,16] Langerin and DC-SIGN share a highly homologous CTLD.[17]

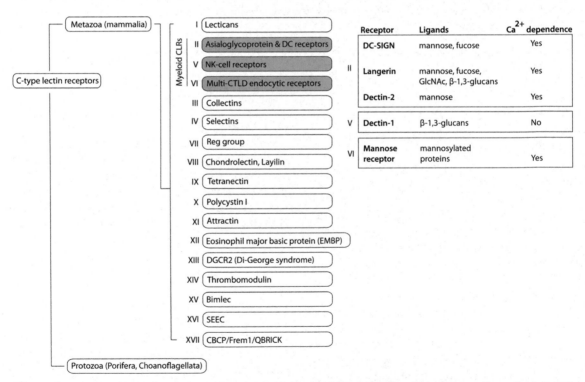

Figure 1. Phylogeny of C-type lectin receptors. Porifera (sponges) and choanoflagellates form the most distant living branch of the animal kingdom[58] and CTLDs have been identified in both types of protozoa,[58,59] indicating CLRs are highly conserved receptors. There are at least 17 subgroups of mammalian CLRs, which are defined by their phylogenetic relationships and domain structures.[7,8] CLRs types II, V, and VI are expressed on myeloid immune cells. Types II and V usually form receptor oligomers on the cell surface (di-, tri-, or tetramers). Langerin and DC-SIGN belong to the type II CLRs, whereas dectin-1 belongs to the type V NK receptor group.[15] CLR, C-type lectin receptor; CTLD, C-type lectin domain; NK, natural killer cell; DC-SIGN, DC-specific ICAM3-grabbing nonintegrin.

Interestingly, langerin and DC-SIGN are present on distinct cell types and function differently, despite having a broad overlap in ligand recognition. Langerin forms trimers on the cell surface of epidermal LCs[18] and recognizes the monosaccharides mannose, fucose, *N*-acetyl-glucosamine (GlcNAc), and the oligosaccharides β-glucan and mannan.[17,19,20] DC-SIGN, on the other hand, is expressed as a tetramer on DCs present in skin and mucosa and recognizes mannose, fucose, GlcNAc, and mannan.[16] Langerin trimerization occurs via a coiled coil structure in the extracellular neck-region, which leads to a more fixed position compared to the related DC-SIGN tetramers. DC-SIGN oligomerizes via its extracellular repeat domains, allowing for more flexibility in ligand binding via its CRDs that are flexibly linked to the neck region.[18]

Typically, Ca^{2+}-dependent CTLDs have four Ca^{2+} binding sites, Ca-1 to Ca-4.[10,17] Binding of Ca^{2+} can have large effect on the tertiary structure of the receptor, and thus can influence the ligand binding of the receptor.[6] Langerin has only the Ca-2 binding site, whereas DC-SIGN has Ca-1–Ca-3, which explains differences in ligand affinity. Furthermore, Chatwel *et al.* identified a second carbohydrate-binding site in langerin that is Ca^{2+}-independent and not present in DC-SIGN.[17] However, the importance of this binding site is not clear, as Ca^{2+}-independent binding of mannose-containing structures by langerin could not be confirmed by other groups.[21]

CTLD of dectin-1

Based on CRD homology, dectin-1 is placed within the type V NK receptor group.[15] Most of the type V receptors express immuno-receptor tyrosine-based inhibitory motifs (ITIM) in their cytoplasmic domains; however, dectin-1 is exceptional in that it contains an activating ITAM-like motif.[22] Dectin-1 is present on LCs, DCs, and macrophages, and binds ligands in a Ca^{2+}-independent manner.[23] Remarkably, the CTLD of dectin-1 does not contain a conserved EPN or QPD motif. Dectin-1 recognizes β-1,3- and β-1,6-glucan carbohydrate structures;[24] the amino acid motif Trp–Ile–His (WIH) has been implicated for β-glucan binding by dectin-1.[11] Murine dectin-1 forms "dimers" upon ligand binding that are bridged intracellularly by spleen tyrosine kinase (SYK).[22,23,25] Although both dectin-

1 and langerin recognize β-glucans,[19] they do not have a highly homologous CTLDs, suggesting that the ability to recognize β-glucans evolved convergently. Recent work has shown that langerin binds β-glucans through the interaction of a single glucose residue within the Ca^{2+} site.[21]

Pathogen recognition, signaling, and the immune response

Antigen presentation and costimulation provided by DCs or LCs to T cells, as well as cytokines secreted by DCs or LCs, determine $CD4^+$ T helper cell differentiation.[3] Cytokine gene transcription by DCs and LCs depends on the activation of the transcription factor NF-κB induced by PRRs, such as the archetypical Toll-like receptors (TLRs) and some CLRs. The NF-κB family consists of five subunits: p65, c-Rel, RelB, p50, and p52. Dimers of NF-κB family members are retained within the cytoplasm but translocate into the nucleus upon PRR signaling to initiate or repress transcription. Signaling by DC-SIGN alone does not lead to NF-κB translocation into the nucleus but enhances activation of certain canonical NF-κB subunits.[26] In contrast, dectin-1 signaling can induce both canonical and noncanonical NF-κB-mediated gene expression independent of other PRRs.[27]

DC-SIGN in mycobacterial and HIV infections

DC-SIGN is a multivalent molecule that interacts with many types of pathogen-associated molecules and self-ligands such as ICAM-3 on T cells.[13,14] Human immunodeficiency virus (HIV), *Mycobacterium* species,[28] *Candida* species,[29] proteins from tick saliva,[30] *Helicobacter pylori*,[26] and helminth structures are among the pathogenic structures bound by DC-SIGN via mannose or fucose moieties.[14,26] Mannose-induced stimulation of DC-SIGN activates the serine/threonine protein kinase RAF1 and thereby induces phosphorylation of NF-κB subunit p65 at serine (Ser) residue 276 (Fig. 2A). The activation of RAF1 by DC-SIGN occurs independently of TLR signaling; however, phosphorylation of p65 requires prior activation of NF-κB, which does depend on TLR signaling.[31] Ser276 phosphorylation of p65 enables binding of the histone acetyl transferases CREB-binding protein (CBP) and p300, which subsequently acetylate p65 (Fig. 2A).[31] This acetylation leads to increased DNA binding affinity of p65 at cytokine genes and

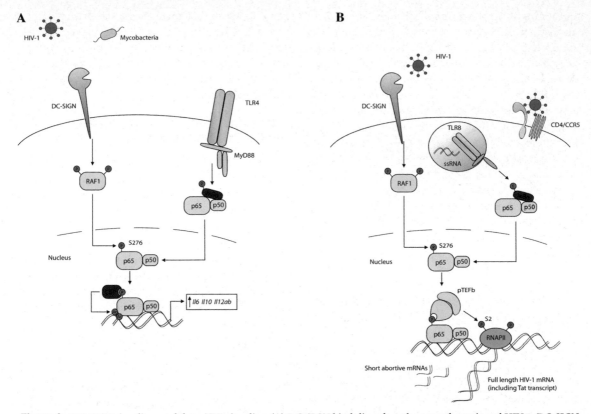

Figure 2. DC-SIGN signaling modulates TLR signaling. (A) DC-SIGN binds ligands such as mycobacteria and HIV-1. DC-SIGN induces RAF1 phosphorylation, which modulates TLR induced NF-κB activation. Upon TLR stimulation the canonical NF-κB subunit p65 is released from its inhibitor and translocates to the nucleus. Phosphorylated RAF1 induces p65 phosphorylation at Ser276 that functions as a binding site for the histone acetylase CBP. Acetylation of p65 induces enhanced and prolonged *Il6*, *Il10* and *Il12ab* transcription. (B) HIV-1 enters the host DC via the coreceptors CD4 and CCR5 leading to viral uncoating, reverse transcription and integration into the host genome (host DNA: black; viral DNA/RNA: red). HIV-1 needs the viral protein Tat for transcription elongation, which is encoded in the integrated viral DNA. Without Tat short abortive mRNAs will be produced. By signaling via TLR8 (ssRNA) and DC-SIGN (gp120) HIV-1 recruits host factors to the transcription initiation complex inducing the first HIV-1 transcripts. TLR8 triggering leads to nuclear translocation of p65. DC-SIGN signaling via gp120 induces RAF1 activation and subsequent p65 phosphorylation at Ser276, which functions as a binding site for pTEFb. pTEFb is recruited to the HIV-1 LTR and phosphorylates RNAPII at Ser2, allowing for transcription elongation and full HIV-1 mRNA transcripts. Once Tat is produced, transcription will be more efficient and enhanced. CBP, CREB binding protein; DC-SIGN, DC-specific ICAM3-grabbing nonintegrin; MyD88, myeloid differentiation primary response protein 88; NF-κB, nuclear factor κ B; IκBα, inhibitor of NF-κBα; pTEFb, positive transcription elongation factor b; RNAPII, RNA polymerase II; TLR, Toll-like receptor.

prolonged nuclear activity, and hence enhanced transcription of *Il6*, *Il10*, *Il12a*, and *Il12b*.[26,31] Thus, while TLR triggering is necessary for NF-κB activation, the additional information provided by a pathogen leads to activation of DC-SIGN and modulation of the TLR trigger, and thus customization of the adaptive immune response to the specific pathogen. This might explain why DC-SIGN recognition of self-ligands, such as adhesion molecules ICAM-2 and ICAM-3, does not lead to DC maturation and cytokine production, as there is no si-

multaneous activation of PRRs (for example, by a pathogen) to induce NF-κB activation.

Mycobacterium tuberculosis is the causative agent of tuberculosis.[32] After the initial immune reactive phase, *M. tuberculosis* infection enters a chronic latent phase, which suggests that *M. tuberculosis* is able to suppress cellular immune responses. Mycobacteria trigger both TLRs and DC-SIGN on DCs via mycobacteria-expressed cell wall structures and mannosylated lipoarabinomannan (ManLAM), respectively.[26,28] Besides ManLAM, DC-SIGN also recognizes mycobacterial α-glucan and phosphatidylinositol mannosides.[33,34] *M. tuberculosis* induces high levels of IL-6, IL-10, and IL-12p70 secretion by DCs, which is dependent on DC-SIGN through the interplay with TLR signaling (Fig. 2A).[26–31] It is likely that *M. tuberculosis* exploits DC-SIGN signaling to evade host immune surveillance by producing both the Th1-repressing cytokine IL-10 and the Th1-promoting cytokine IL-12. This results in IL-10–producing T cells without a bias toward either Th1 or Th2 differentiation, and prevents the host from clearing the pathogen.[26,35] The exact role of DC-SIGN in establishing or retaining the latent phase of tuberculosis has not yet been unraveled. However, a cohort study suggested that decreased levels of DC-SIGN are associated with increased protection against tuberculosis,[36] indicating that mycobacteria have coevolved with the human immune system to evade eradication partly via DC-SIGN.

Another striking example of host–pathogen coevolution is the interaction between DC-SIGN and the human immunodeficiency virus (HIV-1). HIV-1 is a sexual transmitted disease and the causative agent of acquired immunodeficiency syndrome (AIDS). HIV-1 targets $CD4^+$ T cells by fusion to CD4 and chemokine receptors, in particular CCR5 and CXCR4. LCs and DCs line mucosal tissues and are therefore among the first cells to encounter the virus; via DC-SIGN DCs interact with the HIV-1 envelope glycoprotein gp120.[14] Notably, this does not lead to eradication of the virus but promotes HIV-1 transmission and, finally, infection of the host; HIV-1 survives capture by DC-SIGN, and is thereby transported by $DC-SIGN^+$ DCs toward lymph nodes where the virus is subsequently transmitted to $CD4^+$ T cells,[14] the primary target cell of HIV-1. Thus, HIV-1 not only hijacks DCs for transport to lymph nodes but also exploits DC-SIGN for productive infection of DCs.[37] DCs express CD4 and CCR5, which act as coreceptors for HIV-1 (Fig. 2B). Binding results in fusion with the cell membrane, viral uncoating, reverse transcription of HIV-1 single stranded (ss) RNA, and integration of the resulting double stranded DNA into the host genome, where it is subject to transcriptional regulation similar to host genes. For the initiation and elongation of its transcription, HIV-1 is dependent on host as well as viral factors. Host transcription factors such as Sp1 and NF-κB are required to initiate HIV-1 transcription by RNA polymerase II (RNAPII).[38] However, without the viral transcription-elongation factor Tat, RNAPII would detach from DNA after producing only short, abortive mRNAs (Fig. 2B),[37] hence preventing *de novo* synthesis of viral proteins. And while Tat is not included in HIV virions and is not present during the first rounds of transcription initiation, recent work has shown that HIV-1 uses DC-SIGN signaling for the recruitment of host transcription-elongation factors to produce Tat transcripts.[37]

After capture by DC-SIGN on DCs and subsequent internalization, HIV-1 triggers both TLR8, via ssRNA, and DC-SIGN, via gp120, resulting in nuclear translocation of NF-κB and RAF1 activation, respectively (Fig. 2A and 2B). Phosphorylation of NF-κB subunit p65 at Ser276 recruits the host transcription-elongation factor pTEF-b to the HIV-1 transcription complex. pTEF-b phosphorylates RNAPII at Ser2, which promotes transcription elongation and generation of full-length HIV-1 transcripts required for synthesis of new virus particles.[37] Without DC-SIGN signaling and subsequent p65 phosphorylation, pTEFb is not recruited to the initiation site and RNAPII produces short abortive RNAs (Fig. 2B). Therefore, DC-SIGN is indispensible for infection of DC-SIGN$^+$ DCs by HIV-1. This is another example of host–pathogen interactions whereby the pathogen has evolved to use the host's immune system for its own benefit.

Langerin in HIV-1 and fungal infections

In humans, langerin is exclusively expressed by epidermal LCs. Langerin contains an intracellular proline-rich signaling motif[12] that likely functions as a potential docking site for signal transduction proteins.[39] However, not much is known about the role of langerin signaling in the induction of immune responses. Langerin induces the formation of intracellularly located Birbeck granules, which are organelles shaped like tennis racquets. The origin and purpose of Birbeck granules are still poorly understood; however, they are linked with endocytosis.[40] In humans, single nucleotide polymorphisms (SNPs) have been described in the CTLD of langerin that affect carbohydrate recognition. A Trp to Arg mutation at position 264 leads to a lack of Birbeck granule formation.[41] However, no clinical associations or evolutionary benefits or disadvantages have been linked to these SNPs thus far.

While both langerin[42,43] and DC-SIGN bind HIV-1[14] via the glycoprotein gp120, the immunological outcomes of these interactions are very different. First, in contrast to DC-SIGN[+] DCs, LCs are much less susceptible to infection by HIV-1 and they do not transmit the virus to T cells.[43] Second, the virus is internalized via langerin and subsequently degraded.[43] Inhibition of langerin allows infection of LCs, which subsequently transmits HIV-1 to T cells,[43] strongly suggesting that langerin is an important antiviral immune receptor, though it is not clear whether langerin induces signaling processes similar to DC-SIGN. The antiviral function of langerin indicates that the host has evolved this mechanism to prevent HIV-1 infection.

The protective function of langerin against HIV-1 can be abolished by coinfections with other sexual transmitted diseases (STD), such as herpes simplex virus (HSV) or *Candida*.[44] HSV-2 causes genital herpes, which leads to ulcerating and inflamed mucosal tissues, whereas *Candida* can cause genital infections that can be transmitted sexually. Both HSV-2 and *Candida* are able to interact with langerin and inhibit its function, which increases the risk for HIV-1 infection.[43,45] Additionally, HSV-2 is able to infect LCs, which decreases langerin expression and its protective function. Furthermore, HSV-1 and *Candida* infections locally induce the production of TNF-α, which enhances HIV-1 transcription.[44] Thus, coinfections alter the functionality of langerin and abrogate antiviral function of LCs, increasing the risk of acquiring HIV-1 infection and transmission of HIV-1 to T cells.[43-46]

Besides its antiviral role, an antifungal role for langerin has recently been suggested.[19] LCs reside in the epidermis and thus are in close contact with resident fungal species present on human skin. Resident fungi can protect the skin from bacterial infections. However, if the fine balance between resident immune cells and fungi is disturbed, the latter can colonize the skin and cause invasive infections. Opportunistic *Candida* and *Cryptococcus* species are the most common causes of invasive fungal infections in immuno-compromised patients.[47,48] Langerin recognizes β-glucan and mannan structures derived from the fungus *Malassezia furfur* and a variety of *Candida* and *Saccharomyces* species. *Candida* species are internalized by LCs upon binding to langerin,[19] though it remains unknown whether internalization leads to destruction of the fungus.

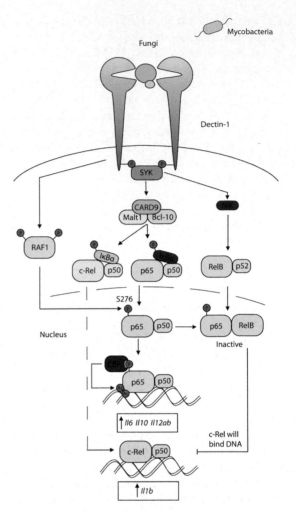

Figure 3. Dectin-1 signaling induces Th-17 cytokine profiles. Dectin-1 ligand binding induces phosphorylation of the tyrosine-based motifs, which subsequently recruits SYK. Activation of SYK leads to the formation of the CARD9/Bcl-10/Malt1 complex ultimately releasing the inhibitor of NF-κB, inducing nuclear translocation of the canonical subunits p65 and c-Rel. Next to that, SYK activation induces NIK resulting in nuclear translocation of the noncanonical NF-κB subunit RelB. RAF1 activation via dectin-1 phosphorylates p65 at serine 276, which functions as a binding site for the histone acetytransferase CBP. Acetylation of p65 leads to enhanced and prolonged *Il6, Il10, Il12a,* and *Il12b* transcription. Furthermore, phosphorylated p65 forms dimers with RelB, disabling the noncanonical subunit from binding to DNA. c-Rel will bind DNA and induce *Il1b* transcription. The net effect is cytokine production skewing naive CD4[+] T cells toward T helper 17 cells. Bcl-10, B cell lymphoma 10; CARD9, caspase recruitment domain family, member 9; CBP, CREB-binding protein; dectin-1, DC-associated C-type lectin 1; IκBα, inhibitor of NF-κBα; MALT-1, mucosa-associated lymphoid tissue lymphoma translocation gene 1; NF-κB, nuclear factor κ B; NIK, NF-κB inducing kinase; SYK, spleen tyrosine kinase.

Interestingly, langerin does not bind structures from *Cryptococcus* species, suggesting no immune recognition of *Cryptococcus* by LCs. Dectin-1 is expressed on immature LCs, although at low levels. It has been shown that langerin is the major antifungal receptor on LCs compared to dectin-1 on LCs;[19] thus, langerin might have evolved to protect skin from invading fungal infections.

Dectin-1 in fungal and mycobacterial infections

Dectin-1 is a unique CLR because it triggers signaling events and cytokine expression without requiring involvement of other PRRs. Dectin-1 recognizes fungal β-glucans and induces anti-fungal Th-17 responses by activation of NF-κB. In contrast to DC-SIGN signaling, dectin-1 does not require additional TLR signaling for the activation of NF-κB. Dectin-1 has a single tyrosine-based motif in the intracellular domain;[25] upon ligand binding, the tyrosine-based motif is phosphorylated and is recognized by SYK via a single Src homology 2 domain (SH2) (Fig. 3). Binding of the two SH2 domains of SYK to separate dectin-1 molecules supposedly induces the formation of a dectin-1 "dimer."[22] SYK-mediated signaling induces the formation of a scaffold complex that consists of CARD9, Bcl-10, and Malt-1 (Fig. 3).[49] This scaffold complex activates the canonical subunits p65 and c-Rel by releasing the NF-κB units from its inhibitor IκBα (Fig. 3).[27,49] Simultaneously, SYK activates NF-κB–inducing kinase (NIK), which subsequently leads to nuclear translocation of the noncanonical subunit RelB[27] and then leads to suppression of *Il12b* and *Il1b* transcription (Fig. 3).

In addition to these NF-κB–activating pathways, dectin-1 activates RAF1, which phosphorylates p65 at Ser276–similar to that which occurs in DC-SIGN signaling (Fig. 3). RAF1-mediated phosphorylation leads to acetylation of p65 and prolonged transcription of certain cytokines (Fig. 3).[27] In contrast to DC-SIGN, dectin-1 by itself activates both p65 and RelB. Ser276 phosphorylation of p65 influences both p65 and RelB activity: phosphorylated p65 dimerizes with RelB, resulting in inactive p65/RelB dimers and attenuation of the respressive function of RelB. Interestingly, p65 activity is not attenuated by p65/RelB dimerization:[27] phosphorylation at Ser276 and subsequent acetylation of p65 increases *Il6*, *Il10*, and *Il12a* transcription, similar

to that described for DC-SIGN signaling. RelB transcriptional suppression of *Il12b* and *Il1b* is reversed by the "capture" of RelB in inactive p65/RelB dimers (Fig. 3). Since RelB is inactivated, p65 and c-Rel are free to bind the promoters of *Il12b* and *Il1b*, respectively, and subsequently induce transcription of typical Th-17-inducing cytokines (Fig. 3).

The net cytokine result of dectin-1 signaling via SYK/CARD9, SYK/NIK, and RAF1 skews T cell differentiation toward Th1 and Th-17 cells, thus providing antifungal immunity. Dectin-1 signaling is a complex pathway that can be influenced by simultaneous PRR activation. Besides activating dectin-1, fungi can simultaneously trigger archetypical TLRs[50] and activation of different NF-κB subunits. Different fungi trigger different patterns of PRRs, which lead to tailored immune responses. A fine balance of signaling will tailor subsequent cytokine production and the type of immune response elicited upon pathogen recognition by dectin-1.

Similar to langerin, dectin-1 recognizes *Candida* species and is involved in immune responses against *Candida* by promoting Th-17 skewing. Notably, an amino acid change present in the human population (Tyr238X) results in defective dectin-1 surface expression.[51] Individuals homozygous for this polymorphism have a higher incidence of mucocutaneous *Candida* infections, which strongly points to the importance of functional dectin-1 for antifungal immune responses.

Dectin-1 is also implicated in antimycobacterial immune responses. Effective immune responses against *Mycobacterium tuberculosis* rely on pathogen recognition by PRRs, such as TLRs and CLRs, as well as NOD-like receptors (NLRs).[52] The immune system has evolved several redundant systems to ensure effective eradication of the pathogen. *M. tuberculosis* leads to signaling via TLR2, TLR4, and TLR9; however, mice deficient for those TLRs or the adaptor molecule MyD88–which transduces signaling by those TLRs–are still able to elicit antimycobacterial immune responses.[53] Similarly, NLR signaling and the complement system are redundant in *M. tuberculosis* infection.[54] This underlines the different back-up systems in the human immune system that have evolved to defend against pathogens. Recently it has been described that mycobacteria trigger CARD9/Bcl-10/Malt1 complexes (Fig. 3) via interaction with dectin-1.[54] Since dectin-1 is a CLR that can induce NF-κB activation, this explains the redundancy of PRRs like TLRs and NLRs. Next to its major antifungal function, CARD9 signaling has been shown to be indispensible for antimycobacterial immune responses.[55] In addition, the dectin-1-related CLR dectin-2 also activates SYK and CARD9/Bcl-10/Malt-1 complexes by recognizing mannan structures derived from fungi. It has recently been shown that activation of dectin-2 induces *Il1b* and *Il23p19* transcription, through c-Rel activation.[56] Thus, dectin-2 activation by fungi promotes the expression of IL-1β and IL-23. Simultaneous triggering of both dectin-1 and dectin-2 thus will boost Th-17-mediated cellular responses.[56] This highlights the evolutionary flexibility, redundancy, and adaptation of the immune system to defeat different pathogens.

Concluding remarks

Host and pathogen are in a continuous race for survival: pathogens evolve to infect the host, whereas the host immune system evolves to counteract pathogen survival. Innate immune cells sense pathogens and steer the immune response toward antipathogen defense mechanisms. CLRs are conserved PRRs present on innate immune cells like DCs, LCs, and macrophages. CLRs recognize sugars and, remarkably, most pathogens contain essential carbohydrates in their cell walls, envelopes, or membranes. CLR triggering by carbohydrates induces cytokine responses that skew T helper cells toward Th1, Th2, or Th-17 cells. In this review, we have outlined the roles of the CLRs DC-SIGN, langerin, and dectin-1 in either effective or defective immune responses against HIV-1, *Candida* species, and mycobacteria.

The immune system utilizes converging signaling pathways that enable a broad spectrum of different responses to different pathogens. The type of antigen, the costimulatory stimulus,[2] and the cytokines provided by antigen presenting cells to naive T cells either elicit or prevent subsequent immune responses. CLRs recognize a diversity of pathogenic patterns through their different CTLDs. Although some CLRs make use of similar signaling pathways, immune responses are fine tuned to defeat pathogens as efficiently as possible. Although a RAF1-mediated pathway is triggered by both DC-SIGN and dectin-1, the immunological outcomes are different. DC-SIGN by itself does not activate NF-κB, whereas dectin-1 triggers p65 as well as RelB and c-Rel. The balance between p65 and RelB activity, and subsequent cytokine production and T helper differentiation, is greatly affected by RAF1 signaling.[26,27] This exemplifies the flexibility and adaptability of the immune system to different conserved patterns on pathogens.

Having complex innate and adaptive immune systems that consist of converging signaling pathways can be beneficial for either the host or the pathogen. For example, a variety of immune signaling pathways provides pathogens with many immune escape possibilities, as interference of a given pathway can occur at may different levels. Mutations in dectin-1, for example, lead to defective immune responses and *Candida* infections,[51] and mutation in the proximal signaling molecule CARD9 leads to propagation of *Candida* infections.[57] Thus, the more steps involved in the induction of immune responses, the more mechanisms by which pathogens can subvert these responses. Yet, on the other hand, the more steps involved in inducing immune responses, the more redundant the induction of immune responses can be. *Mycobacterium tuberculosis* triggers NLRs, TLRs, the complement system, and CLRs like dectin-1, langerin, and DC-SIGN.[28,33,34,50,52,55] Although some PRRs such as dectin-1 seem more important to *M. tuberculosis* infections,[55] it is a clear strength of our immune system that different PRRs provide redundancy in the immune responses to pathogens.

Probably the best example of host–pathogen evolution and adaptation is the difference in outcome of HIV-1 binding to either langerin or DC-SIGN. Langerin on LCs has a protective function against HIV-1 infection,[43] whereas the highly homologous DC-SIGN on DCs is subverted by the virus for viral propagation.[37] By binding to DC-SIGN, HIV-1 hijacks DCs for transport to the lymph nodes to be delivered to the target CD4$^+$ T cell. In addition, the immunological signaling pathway triggered by HIV-1 via DC-SIGN and TLR8 ought to defeat the virus. Nevertheless, this signaling propagates transcription of the indispensable first transcripts of the virus. As the Red Queen from Alice in Wonderland (Lewis Caroll) said: "It takes all the running you can do, to keep in the same place." The one that first stops running, either host or pathogen, will inevitably be seized by the other.

Acknowledgments

We thank the members of the Host Defense group for their valuable input. L.M.vdB. is supported by the Dutch Burns Foundation (grant number 08.109) S.I.G. and T.B.H.G. are supported by the Dutch Scientific Research program (grant number NGI 40-41009-98-8057 to S.I.G.; grant number NWO VICI 918.10.619 to T.B.H.G.).

Conflicts of interest

The authors declare no conflicts of interest.

References

1. Banchereau, J. & R.M. Steinman. 1998. Dendritic cells and the control of immunity. *Nature* **392:** 245–252.
2. Liu, Y. & C.A. Janeway. 1992. Cells that present both specific ligand and costimulatory activity are the most efficient inducers of clonal expansion of normal Cd4 T-cells. *Proc. Natl. Acad. Sci. USA* **89:** 3845–3849.
3. De Jong, E.C., H.H. Smits & M.L. Kapsenberg. 2005. Dendritic cell-mediated T cell polarization. *Springer Semin. Immunopathol.* **26:** 289–307.
4. Kalinski, P., H.H. Smits, J.H.N. Schuitemaker, *et al.* 2000. IL-4 is a mediator of IL-12p70 induction by human Th2 cells: reversal of polarized Th2 phenotype by dendritic cells. *J. Immunol.* **165:** 1877–1881.
5. Dong, C. 2006. Opinion—diversification of T-helper-cell lineages: finding the family root of IL-17-producing cells. *Nat. Rev. Immunol.* **6:** 329–333.
6. Zelensky, A.N. & J.E. Gready. 2005. The C-type lectin-like domain superfamily. *FEBS J.* **272:** 6179–6217.
7. Drickamer, K. 1993. Evolution of Ca2+-dependent animal lectins progress in nucleic acid. *Res. Mol. Biol.* **45:** 207–232.
8. Drickamer, K. & A.J. Fadden. 2002. Genomic analysis of C-type lectins. *Biochem. Soc. Symp.* **69:** 59–72.
9. Drickamer, K. 1999. C-type lectin-like domains. *Curr. Opin. Struct. Biol.* **9:** 585–590.
10. Zelensky, A.N. & J.E. Gready. 2003. Comparative analysis of structural properties of the C-type-lectin-like domain (CTLD). *Protein Struct. Function Genet.* **52:** 466–477.
11. Adachi, Y., T. Ishii, Y. Ikeda, *et al.* 2004. Characterization of beta-glucan recognition site on C-type lectin, dectin 1. *Infect. Immun.* **72:** 4159–4171.
12. Valladeau, J., O. Ravel, C. Dezutter-Dambuyant, *et al.* 2000. Langerin, a novel C-type lectin specific to Langerhans cells, is an endocytic receptor that induces the formation of Birbeck granules. *Immunity* **12:** 71–81.
13. Geijtenbeek, T.B.H., R. Torensma, S.J. Van Vliet, *et al.* 2000. Identification of DC-SIGN, a novel dendritic cell-specific ICAM-3 receptor that supports primary immune responses. *Cell* **100:** 575–585.
14. Geijtenbeek, T.B.H., D.S. Kwon, R. Torensma, *et al.* 2000. DC-SIGN, a dendritic cell-specific HIV-1-binding protein that enhances trans-infection of T cells. *Cell* **100:** 587–597.
15. Cummings, R.D. & R.P. McEver. 2009. C-type lectins. In *Essentials of Glycobiology*, 2nd ed. A. Varki, R.D. Cummings, J.D. Esko, H.H. Freeze, P. Stanley, C.R. Bertozzi, G.W. Hart & M.E. Etzler, Eds.: Chap. 31. Cold Spring Harbor Laboratory Press. Cold Spring Harbor, NY.
16. Figdor, C.G., Y. Van Kooyk & G.J. Adema. 2002. C-type lectin receptors on dendritic cells and Langerhans cells. *Nat. Rev. Immunol.* **2:** 77–84.
17. Chatwell, L., A. Holla, B.B. Kaufer & A. Skerra. 2008. The caxbohydrate recognition domain of Langerin reveals high structural similarity with the one of DC-SIGN but an additional, calcium-independent sugar-binding site. *Mol. Immunol.* **45:** 1981–1994.
18. Feinberg, H., A.S. Powlesland, M.E. Taylor & W.I. Weis. 2010. Trimeric structure of langerin. *J. Biol. Chem.* **285:** 13285–13293.
19. De Jong, M.A.W.P., L.E.M. Vriend, B. Theelen, *et al.* 2010. C-type lectin Langerin is a beta-glucan receptor on human Langerhans cells that recognizes opportunistic and pathogenic fungi. *Mol. Immunol.* **47:** 1216–1225.
20. Stambach, N.S. & M.E. Taylor. 2003. Characterization of carbohydrate recognition by langerin, a C-type lectin of Langerhans cells. *Glycobiology* **13:** 401–410.
21. Feinberg, H., M.E. Taylor, N. Razi, *et al.* 2011. Structural basis for langerin recognition of diverse pathogen and mammalian glycans through a single binding site. *J. Mol. Biol.* **405:** 1027–1039.
22. Rogers, N.C., E.C. Slack, A.D. Edwards, *et al.* 2005. Syk-dependent cytokine induction by dectin-1 reveals a novel pattern recognition pathway for C type lectins. *Immunity* **22:** 507–517.
23. Brown, J., C.A. O'Callaghan, A.S.J. Marshall, *et al.* 2007. Structure of the fungal beta-glucan-binding immune receptor dectin-1: implications for function. *Protein Sci.* **16:** 1042–1052.
24. Brown, G.D. & S. Gordon. 2001. Immune recognition—a new receptor for beta-glucans. *Nature* **413:** 36–37.
25. Osorio, F. & G.R.E. Sousa. 2011. Myeloid C-type lectin receptors in pathogen recognition and host defense. *Immunity* **34:** 651–664.
26. Gringhuis, S.I., J. Den Dunnen, M. Litjens, *et al.* 2009. Carbohydrate-specific signaling through the DC-SIGN signalosome tailors immunity to Mycobacterium tuberculosis, HIV-1 and Helicobacter pylori. *Nat. Immunol.* **10:** 1081–1088.
27. Gringhuis, S.I., J. Den Dunnen, M. Litjens, *et al.* 2009. Dectin-1 directs T helper cell differentiation by controlling noncanonical NF-kappa B activation through Raf-1 and Syk. *Nat. Immunol.* **10:** 203–213.
28. Gcijtenbeek, T.B.H., S.J. Van Vliet, E.A. Koppel, *et al.* 2003. Mycobacteria target DC-SIGN to suppress dendritic cell function. *J. Exp. Med.* **197:** 7–17.
29. Cambi, A., K. Gijzen, J.M. De Vries, *et al.* 2003. The C-type lectin DC-SIGN (CD209) is an antigen-uptake receptor for Candida albicans on dendritic cells. *Eur. J. Immunol.* **33:** 532–538.
30. Hovius, J.W.R., M.A.W.P. De Jong, J. Den Dunnen, *et al.* 2008. Salp15 binding to DC-SIGN inhibits cytokine expression by impairing both nucleosome remodeling and mRNA stabilization. *Plos Pathogens* **4:** 2–14.

31. Gringhuis, S.I., J. Den Dunnen, M. Litjens, *et al.* 2007. C-type lectin DC-SIGN modulates toll-like receptor signaling via Raf-1 kinase-dependent acetylation of transcription factor NF-kappa B. *Immunity* **26:** 605–616.

32. Flynn, J.L. & J. Chan. 2003. Immune evasion by Mycobacterium tuberculosis: living with the enemy. *Curr. Opin. Immunol.* **15:** 450–455.

33. Geurtsen, J., S. Chedammi, J. Mesters, *et al.* 2009. Identification of mycobacterial alpha-glucan as a novel ligand for DC-SIGN: involvement of mycobacterial capsular polysaccharides in host immune modulation. *J. Immunol.* **183:** 5221–5231.

34. Driessen, N.N., R. Ummels, J.J. Maaskant, *et al.* 2009. Role of phosphatidylinositol mannosides in the interaction between mycobacteria and DC-SIGN. *Infect. Immun.* **77:** 4538–4547.

35. Larsen, J.M., C. S.Benn, Y. Fillie, *et al.* 2007. BCG stimulated dendritic cells induce an interleukin-10 producing T-cell population with no T helper 1 or T helper 2 bias in vitro. *Immunology* **121:** 276–282.

36. Vannberg, F.O., S.J. Chapman, C.C. Khor, *et al.* 2008. CD209 genetic polymorphism and tuberculosis disease. *Plos One* **3:** 1–6.

37. Gringhuis, S.I., M. Van Der Vlist, L.M. Van Den Berg, *et al.* 2010. HIV-1 exploits innate signaling by TLR8 and DC-SIGN for productive infection of dendritic cells. *Nat. Immunol.* **11:** U419–U481.

38. Perkins, N.D., N.L. Edwards, C.S. Duckett, *et al.* 1993. A cooperative interaction between Nf-kappa-B and Sp1 is required for HIV-1 enhancer. *Activat. Embo. J.* **12:** 3551–3558.

39. Ren, R.B., B.J. Mayer, P. Cicchetti & D. Baltimore. 1993. Identification of a 10-amino acid proline-rich Sh3 binding-site. *Science* **259:** 1157–1161.

40. Valladeau, J., C. Dezutter-Dambuyant & S. Saeland. 2003. Langerin/CD207 sheds light on formation of birbeck granules and their possible function in langerhans cells. *Immunol. Res.* **28:** 93–107.

41. Ward, E.M., N.S. Stambach, K. Drickamer & M.E. Taylor. 2006. Polymorphisms in human langerin affect stability and sugar binding activity. *J. Biol. Chem.* **281:** 15450–15456.

42. Turville, S.G., P.U. Cameron, A. Handley, *et al.* 2002. Diversity of receptors binding HIV on dendritic cell subsets. *Nat. Immunol.* **3:** 975–983.

43. De Witte, L., A. Nabatov, M. Pion, *et al.* 2007. Langerin is a natural barrier to HIV-1 transmission by Langerhans cells. *Nat. Med.* **13:** 367–371.

44. De Jong, M.A.W.P., L. De Witte, M.J. Oudhoff, *et al.* 2008. TNF-alpha and TLR agonists increase susceptibility to HIV-1 transmission by human Langerhans cells ex vivo. *J. Clin. Invest.* **118:** 3440–3452.

45. De Jong, M.A.W.P., L. De Witte, M.E. Taylor & T.B.H. Geijtenbeek. 2010. Herpes simplex virus type 2 enhances HIV-1 susceptibility by affecting langerhans cell function. *J. Immunol.* **185:** 1633–1641.

46. De Jong, M.A.W.P., L. De Witte, A. Bolmstedt, *et al.* 2008. Dendritic cells mediate herpes simplex virus infection and transmission through the C-type lectin DC-SIGN. *J. Gen. Virol.* **89:** 2398–2409.

47. Pappas, P.G., J.R. Perfect, G.A. Cloud, *et al.* 2001. Cryptococcosis in human immunodeficiency virus-negative patients in the era of effective azole therapy. *Clin. Infect. Dis.* **33:** 690–699.

48. Pappas, P.G., J.H. Rex, J. Lee, *et al.* 2003. A prospective observational study of candidemia: epidemiology, therapy, and influences on mortality in hospitalized adult and pediatric patients. *Clin. Infect. Dis.* **37:** 634–643.

49. Gross, O., A. Gewies, K. Finger, *et al.* 2006. Card9 controls a non-TLR signalling pathway for innate anti-fungal immunity. *Nature* **442:** 651–656.

50. Dennehy, K.M., G. Ferwerda, I. Faro-Trindade, *et al.* 2008. Syk kinase is required for collaborative cytokine production induced through dectin-1 and Toll-like receptors. *Eur. J. Immunol.* **38:** 500–506.

51. Ferwerda, B., G. Ferwerda, T.S. Plantinga, *et al.* 2009. Human dectin-1 deficiency and mucocutaneous fungal infections. *N. Engl. J. Med.* **361:** 1760–1767.

52. Court, N., V. Vasseur, R. Vacher, *et al.* 2010. Partial redundancy of the pattern recognition receptors, scavenger receptors, and C-type lectins for the long-term control of mycobacterium tuberculosis infection. *J. Immunol.* **184:** 7057–7070.

53. Holscher, C., N. Reiling, U.E. Schaible, *et al.* 2008. Containment of aerogenic Mycobacterium tuberculosis infection in mice does not require MyD88 adaptor function for TLR2, -4 and -9. *Eur. J. Immunol.* **38:** 680–694.

54. Marakalala, M.J., L.M. Graham & G.D. Brown. 2010. The role of syk/CARD9-coupled C-type lectin receptors in immunity to Mycobacterium tuberculosis infections. *Clin. Develop. Immunol* **567571:** 1–9.

55. Dorhoi, A., C. Desel, V. Yeremeev, *et al.* 2010. The adaptor molecule CARD9 is essential for tuberculosis control. *J. Exp. Med.* **207:** 777–792.

56. Gringhuis, S.I., B.A. Wevers, T.M. Kaptein, *et al.* 2011. Selective C-Rel activation via Malt1 controls anti-fungal T(H)-17 immunity by dectin-1 and dectin-2. *Plos Pathogens* **7:** 1–11.

57. Glocker, E.O., A. Hennigs, M. Nabavi, *et al.* 2009. A homozygous CARD9 mutation in a family with susceptibility to fungal infections. *N. Engl. J. Med.* **361:** 1727–1735.

58. Harwood, A. & J.C. Coates. 2004. A prehistory of cell adhesion. *Curr. Opin. Cell Biol.* **16:** 470–476.

59. King, N., C.T. Hittinger & S.B. Carroll. 2003. Evolution of key cell signaling and adhesion protein families predates animal origins. *Science* **301:** 361–363.

Ann. N.Y. Acad. Sci. ISSN 0077-8923

ANNALS OF THE NEW YORK ACADEMY OF SCIENCES
Issue: *Glycobiology of the Immune Response*

Integrated approach toward the discovery of glyco-biomarkers of inflammation-related diseases

Takashi Angata, Reiko Fujinawa, Ayako Kurimoto, Kazuki Nakajima, Masaki Kato, Shinji Takamatsu, Hiroaki Korekane, Cong-Xiao Gao, Kazuaki Ohtsubo, Shinobu Kitazume, and Naoyuki Taniguchi

Systems Glycobiology Research Group, Chemical Biology Department, RIKEN Advanced Science Institute, Wako, Saitama, Japan

Address for correspondence: Takashi Angata, Systems Glycobiology Research Group, Chemical Biology Department, RIKEN Advanced Science Institute, 2-1 Hirosawa, Wako, Saitama 351-0198, Japan. angata@riken.jp; Naoyuki Taniguchi, Systems Glycobiology Research Group, Chemical Biology Department, RIKEN Advanced Science Institute, 2-1 Hirosawa, Wako, Saitama 351-0198, Japan. tani52@wd5.so-net.ne.jp

Glycobiology has contributed tremendously to the discovery and characterization of cancer-related biomarkers containing glycans (i.e., glyco-biomarkers) and a more detailed understanding of cancer biology. It is now recognized that most chronic diseases involve some elements of chronic inflammation; these include cancer, Alzheimer's disease, and metabolic syndrome (including consequential diabetes mellitus and cardiovascular diseases). By extending the knowledge and experience of the glycobiology community regarding cancer biomarker discovery, we should be able to contribute to the discovery of diagnostic/prognostic glyco-biomarkers of other chronic diseases that involve chronic inflammation. Future integration of large-scale "omics"-type data (e.g., genomics, epigenomics, transcriptomics, proteomics, and glycomics) with computational model building, or a systems glycobiology approach, will facilitate such efforts.

Keywords: chronic inflammation; glyco-biomarker; cancer; Alzheimer's disease; chronic obstructive pulmonary disease; systems biology

Introduction

It is now widely recognized that identifying biomarkers reflecting disease status is essential not only for reliable diagnosis but also for the development of effective treatments.[1,2] Although any quantifiable trait may serve as a biomarker, biomarkers in bodily fluids that require minimally invasive sampling are most desirable because repeated sampling is required to monitor disease status and/or treatment effectiveness. For this reason, plasma/serum biomarkers are the mainstay for diagnosis and decision making in clinical settings.

Many potential carbohydrate-related biomarkers (glyco-biomarkers) of various cancers have been discovered and some are used in clinical practice (Table 1).[3,24] Many cancer biomarkers are glycoproteins.[22] The tumor microenvironment is rich in inflammatory cells, such as macrophages, natural killer cells, and cytotoxic T cells, which produce proinflammatory stimuli (e.g., proinflammatory cytokines, lipid metabolites such as prostaglandins, and reactive oxygen species) that likely affect the glycosylation patterns of cancer cells and nearby stromal cells. Some, if not all, cancer glyco-biomarkers appear to reflect this inflammatory tumor microenvironment.[25]

In this review, we provide a brief overview of glyco-biomarkers related to the inflammatory aspect of cancer as a guiding principle, and discuss possible ways to discover glyco-biomarkers of other inflammation-related diseases by integrating various approaches relevant to glycobiology.

Glyco-biomarkers of the inflammatory aspect of cancer: a leading paradigm

As mentioned earlier, chronic inflammation is now considered a major factor contributing to the progression of cancer.[26,27] For example,

doi: 10.1111/j.1749-6632.2012.06469.x

Table 1. Cancer biomarkers related to glycans and approved for clinical use

Biomarker	Cancer type	Glycan structure recognized	Approved by Food and Drug Administration (U.S.)[22]	Approved by Ministry of Health, Labor and Welfare (Japan)[23]	Inflammatory conditions in which high values are observed
AFP-L3% (percentage of LCA lectin-reactive α-fetoprotein)	Hepatocellular carcinoma	Core fucose[3]	Yes	Yes	Fulminant hepatitis[4]
CA19–9 (including: SPan-1)	Pancreatic cancer	Sialyl Lewis a[5]	Yes	Yes	Pancreatitis,[6] cholelithiasis,[7] and rheumatoid arthritis[8]
DUPAN-2	Pancreatic cancer	Sialyl Lewis c (sialyl lactotetraose)[9]	No	Yes	Pancreatitis[10]
CSLEX (including: SLX, NCC-ST-439)	Lung, breast, ovarian, and gastrointestinal cancers	Sialyl Lewis x[11]	No	Yes	Various acute and chronic inflammatory conditions[12,13]
CA72–4 (including: STN, CA54/61)	Ovarian cancer	Sialyl Tn[14]	No	Yes	Endometriosis[15]
CA125	Ovarian cancer	(Unidentified carbohydrate epitope on MUC16)[16]	Yes	Yes	Pelvic inflammatory diseases,[17,18] endometriosis,[19] and rheumatoid arthritis[8]
CA15–3	Breast cancer	(Unidentified carbohydrate epitope on MUC1)[20]	Yes	Yes	Pulmonary fibrosis[21] and rheumatoid arthritis[8]

tumor-associated macrophages in solid tumors supply various soluble factors to the microenvironment, such as epidermal growth factor (EGF) to support cancer cell survival and proliferation, transforming growth factor beta (TGF-β) to facilitate epithelial–mesenchymal transition, vascular endothelial growth factor (VEGF) and interleukin-8 (IL-8) to stimulate the development of new blood vessels, and matrix metalloproteinase 9 (MMP-9) to promote tumor dissemination.[28]

Cancer-related glyco-biomarkers currently in clinical use (Table 1) or under development include aberrantly glycosylated glycoproteins (including carbohydrate antigens defined by antibody reactivity such as CA19–9),[3,24,29,30] serum N-glycan profiles,[31,32] and anticarbohydrate autoantibodies.[33] Systematic efforts aiming to discover cancer glyco-biomarkers are underway,[34] including those supported by national projects in Japan (led by Dr. Narimatsu at the National Institute of Advanced Industrial Science and Technology, with funding from the New Energy and Industrial Technology Development Organization)[35,36] and in the United States (the Alliance of Glycobiologists for Detection of Cancer and Cancer Risk in collaboration with the Early Detection Research Network, Glycomics Resource Centers, and the Consortium for Functional Glycomics, all funded by the National Institutes of Health; http://www.nih.gov/news/pr/aug2007/nci-21.htm).

Glyco-biomarkers of particular interest in the context of cancer and inflammation are those carrying fucosylated N-glycans.[37–39] The enzyme fucosyltransferase-8 (Fut8) catalyzes the synthesis of the core fucose structure, and several α1–3 fucosyltransferases (Fut3–7 and 9) catalyze the synthesis of the terminally fucosylated type II structures, such as the Lewis x structure (Galβ1–4[Fucα1–3]GlcNAcβ1-) (Fig. 1). Fut3 is solely responsible for the synthesis of fucosylated type I structures, such as the Lewis a structure (Galβ1–3[Fucα1–4]GlcNAcβ1-).[40,41]

Chronic hepatitis B or C virus infection underlies most hepatocellular carcinoma (HCC) cases in industrialized countries. The disease progresses from chronic hepatitis through liver fibrosis to liver cirrhosis and eventually to HCC.[42,43] It is therefore natural to assume that chronic inflammation forms the basis of HCC and that HCC biomarkers reflect this inflammatory condition. The oncofetal glycoprotein alpha-fetoprotein (AFP) serves as a classic marker of HCC. The specificity of HCC diagnosis by AFP was improved by introducing glycan information: the percentage of core-fucosylated AFP relative to total AFP (i.e., AFP-L3% assay) is now used as a more specific marker of HCC.[3,4,44] Increased fucosylation of AFP in HCC appears to reflect a complex array of events including the induction of Fut8,[45] enhanced GDP–fucose synthesis and transportation into the Golgi apparatus,[46,47] and loss of transportation/secretion polarity.[48] A recently discovered potential liver fibrosis/cirrhosis biomarker that indicates increased core fucosylation of alpha1-acid glycoprotein[49] may also reflect these inflammation-induced events but at an earlier phase of disease progression (i.e., liver fibrosis and cirrhosis). Glycan information may also improve the predictive value of prostate-specific antigen (PSA). The glycan structure of PSA differs between prostate cancer and benign prostate hyperplasia;[50,51] thus altered PSA glycosylation may reflect local inflammation in prostate cancer tissue.[52,53]

A recent genome-wide association study examining correlations between genetic variations and N-glycosylation patterns of serum glycoproteins identified transcription factors involved in various aspects of fucosylation.[54] Hepatic nuclear factor 1α (HNF1α), the key transcription factor identified in the study, regulates glucose homeostasis. The study demonstrated that HNF1α directly and indirectly (via HNF1α, another transcription factor downstream in the regulatory cascade) regulates the expression of multiple genes involved in fucosylation, such as fucosyltransferases (e.g., suppression of Fut8 and induction of Fut3, 5, and 6) and a key enzyme involved in the biosynthesis of GDP-fucose (GDP-mannose 4,6-dehydratase, encoded by *GMDS*). HNF1α polymorphism is associated with susceptibility to coronary heart diseases[55] and type 2 diabetes,[56] implying that fucosylated serum glycoproteins may also serve as biomarkers for susceptibility to and/or prognosis of metabolic syndrome-associated diseases.

Another candidate cancer biomarker associated with inflammation is fucosylated haptoglobin, a recently identified potential biomarker of pancreatic

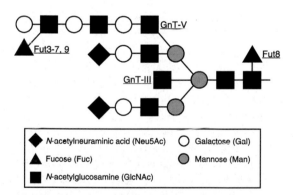

Figure 1. Glycosyltransferases and the sugar residues they attach to N-glycans. Fut8 (encoded by *FUT8*) is responsible for the synthesis of the "core fucose" attached in α1–6 linkage to the GlcNAc at the reducing end of N-glycan. Fut3–7 and 9 (encoded by *FUT3, FUT4, FUT5, FUT6, FUT7*, and *FUT9*) attach a fucose in α1–3 linkage to the GlcNAc in Galβ1–4GlcNAc (type II) structure. Fut3 can also transfer a fucose in α1–4 linkage to the GlcNAc in Galβ1–3GlcNAc (type I) structure (not shown). GnT-III (encoded by *MGAT3*) and GnT-V (encoded by *MGAT5*) catalyze the attachment of a GlcNAc to the mannose residue of Manβ1–4GlcNAc in β1–4 linkage and to the mannose residue (underlined) of GlcNAcβ1-2Manα1-6Man in β1-6 linkage, respectively, in the N-glycan core. Symbols for sugar residues are in accordance with those proposed by the Consortium for Functional Glycomics (http://www.functionalglycomics.org/static/consortium/Nomenclature.shtml).

cancer.[57,58] Haptoglobin is a glycoprotein synthesized primarily in the liver. Its glycan structure (both terminal and core fucosylation) is strongly influenced by the inflammatory mediator, IL-6, which is released into circulation by the inflamed pancreas.[59] Activation of the IL-6 receptor amplifies liver gene expression via HNF1α,[60] which appears relevant to the central role of HNF1α as regulator of fucosylation in the liver. In this case, a mediator released from the inflammatory tumor microenvironment influences the glycan structure of abundant glycoprotein synthesized in the liver. The mechanism by which the glycan structure of haptoglobin is specifically targeted is worth investigating.

In short, prototypical glyco-biomarkers related to the inflammatory aspect of cancer fall into the following two categories: (1) tissue-specific glycoproteins in which the glycan structures are affected by the inflammatory tumor microenvironment, and (2) abundant serum glycoproteins with glycosylation patterns that reflect inflammation in another tissue. Tissue-specific glycoproteins are more specific to the disease, whereas serum glycoproteins may be more sensitive markers of inflammation. However, inflammatory diseases other than cancer may also influence the glycan structures of serum glycoproteins (most of which are synthesized in the liver) via a similar mechanism. Additional methods should be employed to diagnose particular diseases. In addition, some *FUT3* gene polymorphisms result in the absence of Fut3 enzyme expression and subsequent loss of fucosylated type I structures including CA19–9.[61,62] Thus, human genetic polymorphisms should also be taken into account in the advanced stages of biomarker validation.

Road to discovery of possible glyco-biomarkers of other inflammation-related diseases

By following a similar logic (Fig. 2), we should be able to identify glyco-biomarkers of other inflammation-related diseases. Two examples under investigation by our group are discussed below.

Alzheimer's disease

Alzheimer's disease is a leading cause of senile dementia, and one in every three persons is predicted to develop the disease after the age of 85.[63] The most popular hypothesis regarding the disease etiology is as follows: short proteolytic fragments of amyloid

1. Clinical samples
2. Disease model-derived samples
   (animal models, cellular models)

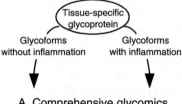

Figure 2. A strategy to identify a tissue-specific glyco-biomarker reflecting inflammation. A tissue-specific glycoprotein (found in serum/plasma) is extracted from case and control samples (from clinical samples or disease models) and their glycan structures compared. A comprehensive glycomics approach is feasible only when samples are available in sufficient quantities. An alternative approach in the form of focused glycomics is necessary when sample quantities are limited. Both empirical (based on previous knowledge) and predictive (based on *in silico* simulations) approaches should be considered for identifying candidate glycan epitopes likely altered by the disease. See Figure 3 for data integration regarding construction of an *in silico* model of glycosylation machinery.

precursor protein (APP) generated by the sequential cleavage by β- and γ-secretases accumulate in the brain parenchyma; this fragment (known as Aβ42) forms oligomers that are neurotoxic, resulting in neuronal death. Definitive diagnosis of Alzheimer's disease still requires postmortem biopsy and confirmation of amyloid plaques (formed by Aβ42) and neurofibrillary tangles (formed by phosphorylated tau protein) in the brain of affected individuals. Current diagnoses to find probable cases of Alzheimer's disease are based on a set of physical and neurological examinations and somewhat subjective criteria regarding cognitive skills, requiring skilled medical professionals. Imaging of amyloid plaques using positron emission tomography (PET) is considered a promising approach for objective diagnosis,[64] but the high cost of PET scans presents an obstacle in making such diagnostics routine. Objective biomarkers of Alzheimer's disease in bodily fluids are thus eagerly sought.[65,66] At present, Aβ42 and total and phosphorylated tau protein in cerebrospinal fluid are leading candidates for such objective diagnostic markers.[67,68]

Inflammation plays a major role in the development of Alzheimer's disease.[63] The inflammation may be confined primarily to inside the blood–brain barrier; therefore, the inflammatory signature molecules may not easily reach the bloodstream. However, in a clinical condition known as cerebral amyloid angiopathy that coincides with over 80% of Alzheimer's disease cases, Aβ42 deposited in blood vessels likely includes those of both neuronal and endothelial origins.[69,70]

A previous study reported the induction of N-acetylglucosaminyltransferase III (GnT-III), which is responsible for the synthesis of bisecting GlcNAc structure (Fig. 1), in macrophages stimulated with Aβ42.[71] This overexpression may be reflected by the extreme increase of bisecting GlcNAc structure–or reduction of structures that are preempted by the action of GnT-III on the common substrate (e.g., GlcNAcβ1–6Manα1–6 branch formed by GnT-V or core fucose transferred by Fut8[37,72,73])–in glycoproteins secreted by or expressed on macrophages in the vicinity of amyloid depositions in blood vessels; however, formal demonstration is still lacking. Our preliminary study suggested that overexpression of GnT-III in macrophages indeed produced altered glycosylation of cell-surface glycoproteins and also altered phagocytic properties of the macrophages (Takamatsu *et al.*, unpublished observation). One may argue that glycosylation changes in macrophages may be of little use in the search for serum/plasma biomarkers; nevertheless, such information may be eventually used to develop probes for *in vivo* imaging. In addition, elevated expression of GnT-III in the brains of Alzheimer's disease patients has been reported,[74] raising the possibility that glycan structural alterations may not be limited to those produced by macrophages. Other candidate biomarkers of Alzheimer's disease include aberrantly glycosylated glycoproteins, such as acetylcholine esterase,[75,76] transferrin,[77] and APP itself,[78] although it is not known whether inflammation plays a role in the generation of these glycoforms.

On a separate note, recent genome-wide association studies have reported an association between late-onset (i.e., nonfamilial) Alzheimer's disease and a single-nucleotide polymorphism in the CD33/Siglec-3 promoter region.[79,80] CD33/Siglec-3 is a lectin expressed on myeloid cells that recognizes glycans terminated with sialic acids.[81,82] The mechanism by which CD33/Siglec-3 is involved in Alzheimer's disease is neither known

nor easily predicted; for example, it is not even known whether CD33/Siglec-3 is expressed on brain microglia. However, it is tempting to speculate that CD33/Siglec-3 expressed on circulating monocytes and tissue macrophages (possibly including microglia) modulates the immune response of these cells toward amyloid deposits and/or other molecular signatures of cellular damage. In this regard, CD33 was reported to repress the inflammatory response of monocytes through recognition of sialic acids.[83] Analysis of glycan structures in the vicinity of cerebral amyloid angiopathy may provide additional clues regarding progression mechanisms of Alzheimer's disease and may lead to the development of a novel intervention strategy to stop or slow the process.

Chronic obstructive pulmonary disease

Chronic obstructive pulmonary disease (COPD) refers to lung disease characterized by airflow obstruction that is not fully reversible, unlike asthma, which can be fully reversed by appropriate treatment.[84] COPD is an umbrella term for chronic bronchitis and emphysema, a combination of which typically occurs in patients with differing proportions. It is projected to be the third leading cause of mortality worldwide by the year 2030.[85] A major cause of COPD in industrialized countries is tobacco smoking, and the chronic inflammatory responses elicited by smoking seem to play significant role in the development of the disease.[86] Exacerbation of COPD, which often leads to emergency department visits and hospitalization, is typically caused by acute airway inflammation in response to viral or bacterial infection.[87] COPD is objectively diagnosed by a pulmonary function test (spirometry), but no established objective diagnostic criteria currently exist for COPD exacerbation. Despite its impact on the health of senior smokers and ex-smokers and the health care system as a whole, COPD is grossly underdiagnosed in some countries, including Japan.[88] Therefore, introducing a serum/plasma marker in routine health examinations may improve diagnosis and subsequent care.

Our previous studies have shown that Fut8-null mice spontaneously develop emphysema,[89] and that heterozygous mice are prone to develop emphysema from chronic tobacco smoke exposure (Gao *et al.*, unpublished observation). A recent report regarding the correlation between a single nucleotide polymorphism of human *FUT8* and pulmonary emphysema appears to corroborate the relevance of Fut8 in human COPD pathology.[90] Thus, differences in glycosylation patterns among individuals (including, but not limited to, core fucosylation of N-glycans) likely influence individual susceptibility to COPD. The development and/or progression of COPD may in turn alter the glycosylation patterns of some proteins found in the serum/plasma.

Glycoproteins synthesized specifically in the lung, such as surfactant protein-D (SP-D), can be detected in the blood of healthy individuals as well as those with COPD. The difference in SP-D concentrations of healthy individuals and COPD patients is not sufficient to establish a clear cut-off value, and smoking status also affects its blood concentration.[91–93] However, changes in the SP-D glycosylation pattern due to chronic inflammation associated with COPD may be observed, as is the case with some cancer glyco-biomarkers. Our preliminary data suggested that an N-glycan structural motif on SP-D was indeed altered in COPD patients compared with that of long-term smokers who have not developed COPD (unpublished observation). In addition, acute inflammation during COPD exacerbation may also influence the glycan structures of lung glycoproteins. These pieces of information may be integrated for the identification of biomarkers for the diagnosis of COPD and its exacerbation.

Omics data integration for glyco-biomarker discovery

Although much has been learned regarding how glycans are synthesized by cooperation and competition of a number of glycosyltransferases, it is still not possible to reliably predict how cellular glycosylation patterns change in response to environmental stimuli, such as inflammatory mediators. To facilitate discovery of glyco-biomarkers related to inflammation, it is desirable to develop a computational model that can predict glycosylation changes in response to environmental stimuli, by integrating data relevant to glycobiology. Integration of comprehensive data sets at different levels of measurements (e.g., signal transduction network that connects cell surface events to the transcription factor network, which in turn determines transcriptomic response, and eventually the glycomic response as an output) is needed to achieve this goal (Fig. 3). A recent review has provided an excellent overview of the

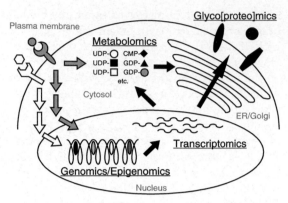

Figure 3. Data integration for glyco-biomarker discovery. Signals received at the cell surface are sent to the nucleus via signal transduction cascades, which mobilize transcription factor cascades. These signals may also directly modify the cellular metabolome relevant to glycosylation (e.g., concentration of nucleotide sugars), which should be determined using a focused metabolomics approach. The transcription factor cascade determines the level at which genes involved in glycan biosynthesis (glycogenes) are expressed, and this information may be comprehensively acquired by DNA microarray or quantitative RT-PCR based approaches (transcriptomics). Glycogene expression is also defined at the genomic level (by genetic polymorphism) and epigenomic level (histone and DNA modifications), which should be also taken into account. Changes in glycogene expression, by modifying various parameters in the glycan biosynthesis pathway, are reflected in altered glycosylation of glycoproteins (and glycolipids), which should be analyzed by glycomics and/or glycoproteomics approaches. Changes in the expression of glycoproteins, which serve as carriers of glycan information, should also be measured at both the transcriptomic and proteomic levels. Finally, a computational (*in silico*) model should eventually integrate these different layers of information relevant to glycobiology to serve as a platform to facilitate biomarker discovery and development of biological hypotheses.

current status of this data integration toward systems glycobiology.[94] Recent efforts toward systems glycobiology modeling primarily involve integrating the glycogene transcriptome (i.e., expression of glycosyltransferases, sulfotransferases, and nucleotide sugar transporters), glycosylation pathway, and quasi-quantitative glycomics data obtained by mass spectrometry and/or lectin microarray analyses of cellular glycoproteins. Integration of these "glyco-centered" data alone is already a formidable challenge; however, serious consideration must be given at the outset regarding how to integrate this glycan-related module into larger *in silico* biological models that display systems behavior.[95]

Some shortcomings in the current omics-type data acquisition approaches should also be addressed. For example, glycogene expression changes are often missed with regular DNA microarrays, possibly because of their low transcript abundance or poor representation in commercial DNA microarrays.[96] Information regarding how glycogene expression is regulated by external stimuli via signal transduction and transcription factor cascades may be more easily and accurately obtained with a dedicated glycogene DNA array[97,98] and/or quantitative reverse transcription-polymerase chain reaction (RT-PCR).[35,99] Epigenetic modification of glycogene expression[100] is another field to be integrated into *in silico* models in the future. In addition, complementary approaches that connect environmental stimuli directly with cellular metabolomic responses relevant to glycosylation, such as nucleotide sugar concentrations,[101] will be necessary.

Challenges in glyco-biomarker development beyond discovery phase

We propose that data integration of tissue-specific glycoproteins found in serum/plasma and of inflammation-induced changes in their glycan structures may identify a pool of candidate biomarkers for diseases in which inflammation plays a role. It is important to emphasize that biochemical characterization of glycosylation changes using model cell lines/animals and clinical samples are essential, but the sensitivity of glycan structural analysis techniques, even with the most advanced mass spectrometry, is still insufficient, in many cases, for clinical sample analysis. Further improvements in instrument sensitivity, as well as in sample preparation and derivatization protocols are anticipated.

Although this paper is focused on the discovery phase of candidate glyco-biomarker understanding and usage, it is obvious that candidate biomarkers must be verified, validated, and authorized before clinical use.[1] It is a long and arduous process, in which many candidates fail.[102] In the case of glyco-biomarkers, a diagnostic kit likely consists of an antibody that recognizes the protein backbone and a reagent (either a lectin or an antibody) that recognizes the glycan epitope. Major and unique obstacles in glyco-biomarker detection development may include the difficulties in developing antibodies against glycans and the inherent low affinity of lectins.[103] Recent advancements in the generation of

glycan-detection reagents, such as antibodies from a phage display library[104] and lectins from a genetically modified lectin library,[105,106] may eventually overcome these obstacles. Nevertheless, it should be noted that the proper choice of a reagent recognizing a glycan epitope and an appropriate assay format has enabled the development of a useful cancer biomarker (i.e., AFP-L3%) assay that has been approved for clinical use,[3] and the assay's modification may allow more widespread use without compromising its specificity.[107]

Conclusions

Most cancer biomarkers have been developed by screening specific monoclonal antibodies against tumor tissues or by focusing on tumor tissue oncofetal properties. Detailed analyses of these cancer biomarkers have revealed that their structural determinants are often glycans. At least some cancer-associated glycosylation changes appear to reflect the inflammatory tumor microenvironment; other chronic diseases in which inflammation plays a part may also result in glycosylation changes of the affected and/or surrounding cells/tissues. Thus, chronic inflammation and glycosylation changes associated with the inflammatory environment may serve as a foundation for the development of biomarkers of various chronic diseases.

Although functional studies that focus on a particular molecule of interest still hold value in biology, it appears inevitable that there will be some shift toward large-scale factory-type research in the field. Various levels/modes of comprehensive data (genomic, epigenomic, transcriptomic, metabolomic, and glycomic data) should eventually be integrated into an *in silico* model of the cellular glycosylation system. Determining how glycosylation responds to environmental stimuli will also be essential in the future development of clinical diagnosis and therapeutics involving glycoproteins. Omics-type data acquisition and systematic integration will take years if not decades of work along with dedicated resources and close collaboration between computational and experimental biologists, as is the case with any systems biology endeavors. However, a comprehensive picture of how glycosylation is regulated will be a tremendously useful resource for the development of clinical diagnostics, optimization of glycoprotein biologics/therapeutics, and a better understanding of glycan functions.

Acknowledgments

This work was supported by the Global COE program "Frontier Biomedical Science Underlying Organelle Network" from the Ministry of Education, Culture, Sports, Science and Technology of Japan to Osaka University (T.A. and N.T.), the Naito Foundation Subsidy for Promotion of Specific Research Projects (N.T.), Grant-in-Aid for Scientific Research (A), No. 20249018 (N.T.), and the Program for Promotion of Fundamental Studies in Health Sciences of the National Institute of Biomedical Innovation (NIBIO), Japan (N.T.).

Conflicts of interest

The authors declare no conflicts of interest.

References

1. Rifai, N., M.A. Gillette & S.A. Carr. 2006. Protein biomarker discovery and validation: the long and uncertain path to clinical utility. *Nat. Biotechnol.* **24:** 971–983.
2. Sawyers, C.L. 2008. The cancer biomarker problem. *Nature* **452:** 548–552.
3. Aoyagi, Y., Y. Tamura & T. Suda. 2011. History and recent progress in evaluation of the fucosylated alpha-fetoprotein fraction. *J. Gastroenterol. Hepatol.* **26:** 615–616.
4. Taketa, K., Y. Endo, C. Sekiya, *et al.* 1993. A collaborative study for the evaluation of lectin-reactive alpha-fetoproteins in early detection of hepatocellular carcinoma. *Cancer Res.* **53:** 5419–5423.
5. Magnani, J.L., B. Nilsson, M. Brockhaus, *et al.* 1982. A monoclonal antibody-defined antigen associated with gastrointestinal cancer is a ganglioside containing sialylated lacto-N-fucopentaose II. *J. Biol. Chem.* **257:** 14365–14369.
6. Safi, F., R. Roscher, R. Bittner, *et al.* 1987. High sensitivity and specificity of CA 19–9 for pancreatic carcinoma in comparison to chronic pancreatitis. Serological and immunohistochemical findings. *Pancreas* **2:** 398–403.
7. Sawabu, N., Y. Takemori, D. Toya, *et al.* 1986. Factors affecting serum levels of CA 19–9 with special reference to benign hepatobiliary and pancreatic diseases. *Gastroenterol. Jpn.* **21:** 491–498.
8. Szekanecz, E., Z. Sandor, P. Antal-Szalmas, *et al.* 2007. Increased production of the soluble tumor-associated antigens CA19–9, CA125, and CA15–3 in rheumatoid arthritis: potential adhesion molecules in synovial inflammation? *Ann. N.Y. Acad. Sci.* **1108:** 359–371.
9. Kawa, S., M. Tokoo, H. Oguchi, *et al.* 1994. Epitope analysis of SPan-1 and DUPAN-2 using synthesized glycoconjugates sialyllact-N-fucopentaose II and sialyllact-N-tetraose. *Pancreas* **9:** 692–697.
10. Mahvi, D.M., H.F. Seigler, W.C. Meyers, *et al.* 1988. DUPAN-2 levels in serum and pancreatic ductal fluids of patients with benign and malignant pancreatic disease. *Pancreas* **3:** 488–493.
11. Kumamoto, K., C. Mitsuoka, M. Izawa, *et al.* 1998. Specific detection of sialyl Lewis X determinant carried on the mucin GlcNAcbeta1->6GalNAcalpha core structure as a tumor-associated antigen. *Biochem. Biophys. Res. Commun.* **247:** 514–517.
12. De Graaf, T.W., M.E. Van der Stelt, M.G. Anbergen & W. van Dijk. 1993. Inflammation-induced expression of sialyl Lewis X-containing glycan structures on alpha 1-acid glycoprotein (orosomucoid) in human sera. *J. Exp. Med.* **177:** 657–666.
13. van Dijk, W., E.C. Havenaar & E.C. Brinkman-van der Linden. 1995. Alpha 1-acid glycoprotein (orosomucoid): pathophysiological changes in glycosylation in relation to its function. *Glycoconj. J.* **12:** 227–233.
14. Kjeldsen, T., H. Clausen, S. Hirohashi, *et al.* 1988. Preparation and characterization of monoclonal antibodies directed to the tumor-associated O-linked sialosyl-2–6 alpha-N-acetylgalactosaminyl (sialosyl-Tn) epitope. *Cancer Res.* **48:** 2214–2220.
15. Nozawa, S., D. Aoki, M. Yajima, *et al.* 1992. CA54/61 as a marker for epithelial ovarian cancer. *Cancer Res.* **52:** 1205–1209.
16. Hanisch, F.G., G. Uhlenbruck, C. Dienst, *et al.* 1985. Ca 125 and Ca 19–9: two cancer-associated sialylsaccharide antigens on a mucus glycoprotein from human milk. *Eur. J. Biochem.* **149:** 323–330.
17. Halila, H., U.H. Stenman & M. Seppala. 1986. Ovarian cancer antigen CA 125 levels in pelvic inflammatory disease and pregnancy. *Cancer* **57:** 1327–1329.
18. Daoud, E. & G. Bodor. 1991. CA-125 concentrations in malignant and nonmalignant disease. *Clin. Chem.* **37:** 1968–1974.
19. Patton, P.E., C.S. Field, R.W. Harms & C.B. Coulam. 1986. CA-125 levels in endometriosis. *Fertil. Steril.* **45:** 770–773.
20. Hinoda, Y., H. Kakiuchi, N. Nakagawa, *et al.* 1992. Circulating tumor-associated antigens detected by monoclonal antibodies against the polypeptide core of mucin–comparison of antigen MUSE11 with CA15–3. *Gastroenterol. Jpn.* **27:** 390–395.
21. Ricci, A., S. Mariotta, E. Bronzetti, *et al.* 2009. Serum CA 15–3 is increased in pulmonary fibrosis. *Sarcoidosis Vasc. Diffuse. Lung Dis.* **26:** 54–63.
22. Ludwig, J.A. & J.N. Weinstein. 2005. Biomarkers in cancer staging, prognosis and treatment selection. *Nat. Rev. Cancer* **5:** 845–856.
23. Kannagi, R. 2009. List of tumor markers. *J. Jpn. Med. Assoc.* **138** (Special issue 1): S346–S349 (in Japanese).
24. Steinberg, W. 1990. The clinical utility of the CA 19–9 tumor-associated antigen. *Am. J. Gastroenterol.* **85:** 350–355.
25. Arnold, J.N., R. Saldova, U.M. Hamid & P.M. Rudd. 2008. Evaluation of the serum N-linked glycome for the diagnosis of cancer and chronic inflammation. *Proteomics* **8:** 3284–3293.
26. Mantovani, A., P. Allavena, A. Sica & F. Balkwill. 2008. Cancer-related inflammation. *Nature* **454:** 436–444.
27. Grivennikov, S.I., F.R. Greten & M. Karin. 2010. Immunity, inflammation, and cancer. *Cell* **140:** 883–899.
28. Weinberg, R.A. 2006. *The Biology of Cancer*. Garland Science. New York.

29. Brockhausen, I. 2006. Mucin-type O-glycans in human colon and breast cancer: glycodynamics and functions. *EMBO Rep.* **7:** 599–604.

30. Hakomori, S. 2001. Tumor-associated carbohydrate antigens defining tumor malignancy: basis for development of anti-cancer vaccines. *Adv. Exp. Med. Biol.* **491:** 369–402.

31. Callewaert, N., H. Van Vlierberghe, A. Van Hecke, *et al.* 2004. Noninvasive diagnosis of liver cirrhosis using DNA sequencer-based total serum protein glycomics. *Nat. Med.* **10:** 429–434.

32. Goldman, R., H.W. Ressom, R.S. Varghese, *et al.* 2009. Detection of hepatocellular carcinoma using glycomic analysis. *Clin. Cancer Res.* **15:** 1808–1813.

33. Jacob, F., D.R. Goldstein, N.V. Bovin, *et al.* 2012. Serum antiglycan antibody detection of nonmucinous ovarian cancers by using a printed glycan array. *Int. J. Cancer* **130:** 138–146.

34. Packer, N.H., C.W. von der Lieth, K.F. Aoki-Kinoshita, *et al.* 2008. Frontiers in glycomics: bioinformatics and biomarkers in disease. An NIH white paper prepared from discussions by the focus groups at a workshop on the NIH campus, Bethesda MD (September 11–13, 2006). *Proteomics* **8:** 8–20.

35. Ito, H., A. Kuno, H. Sawaki, *et al.* 2009. Strategy for glycoproteomics: identification of glyco-alteration using multiple glycan profiling tools. *J. Proteome Res.* **8:** 1358–1367.

36. Narimatsu, H., H. Sawaki, A. Kuno, *et al.* 2010. A strategy for discovery of cancer glyco-biomarkers in serum using newly developed technologies for glycoproteomics. *FEBS J.* **277:** 95–105.

37. Taniguchi, N., A. Ekuni, J.H. Ko, *et al.* 2001. A glycomic approach to the identification and characterization of glycoprotein function in cells transfected with glycosyltransferase genes. *Proteomics* **1:** 239–247.

38. Miyoshi, E., K. Moriwaki & T. Nakagawa. 2008. Biological function of fucosylation in cancer biology. *J. Biochem.* **143:** 725–729.

39. Moriwaki, K. & E. Miyoshi. 2010. Fucosylation and gastrointestinal cancer. *World J. Hepatol.* **2:** 151–161.

40. Nishihara, S., H. Narimatsu, H. Iwasaki, *et al.* 1994. Molecular genetic analysis of the human Lewis histo-blood group system. *J. Biol. Chem.* **269:** 29271–29278.

41. Mollicone, R., A. Cailleau & R. Oriol. 1995. Molecular genetics of H, Se, Lewis and other fucosyltransferase genes. *Transfus. Clin. Biol.* **2:** 235–242.

42. El-Serag, H.B. & K.L. Rudolph. 2007. Hepatocellular carcinoma: epidemiology and molecular carcinogenesis. *Gastroenterology* **132:** 2557–2576.

43. Farazi, P.A. & R.A. DePinho. 2006. Hepatocellular carcinoma pathogenesis: from genes to environment. *Nat. Rev. Cancer* **6:** 674–687.

44. Aoyagi, Y., Y. Suzuki, M. Isemura, *et al.* 1988. The fucosylation index of alpha-fetoprotein and its usefulness in the early diagnosis of hepatocellular carcinoma. *Cancer* **61:** 769–774.

45. Noda, K., E. Miyoshi, N. Uozumi, *et al.* 1998. Gene expression of alpha1–6 fucosyltransferase in human hepatoma tissues: a possible implication for increased fucosylation of alpha-fetoprotein. *Hepatology* **28:** 944–952.

46. Noda, K., E. Miyoshi, J. Gu, *et al.* 2003. Relationship between elevated FX expression and increased production of GDP-L-fucose, a common donor substrate for fucosylation in human hepatocellular carcinoma and hepatoma cell lines. *Cancer Res.* **63:** 6282–6289.

47. Moriwaki, K., K. Noda, T. Nakagawa, *et al.* 2007. A high expression of GDP-fucose transporter in hepatocellular carcinoma is a key factor for increases in fucosylation. *Glycobiology* **17:** 1311–1320.

48. Nakagawa, T., N. Uozumi, M. Nakano, *et al.* 2006. Fucosylation of N-glycans regulates the secretion of hepatic glycoproteins into bile ducts. *J. Biol. Chem.* **281:** 29797–29806.

49. Kuno, A., Y. Ikehara, Y. Tanaka, *et al.* 2011. Multilectin assay for detecting fibrosis-specific glyco-alteration by means of lectin microarray. *Clin. Chem.* **57:** 48–56.

50. Dwek, M.V., A. Jenks & A.J. Leathem. 2010. A sensitive assay to measure biomarker glycosylation demonstrates increased fucosylation of prostate specific antigen (PSA) in patients with prostate cancer compared with benign prostatic hyperplasia. *Clin. Chim. Acta* **411:** 1935–1939.

51. Fukushima, K., T. Satoh, S. Baba & K. Yamashita. 2010. Alpha1,2-fucosylated and beta-N-acetylgalactosaminylated prostate-specific antigen as an efficient marker of prostatic cancer. *Glycobiology* **20:** 452–460.

52. De Marzo, A.M., E.A. Platz, S. Sutcliffe, *et al.* 2007. Inflammation in prostate carcinogenesis. *Nat. Rev. Cancer* **7:** 256–269.

53. Radhakrishnan, P., V. Chachadi, M.F. Lin, *et al.* 2011. TNFalpha enhances the motility and invasiveness of prostatic cancer cells by stimulating the expression of selective glycosyl- and sulfotransferase genes involved in the synthesis of selectin ligands. *Biochem. Biophys. Res. Commun.* **409:** 436–441.

54. Lauc, G., A. Essafi, J.E. Huffman, *et al.* 2010. Genomics meets glycomics-the first GWAS study of human N-Glycome identifies HNF1alpha as a master regulator of plasma protein fucosylation. *PLoS Genet.* **6:** e1001256.

55. Reiner, A.P., M.D. Gross, C.S. Carlson, *et al.* 2009. Common coding variants of the HNF1A gene are associated with multiple cardiovascular risk phenotypes in community-based samples of younger and older European-American adults: the Coronary Artery Risk Development in Young Adults Study and The Cardiovascular Health Study. *Circ. Cardiovasc. Genet.* **2:** 244–254.

56. Voight, B.F., L.J. Scott, V. Steinthorsdottir, *et al.* 2010. Twelve type 2 diabetes susceptibility loci identified through large-scale association analysis. *Nat. Genet.* **42:** 579–589.

57. Okuyama, N., Y. Ide, M. Nakano, *et al.* 2006. Fucosylated haptoglobin is a novel marker for pancreatic cancer: a detailed analysis of the oligosaccharide structure and a possible mechanism for fucosylation. *Int. J. Cancer* **118:** 2803–2808.

58. Nakano, M., T. Nakagawa, T. Ito, *et al.* 2008. Site-specific analysis of N-glycans on haptoglobin in sera of patients with pancreatic cancer: a novel approach for the development of tumor markers. *Int. J. Cancer* **122:** 2301–2309.

59. Narisada, M., S. Kawamoto, K. Kuwamoto, *et al.* 2008. Identification of an inducible factor secreted by pancreatic cancer cell lines that stimulates the production of fucosylated haptoglobin in hepatoma cells. *Biochem. Biophys. Res. Commun.* **377:** 792–796.

60. Leu, J.I., M.A. Crissey, J.P. Leu, *et al.* 2001. Interleukin-6-induced STAT3 and AP-1 amplify hepatocyte nuclear factor 1-mediated transactivation of hepatic genes, an adaptive response to liver injury. *Mol. Cell Biol.* **21:** 414–424.

61. Narimatsu, H., H. Iwasaki, F. Nakayama, *et al.* 1998. Lewis and secretor gene dosages affect CA19–9 and DU-PAN-2 serum levels in normal individuals and colorectal cancer patients. *Cancer Res.* **58:** 512–518.

62. Narimatsu, H., H. Iwasaki, S. Nishihara, *et al.* 1996. Genetic evidence for the Lewis enzyme, which synthesizes type-1 Lewis antigens in colon tissue, and intracellular localization of the enzyme. *Cancer Res.* **56:** 330–338.

63. Querfurth, H.W. & F.M. LaFerla. 2010. Alzheimer's disease. *N. Engl. J. Med.* **362:** 329–344.

64. Klunk, W.E., H. Engler, A. Nordberg, *et al.* 2004. Imaging brain amyloid in Alzheimer's disease with Pittsburgh Compound-B. *Ann. Neurol.* **55:** 306–319.

65. Williams, R. 2011. Biomarkers: warning signs. *Nature* **475:** S5–S7.

66. Buckholtz, N.S. 2011. Perspective: in search of biomarkers. *Nature* **475:** S8.

67. Humpel, C. 2011. Identifying and validating biomarkers for Alzheimer's disease. *Trends Biotechnol.* **29:** 26–32.

68. Blennow, K., H. Hampel, M. Weiner & H. Zetterberg. 2010. Cerebrospinal fluid and plasma biomarkers in Alzheimer disease. *Nat. Rev. Neurol.* **6:** 131–144.

69. Calhoun, M.E., P. Burgermeister, A.L. Phinney, *et al.* 1999. Neuronal overexpression of mutant amyloid precursor protein results in prominent deposition of cerebrovascular amyloid. *Proc. Natl. Acad. Sci. U. S. A.* **96:** 14088–14093.

70. Nicoll, J.A., M. Yamada, J. Frackowiak, *et al.* 2004. Cerebral amyloid angiopathy plays a direct role in the pathogenesis of Alzheimer's disease. Pro-CAA position statement. *Neurobiol. Aging* **25:** 589–597; discussion 603–604.

71. Fiala, M., P.T. Liu, A. Espinosa-Jeffrey, *et al.* 2007. Innate immunity and transcription of MGAT-III and Toll-like receptors in Alzheimer's disease patients are improved by bisdemethoxycurcumin. *Proc. Natl. Acad. Sci. U. S. A.* **104:** 12849–12854.

72. Taniguchi, N., E. Miyoshi, J.H. Ko, *et al.* 1999. Implication of N-acetylglucosaminyltransferases III and V in cancer: gene regulation and signaling mechanism. *Biochim. Biophys. Acta* **1455:** 287–300.

73. Ferrara, C., P. Brunker, T. Suter, *et al.* 2006. Modulation of therapeutic antibody effector functions by glycosylation engineering: influence of Golgi enzyme localization domain and co-expression of heterologous beta1, 4-N-acetylglucosaminyltransferase III and Golgi alphamannosidase II. *Biotechnol. Bioeng.* **93:** 851–861.

74. Akasaka-Manya, K., H. Manya, Y. Sakurai, *et al.* 2010. Protective effect of N-glycan bisecting GlcNAc residues on beta-amyloid production in Alzheimer's disease. *Glycobiology* **20:** 99–106.

75. Saez-Valero, J., G. Sberna, C.A. McLean, *et al.* 1997. Glycosylation of acetylcholinesterase as diagnostic marker for Alzheimer's disease. *Lancet* **350:** 929.

76. Saez-Valero, J., G. Sberna, C.A. McLean & D.H. Small. 1999. Molecular isoform distribution and glycosylation of acetylcholinesterase are altered in brain and cerebrospinal fluid of patients with Alzheimer's disease. *J. Neurochem.* **72:** 1600–1608.

77. Taniguchi, M., Y. Okayama, Y. Hashimoto, *et al.* 2008. Sugar chains of cerebrospinal fluid transferrin as a new biological marker of Alzheimer's disease. *Dement. Geriatr. Cogn. Disord.* **26:** 117–122.

78. Halim, A., G. Brinkmalm, U. Ruetschi, *et al.* 2011. Site-specific characterization of threonine, serine, and tyrosine glycosylations of amyloid precursor protein/amyloid beta-peptides in human cerebrospinal fluid. *Proc. Natl. Acad. Sci. U. S. A.* **108:** 11848–11853.

79. Hollingworth, P., D. Harold, R. Sims, *et al.* 2011. Common variants at ABCA7, MS4A6A/MS4A4E, EPHA1, CD33 and CD2AP are associated with Alzheimer's disease. *Nat. Genet.* **43:** 429–435.

80. Naj, A.C., G. Jun, G.W. Beecham, *et al.* 2011. Common variants at MS4A4/MS4A6E, CD2AP, CD33 and EPHA1 are associated with late-onset Alzheimer's disease. *Nat. Genet.* **43:** 436–441.

81. Crocker, P.R., J.C. Paulson & A. Varki. 2007. Siglecs and their roles in the immune system. *Nat. Rev. Immunol.* **7:** 255–266.

82. Varki, A. & T. Angata. 2006. Siglecs–the major subfamily of I-type lectins. *Glycobiology* **16:** 1R–27R.

83. Lajaunias, F., J.M. Dayer & C. Chizzolini. 2005. Constitutive repressor activity of CD33 on human monocytes requires sialic acid recognition and phosphoinositide 3-kinase-mediated intracellular signaling. *Eur. J. Immunol.* **35:** 243–251.

84. Rabe, K.F., S. Hurd, A. Anzueto, *et al.* 2007. Global strategy for the diagnosis, management, and prevention of chronic obstructive pulmonary disease: GOLD executive summary. *Am. J. Respir. Crit. Care Med.* **176:** 532–555.

85. World Health Organization. 2008. *World Health Statistics 2008.* WHO Press. Geneva.

86. Barnes, P.J. 2000. Chronic obstructive pulmonary disease. *N. Engl. J. Med.* **343:** 269–280.

87. Sethi, S. & T.F. Murphy. 2008. Infection in the pathogenesis and course of chronic obstructive pulmonary disease. *N. Engl. J. Med.* **359:** 2355–2365.

88. Fukuchi, Y., M. Nishimura, M. Ichinose, *et al.* 2004. COPD in Japan: the Nippon COPD Epidemiology study. *Respirology* **9:** 458–465.

89. Wang, X., S. Inoue, J. Gu, *et al.* 2005. Dysregulation of TGF-beta1 receptor activation leads to abnormal lung development and emphysema-like phenotype in core fucose-deficient mice. *Proc. Natl. Acad. Sci. U. S. A.* **102:** 15791–15796.

90. Yamada, M., T. Ishii, S. Ikeda, *et al.* 2011. Association of fucosyltransferase 8 (FUT8) polymorphism Thr267Lys with pulmonary emphysema. *J. Hum. Genet.* **56:** 857–860.

91. Sin, D.D., R. Leung, W.Q. Gan & S.P. Man. 2007. Circulating surfactant protein D as a potential lung-specific biomarker of health outcomes in COPD: a pilot study. *BMC Pulm. Med.* **7:** 13.

92. More, J.M., D.R. Voelker, L.J. Silveira, *et al.* 2010. Smoking reduces surfactant protein D and phospholipids in patients with and without chronic obstructive pulmonary disease. *BMC Pulm. Med.* **10:** 53.

93. Winkler, C., E.N. Atochina-Vasserman, O. Holz, *et al.* 2011. Comprehensive characterisation of pulmonary and serum surfactant protein D in COPD. *Respir. Res.* **12:** 29.

94. Neelamegham, S. & G. Liu. 2011. Systems glycobiology: biochemical reaction networks regulating glycan structure and function. *Glycobiology* **21:** 1541–1553.

95. Katayama, T., K. Arakawa, M. Nakao, *et al.* 2010. The DB-CLS BioHackathon: standardization and interoperability for bioinformatics web services and workflows. The DB-CLS BioHackathon Consortium*. *J. Biomed. Semantics* **1:** 8.

96. Comelli, E.M., M. Amado, S.R. Head & J.C. Paulson. 2002. Custom microarray for glycobiologists: considerations for glycosyltransferase gene expression profiling. *Biochem. Soc. Symp.* **69:** 135–142.

97. Comelli, E.M., S.R. Head, T. Gilmartin, *et al.* 2006. A focused microarray approach to functional glycomics: transcriptional regulation of the glycome. *Glycobiology* **16:** 117–131.

98. Yamamoto, H., H. Takematsu, R. Fujinawa, *et al.* 2007. Correlation index-based responsible-enzyme gene screening (CIRES), a novel DNA microarray-based method for enzyme gene involved in glycan biosynthesis. *PLoS One* **2:** e1232.

99. Nairn, A.V., W.S. York, K. Harris, *et al.* 2008. Regulation of glycan structures in animal tissues: transcript profiling of glycan-related genes. *J. Biol. Chem.* **283:** 17298–17313.

100. Kizuka, Y., S. Kitazume, M. Yoshida & N. Taniguchi. 2011. Brain-specific expression of N-acetylglucosaminyltransferase IX (GnT-IX) is regulated by epigenetic histone modifications. *J. Biol. Chem.* **286:** 31875–31884.

101. Nakajima, K., S. Kitazume, T. Angata, *et al.* 2010. Simultaneous determination of nucleotide sugars with ion-pair reversed-phase HPLC. *Glycobiology* **20:** 865–871.

102. Gutman, S. & L.G. Kessler. 2006. The US Food and Drug Administration perspective on cancer biomarker development. *Nat. Rev. Cancer* **6:** 565–571.

103. Cummings, R.D. & M.E. Etzler. 2009. Antibodies and lectins in glycan analysis. In *Essentials of Glycobiology*. Chapter 45. 2nd ed. A. Varki, *et al.*, Eds.: 633–647. Cold Spring Harbor Laboratory Press. New York.

104. Schoonbroodt, S., M. Steukers, M. Viswanathan, *et al.* 2008. Engineering antibody heavy chain CDR3 to create a phage display Fab library rich in antibodies that bind charged carbohydrates. *J. Immunol.* **181:** 6213–6221.

105. Yim, M., T. Ono & T. Irimura. 2001. Mutated plant lectin library useful to identify different cells. *Proc. Natl. Acad. Sci. U. S. A.* **98:** 2222–2225.

106. Yabe, R., Y. Itakura, S. Nakamura-Tsuruta, *et al.* 2009. Engineering a versatile tandem repeat-type alpha2–6sialic acid-binding lectin. *Biochem. Biophys. Res. Commun.* **384:** 204–209.

107. Korekane, H., T. Hasegawa, A. Matsumoto, *et al.* 2012. Development of an antibody-lectin enzyme immunoassay for fucosylated alpha-fetoprotein. *Biochim. Biophys. Acta.* In press.

Ann. N.Y. Acad. Sci. ISSN 0077-8923

ANNALS OF THE NEW YORK ACADEMY OF SCIENCES
Issue: *Glycobiology of the Immune Response*

Novel roles for the IgG Fc glycan

Robert M. Anthony, Fredrik Wermeling, and Jeffrey V. Ravetch

Leonard Wagner Laboratory of Molecular Genetics and Immunology, The Rockefeller University, New York, New York

Address for correspondence: Robert M. Anthony or Jeffrey V. Ravetch, The Laboratory of Molecular Genetics and Immunology, The Rockefeller University, 1230 York Avenue, Box 98, New York, NY 10065. ranthony01@rockefeller.edu, ravetch@rockefeller.edu

IgG antibodies trigger leukocyte activation and inflammation by forming immune complexes that crosslink activating Fcγ receptors (FcγRs). This is essential to combat infection, but detrimental if antibodies target or cross-react with autoantigens. The high specificity and long serum half-life of IgG antibodies confers tremendous therapeutic potential. Indeed, antibodies have been successfully employed to target cancers, autoreactive B cells, and pro-inflammatory cytokines. Conversely, IgG antibodies can also initiate anti-inflammatory responses. In the form of intravenous immunoglobulin (IVIG), IgGs are routinely administered to treat inflammatory autoimmune diseases. Importantly, the N-linked glycans on the IgG Fc are absolutely required for initiating these IgG effector functions. In fact, the Fc glycan composition dictates IgG affinity to individual FcγRs, and in a broader sense, binding to different FcγRs classes: activating, inhibitory, and anti-inflammatory (dendritic cell-specific ICAM-3 grabbing nonintegrin, DC-SIGN). The Fc glycan requirements to initiate and suppress inflammation will be discussed herein.

Keywords: DC-SIGN; Fcγ receptor; inflammation; autoimmune disease; sialylation

Introduction

IgG antibodies are important mediators of inflammation. These molecules are responsible for the antitoxin activity described by von Behring and Kitasato in the late 19th century used to treat diptheria, for horror autotoxicus described by Ehrlich in the early 20th century, and are the basis for a number of therapeutics currently in use.[1,2] IgG antibodies have tremendous therapeutic potential because they are highly specific, have a long serum half-life, and are well tolerated by patients. In fact, monoclonal IgG antibody drugs are among the most successful therapeutics developed in the last 15 years. They have been used to effectively treat breast cancer and autoimmune disease by targeting surface antigens, leading to specific clearance of pathogenic cells, and attenuate inflammation by blocking inflammatory cytokines.[3–10]

IgG antibodies are the predominant antibody class in circulation and comprise two identical light chains and heavy chains, which couple to form a "Y-shaped" structure.[11] The two domains of IgG antibodies that are responsible for their *in vivo* properties are the Fab (antigen binding fragment) and the Fc (crystalizeable fragment) (Fig. 1A).[12] The Fab portion binds its targets with high affinity, leaving the Fc to interact with FcγRs expressed by leukocytes in a low affinity, high avidity interaction. Four distinct IgG classes, which differ in their heavy chains (and, consequentially, Fc), exist in humans (hIgG1-4) and in mice (mIgG1, 2a, 2b, 3). To initiate inflammation, IgG antibodies bind FcγRs, which are classically described as activating FcγRs or inhibitory FcγRs, signaling through immunoreceptor tyrosine activation motifs (ITAMs) or immunoreceptor tyrosine inhibitory motifs (ITIMs), respectively. The relative affinity of IgG Fcs for respective FcγRs, as well as the expression levels of activating and inhibitory FcγRs, ultimately dictates the ensuing inflammatory response type, which has been reviewed extensively.[13–17]

The Fc glycan

A single N-linked glycan is attached to each heavy chain in the Fc portion asparagine-297 (Asn297, Fig. 1A).[18] The glycan has a complex biantennary

doi: 10.1111/j.1749-6632.2011.06305.x

Ann. N.Y. Acad. Sci. 1253 (2012) 170–180 © 2012 New York Academy of Sciences.

A

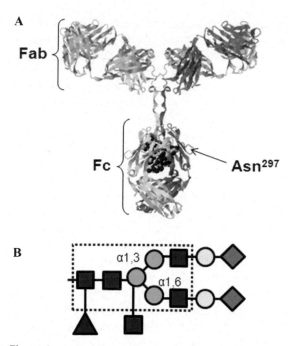

B

Figure 1. IgG and the Fc glycan structure. (A) The Y-shaped structure of human IgG1 antibody b12 (PDB number 1HZH displayed using FirstGlance in Jmol), with the protein backbone displayed in ribbon, with space filling depiction of the glycan.[96] The IgG protein heavy and light chains combine to form the antigen-binding Fab portion, and the heavy chains extend to the Fc portion, which is responsible for initiating effector functions. (B) The fully processed Fc glycan has a complex, biantennary structure. The core structure within the box is composed of *N*-acetylglucosamine (blue squares) and mannose (green circles). The core glycan can be modified by the addition of fucose (red triangle), bisecting *N*-acetylglucosamine, and the two arms, defined by $\alpha 1,3$ and $\alpha 1,6$ mannose linkages, can be extended by the addition of galactose (yellow circles) and sialic acid (pink diamonds).

structure (Fig. 1B), and can vary by the addition of sugar residues to specific parts of the core structure. There is tremendous heterogeneity in the IgG Fc glycan, with over 30 distinct glycans detected on circulated IgG antibodies in healthy individuals.[19] In addition, approximately 20% of IgG Fabs are glycosylated, which results from introduction of a glycosylation site during somatic hypermutation.[18] Glycosylation of the Fab can be important for the binding to antigens, as selected by affinity maturation, and the presence of a Fab glycan can confound analysis of the Fc glycan composition, because the Fab glycan is more sterically accessible.

The Fc glycans are positioned facing toward the center of the IgG molecule, with the $\alpha 1,3$ arm protruding into the cavity between heavy chains, and

the $\alpha 1,6$ arm extending along the heavy chain backbone.[15,18] The Fc glycan is an absolute requirement for binding of wild type IgG to FcγRs, as the interaction is lost by deglycosylating IgG.[18] The Fc glycan is thought to maintain an open confirmation of the Fc heavy chains required for interactions with FcγRs. The structure of aglycosylated Fcs supports this, as the two heavy chains form a closed conformation, preventing formation of the FcγR binding pocket.[20] However, mutations in the Fc backbone can be introduced that enable aglycosylated Fcs to bind FcγRs, which presumably mimic the appropriate conformation folding.[21] This indicates the Fc glycan primarily affects protein–protein interactions by altering IgG backbone conformation.

In the lumen of the endoplasmic reticulum, a 14 monosaccharide glycan (glucose3 mannose9N-acetylglucosamine2, (Glc3Man9GlcNAc2)) is transferred to asparagine-297 (Asn297, Fig. 1A) on each IgG heavy chain by the enzyme oligosaccharyl-transferase (Fig. 2).[22] Next, the glycans are trimmed to a high-mannose structure (Man8–9GlcNAc2) by exoglycosidases, the IgG heavy and light chains are assembled together, and the intact IgG molecule is transported to the Golgi. The glycan structure is further processed throughout the secretory pathway. In the cis-Golgi, the mannose residues are trimmed by $\alpha 1,2$ mannosidase-I to yield Man5GlcNAc2. In the medial-Golgi, *N*-acetylglucosamine is added by $\beta 1,2$-*N*-acetylglucosaminyltransferase-I forming GlcNAc1Man5GlcNAc2, and $\alpha 1,2$ mannosidase-II further removes mannose residues forming the hybrid glycan structure GlcNAc1Man3GlcNAc2. Next, the core IgG glycan (GlcNAc3Man3GlcNAc2) is generated by the transfer of *N*-acetylglucosasmine by $\beta 1,2$-*N*-acetylglucosaminyltransferase-II. Here, the core *N*-acetylglucosamine is available for fucosylation by $\alpha 1,6$-fucosyltransferase. Bisecting *N*-acetylglucosamine is attached to the core by *N*-acetylglucosaminyltransferase-III. As the IgG antibody progresses along the secretory pathway, the glycan can be further modified in the trans-Golgi by the addition of galactose and sialic acid to the arms by $\beta 1,4$ galactosyltransferase and $\alpha 2,6$ sialyltransferase, respectively.

The Fc glycan in inflammation

As mentioned previously, the Fc glycan provides tremendous heterogeneity to IgG antibodies, with the variable addition of the bisecting

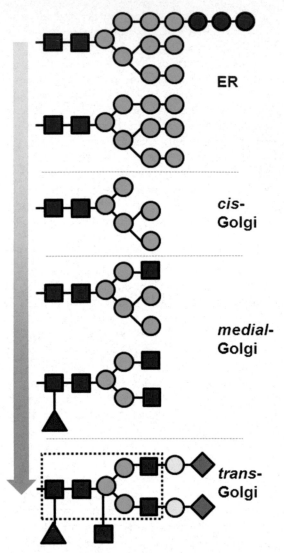

Figure 2. Processing of the Fc glycan. The precursory N-linked glycan (Glc3Man9GlcNAc2) is transferred to Asn-297 on the IgG heavy chain by oligosaccharyltransferase in the ER. Next, the glucose residues are trimmed to form a high-mannose structure (Man8-9GlcNAc2), the IgG heavy and light chains are assembled, and the complex is transported to the Golgi. In the cis-Golgi, the mannose residues are trimmed to yield Man5GlcNAc2. In the medial-Golgi, N-acetylglucosamine is added forming GlcNAc1Man5GlcNAc2, and more mannose residues are removed forming the hybrid glycan structure Glc-NAc1Man3GlcNAc2. Next, the core structure is formed by transfer of N-acetylglucosasmine. The core N-acetylglucosamine is available for fucosylation, and bisecting N-acetylglucosamine is added. The glycan can be further modified in the trans-Golgi by the addition of galactose and sialic acid to the arms. Glucose (blue circles), N-acetylglucosamine (blue squares), mannose (green circles), fucose (red triangles), galactose (yellow circles), sialic acid (pink diamonds).

N-acetylglucosamine, fucose to the core, as well as galactose and sialic acid to the arms of the biantennary structure (Fig. 1B). Interestingly, the composition of the IgG Fc glycan appears to be regulated by the immunological mileu, which feeds back by contributing to either maintenance of homeostasis or by enhancing inflammation (Fig. 3).

Increased levels of fucosylation have been observed in rheumatoid arthritis patients, follow analysis of all circulating proteins, as well as specifically on the IgG heavy chain.[23,24] Murine experiments have demonstrated that repeated immunizations resulted in increased fucosylation of antigen-specific IgG, however, no changes in fucosyltransferase expression in antigen-specific B cells were observed.[25] Attachment of fucose to the IgG core glycan negatively affects antibody effector functions, and afucosylated IgG1 antibodies have a 50- to 100-fold increase in affinity to FcγRIIIa, thought to be the result of interactions between the Fc glycan and the FcγR glycan (Fig. 3).[26–31]

The bisecting N-acetylglucosamine residue also affects IgG interactions with FcγRs, as the presence of this residue increases affinity to FcγRIII (Fig. 3).[32] Consequentially, IgGs with bisecting N-acetylglucosamine display more potent ADCC than N-acetylglucosaminated controls. Increases in addition to bisecting N-acetylglucosamine have been reported in Lambert–Eaton myasthenic symdrome, but were unchanged in Myasthenia gravis patients.[33]

The presence of terminal sialic acid on the Fc glycan reduces FcγR affinity 10-fold, and results in less potent IgG antibodies *in vivo* (Fig. 3).[34] Consistent with this, sialylated IgGs were significantly less efficient at antibody-dependent cytotoxicity (ADCC) compared to asialylated control IgGs.[35] The reduction in FcγR affinity caused by sialylation is independent of sialic acid linkage, as seen in both α2,3 and α2,6 sialic acid attachments.[34] Importantly, reductions in circulating siaylated IgGs are reported during inflammation. Following immunization, sialylation was markedly reduced on Fc glycans in mice.[19,36] Consistent with this notion, rheumatoid arthritis patients and Wegener's granulomatosis patients are reported to produce decreased levels of antigen-specific sialylated IgG for citrillated proteins and proteinase-3, respectively.[37,38] Interestingly, an increase in sialylated autoantigen-specific IgG has been observed during remission of these

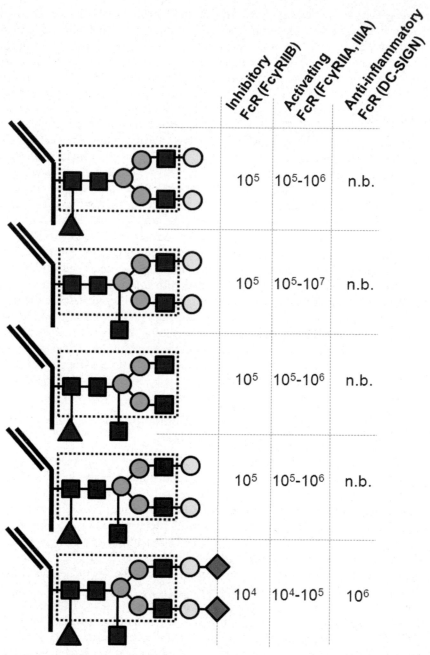

Figure 3. Fc glycan composition dictates FcγR affinity. Approximate association constant ranges (K_a in M-1) of activating, inhibitory, anti-inflammatory FcγRs for various human IgG1 glycoforms. *N*-acetylglucosamine (blue squares), mannose (green circles), fucose (red triangles), galactose (yellow circles), sialic acid (pink diamonds).

diseases, consistent with tight regulation of α2,6 sialyltransferase.[37,38]

Initial reports described reduction in terminal galactose moieties on IgG in patients suffering from various inflammatory diseases, including rheumatoid arthritis, and suggested that galactose might play a regulatory role. However, enzymatic removal of terminal galactose has no effect on FcγR affinity, serum half-life, or induction of inflammation in induced arthritis and ITP models.[35,39] These studies suggest that galactose itself does not directly attribute to IgG effector properties. Therefore, the enhanced inflammatory activity of these antibodies is likely explained by the reduced levels of sialylated IgG Fc glycans, of which a consequence is exposure of terminal galactose residues.

Anti-inflammatory IgG

The immunoregulatory potential of IgG antibodies was first described in 1981 following successful treatment of pedritatric immune-mediated thrombocytopenia (ITP), an autoimmune disease in which platelets are targeted by autoantibodies.[40] Having exhausted other treatment avenues, patients were administered high doses of IVIG out of desperation, which had surprising results. Platelet numbers rebounded in 10 days after treatment. A further insight described in this manuscript indicated the F(ab)2 fragments generated by pepsin digestion of IVIG were ineffective as rescuing platelet numbers. These results were supported by another study that successfully treated pediatric ITP by infusion of IVIG-derived Fcs.[41] These studies confirmed that the general anti-inflammatory activity of IgG antibodies was a function of the Fc, consistent with all effector functions of antibodies.

To explain the observations that high doses of IgG antibodies can act to suppress inflammation, three hypotheses have been put forth.[42,43] One suggests the high dose of administered antibodies saturate the neonatal Fc receptor (FcRn), promoting increased catabolism of pathogen autoantibodies. Another suggests that high doses of IgG antibodies bind to activating FcγRs on inflammatory cells, and prevent binding of pathogenic antibodies to FcγRs and subsequent autoantibody induced inflammation. A third proposes that the high dose of IgG antibodies alters the ratio of FcγR types on inflammatory cells by increasing expression of the inhibitory FcγRIIB, thereby forcing inhibitory signaling following interaction of inflammatory cells and anti-antibody immune complexes.

A number of experimental systems have shed light of these hypotheses. These include the passive transfer of K/BxN serum inducing the effector stage of rheumatoid arthritis, passive transfer of antiplatelet antibodies modeling ITP, and an active immunization nephrotoxic nephritis model.[36,44–47] Importantly, these models mimic human antibody-mediated autoimmune disease, which are prevented by the clinical dose of IVIG. Studies from these models have suggested that IVIG anti-inflammatory activity is resulting from the Fc portion, requires the Fc glycan, and the inhibitory Fc receptor (FcγRIIB) (reviewed in Refs.[15,42,43,48], and.[49] FcRn interactions with IgG antibodies are independent of the Fc glycan, and the glycan requirement of IVIG indicates that FcRn is dispensable for IVIG anti-inflammatory activity. Furthermore, FcγR binding has been well described as a low affinity, high avidity interaction.[13,14,16,17] Therefore, it stands to reason that monomeric IgG antibodies, with low affinity and single valency, will not prevent immune complexes with multiple Fc from the binding to FcγRs.

Importantly, increased FcγRIIB expression in patients suffering from chronic inflammatory demyelinating polyneuropathy (CIDP) was associated with improved responsiveness to IVIG treatment.[50,51] Further, increased expression of FcγRIIB following IVIG administration has been observed in a number of animal models, and mice lacking this receptor are unresponsive to IVIG.[19,36,43,45,46,52] These observations strongly support a role for increased expression of FcγRIIB in the anti-inflammatory activity of IVIG.

Anti-inflammatory activity of the IgG Fc glycan

The high dose requirement of IVIG to initiate the anti-inflammatory activity of IgG antibodies suggested that a minor component of polyclonal IVIG preparations was responsible for this activity. As mentioned above, deglycosylated IVIG lost all anti-inflammatory activity, confirming an essential role of the Fc glycan.[19] Next, IVIG preparations were treated with neuraminidase to remove terminal sialic acid from the Fc glycan.[19] Similar to the deglycosylated IVIG, asialylated IVIG exhibited no anti-inflammatory activity *in vivo*. Further, enrichment for sialylated IgG antibodies yielded 10% of the IgG antibodies in an IVIG preparation. This sialylated IVIG preparation was effective at suppressing induced arthritis and nephrotoxic nephritis at a 10-fold lower dose than whole IVIG[19] (Ravetch and Kaneko, unpublished data). Furthermore, sialylated IgG Fcs (sFcs), generated either by sialic acid-specific lectin enrichment or *in vitro* sialylation, suppressed inflammation at a 30-fold lower dose than IVIG.[19,35] This anti-inflammatory activity required the attachment of sialic acid in an α2,6 linkage, as α2,3 sialylated Fcs were unable to suppress induced arthritis.[35] These results confirmed an essential role for sialylated IgG Fc glycans in the anti-inflammatory activities of IgG antibodies.

As mentioned previously, the addition of sialic acid to the Fc glycan results in reduced affinity for FcγRs (Fig. 3). Therefore, IgGs with sialylated glycans in IVIG would not bind FcγRs, indicating that high dose IVIG would not suppress inflammation by saturating FcγRs, as the lower affinity IgG antibodies would not replace high affinity, high avidity autoimmunecomplexes. Further, the interaction of IgG antibodies with the FcRn is independent of the Fc glycan, and the glycan requirement of IVIG further supports no direct role for FcRn in this anti-inflammatory pathway.

Anti-inflammatory activity of SIGN receptors

Upon the description of the anti-inflammatory activity of sialylated IgG Fcs, studies were commenced to identify the cellular receptor responsible for their anti-inflammatory activity. Sialylation of the Fc glycan results in reduced affinity for FcγRs, indicating an additional receptor triggered by this ligand induced a novel pathway. A screen of genetically modified mice with defined defects in the immune system shed light on the localization of targeted cells. Mice deficient in CD4[+] T cells or B cells were protected by IVIG from induced arthritis.[53] However, op/op mice, deficient in CSF and consequentially specific macrophage populations, Rag1[−/−] mice, and splenectomized mice were not protected.[52,53] These results indicated a splenic population that required anatomical architecture of secondary lymphoid organs (which is severely disrupted in Rag1[−/−] mice, which lack both T and B cells[54]) was required by IVIG, strongly implicated macrophages in the splenic marginal zone.

In the mouse, two populations of splenic marginal zone macrophages (MZMΦs) are defined by expression of Siglec-1 or MARCO.[55–57] Siglec-1[+] metallophillic MZMΦs encircle the white pulp, and are themselves encircled by a ring of MARCO[+] MZMΦs.[58–66] Blocking antibodies to Siglec-1 or MARCO had no effect on the anti-inflammatory activity of IVIG.[53] However, blockade of another receptor expressed by MARCO[+] MZMΦs, specific ICAM3-grabbing non-

integrin, related 1 (SIGN-R1), completely ablated the ability of IVIG to suppress autoantibody induced arthritis.[67–73] Similarly, SIGN-R1–deficient mice (SIGN-R1[−/−]) were not protected from induced arthritis by IVIG or sialylated Fc fragments (sFc).[53,74]

The human orthologue of SIGN-R1 is DC-SIGN (hDC-SIGN), a well-documented lectin recognizing high-mannose glycans.[75–84] Indeed, α2,6 sialylation of Fcs resulted in the ability to bind SIGN-R1 and hDC-SIGN, in addition to reducing FcγR affinity.[53] SIGN receptor binding was not observed with α2,3 sialylated Fcs, consistent with the linkage requirements of the anti-inflammatory activity.[35] These results suggest α2,6 sialylation of the Fc results in a conformation change in the protein backbone that conveys SIGN receptor binding. SIGN-R1[−/−] mice engineered to express hDC-SIGN were responsive to IVIG and sFc, indicating that functionally, human DC-SIGN could replace SIGN-R1 in the IVIG induced anti-inflammatory pathway.[85,86] Furthermore, macrophages cultured from the bone marrow of hDC-SIGN[+] mice, when pulsed *in vitro* with sFc, could transfer anti-inflammatory activity and suppress induced arthritis.

Characterization of this response demonstrated that an innate Th2 response was triggered by sFc through DC-SIGN. Induction of the cytokine IL-33 in the spleen was observed after administration of sFc or IVIG, and blockade of the IL-33 receptor ST2 prevented sFc suppression of induced arthritis. Furthermore, exogenous IL-33 was sufficient to suppress the inflammation. IL-33 has been reported to induce an innate Th2 response, triggering IL-13 production by nuocytes, and the anti-inflammatory activity of IL-33 suggested Th2 cytokines might be involved in this response.[87]

Indeed, IL-4[−/−], IL-4R[−/−], and Stat6[−/−] mice were all unresponsive to sFc, confirming a role for a Th2 response in the anti-inflammatory activity of IVIG. However, these results suggested that IL-13, a cytokine closely related to IL-4, could not act to replace IL-4 in this pathway. However, administration of IL-4 or IL-13 was effective at inducing the pathway, indicating signaling through the IL-4 receptor was the important step triggered. Further, these cytokines potentially upregulated surface expression of FcγRIIB, consistent with the *in vivo* requirements of IVIG (Fig. 4).

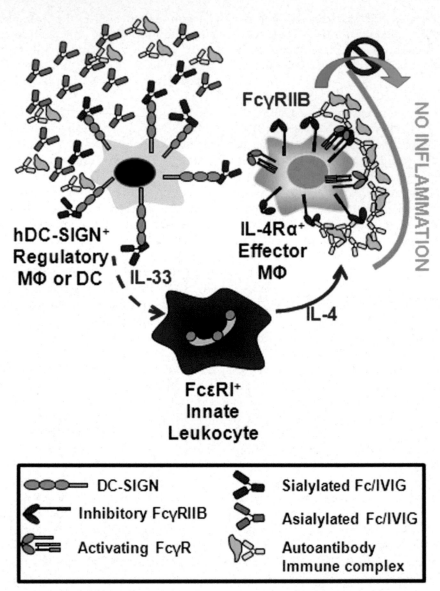

Figure 4. sFc suppresses autoantibody inflammation by inducing an innate Th2 response. Autoantibody immune complexes crosslink activating FcγRs promoting activation of macrophages, and inflammation associated with autoantibody-mediated autoimmune disease. Following administration of IVIG, antibodies with sialylated IgG Fcs bind DC-SIGN+ macrophages (MΦs) or dendritic cells (DCs), promoting IL-33 expression, which activated FcεRI+ innate leukocytes to produce IL-4. This cytokine promotes upregulation of FcγRIIB on macrophages, thereby increasing the activation threshold required to trigger inflammation.

Conclusions and perspectives

The IgG Fc glycan is essential for the structural integrity of IgG, an absolute requirement for FcγR interactions, which dictates the type of Fc receptor that is ligated. This level of contribution to a protein's biology is rather unusual for a glycan. While variations on the IgG Fc glycans directly contribute to the effector functions of antibodies, and specific glycan forms are associated with immunological status, little is known about the regulation of glycosyltransferases, and more investigation is required. As noted, increases in fucosylation and decreased siaylation and galactosylation on the Fc glycan are observed during inflammatory conditions. However, it is not clear whether the expression of glycosyltransferases is regulated or whether other regulatory mechanisms are involved. Further, the specific inflammatory cytokines or coreceptor interactions that contribute to Fc glycan regulation are only beginning to be understood. IL-21 has been shown to increase galactosylation and sialylation on IgG produced by cultured B cells, and Th2 cytokines have been demonstrated to influence IgA glycosylation.[88,89]

The immunoregulatory potential of the IgG Fc glycan demonstrated by IVIG treatment, suggests that this anti-inflammatory pathway, triggered by an endogenous ligand, sFc, through a pattern-recognition receptor is important for the maintenance of homeostasis. Consistent with this notion, sialylation levels are decreased during inflammation, indicating this regulatory pathway is not triggered in patients suffering from chronic inflammatory diseases, and hence the effectiveness of increased sialylated Fc levels by high dose IVIG. Further characterization of the interaction of sFc and DC-SIGN, and examination of the downstream signaling events may help in the development of more effective anti-inflammatory therapies. The effectiveness of IVIG in Alzheimer's disease is currently being explored.[90–95] While preliminary studies have reported encouraging results, it is not clear whether sialylation of the Fc glycan is required for this activity.

Lastly, the Fc glycan composition is an important consideration for rational design of therapeutic antibodies. To potently trigger inflammation and ADCC, IgG Fc glycans should be afucosylated and asialyated, but contain bisecting *N*-acetylglucosamine. In contrast, anti-inflammatory antibodies should ideally be sialylated and fucosylated, and lacking *N*-acetylglucosamine. Further studies are necessary to dissect the regulation of glycosylation *in vivo* and to ensure immunizations and vaccinations result in the most efficient and appropriate IgG effector functions.

Conflicts of interest

The authors declare no conflicts of interest.

References

1. von Behring, E. & S. Kitasato. 1991. The mechanism of diphtheria immunity and tetanus immunity in animals. 1890. *Mol. Immunol.* **28:** 1317, 9–20.

2. Speiser, P. 1957. Observation of a temporary exception to Ehrlich's rule of horror autotoxicus in idiopathic hemolytic anemia (presence of a specific auto-Rh-antibody (D) in a case of acquired hemolytic anemia during a crisis and its disappearance during hormonal therapy. *Wien. Klin. Wochenschr.* **69:** 149–154.

3. Chan, A.C. & P.J. Carter. 2010. Therapeutic antibodies for autoimmunity and inflammation. *Nat. Rev. Immunol.* **10:** 301–316.

4. Beck, A., T. Wurch, C. Bailly & N. Corvaia. 2010. Strategies and challenges for the next generation of therapeutic antibodies. *Nat. Rev. Immunol.* **10:** 345–352.

5. Elliott, M.J., R.N. Maini, M. Feldmann, *et al.* 1994. Randomised double-blind comparison of chimeric monoclonal antibody to tumour necrosis factor alpha (cA2) versus placebo in rheumatoid arthritis. *Lancet* **344:** 1105–1110.

6. van Dullemen, H.M., S.J. van Deventer, D.W. Hommes, *et al.* 1995. Treatment of Crohn's disease with anti-tumor necrosis factor chimeric monoclonal antibody (cA2). *Gastroenterology* **109:** 129–135.

7. Maloney, D.G., A.J. Grillo-Lopez, D.J. Bodkin, *et al.* 1997. IDEC-C2B8: results of a phase I multiple-dose trial in patients with relapsed non-Hodgkin's lymphoma. *J. Clin. Oncol.* **15:** 3266–3274.

8. Maloney, D.G., A.J. Grillo-Lopez, C.A. White, *et al.* 1997. IDEC-C2B8 (Rituximab) anti-CD20 monoclonal antibody therapy in patients with relapsed low-grade non-Hodgkin's lymphoma. *Blood* **90:** 2188–2195.

9. Hudziak, R.M., G.D. Lewis, M. Winget, *et al.* 1989. p185HER2 monoclonal antibody has antiproliferative effects in vitro and sensitizes human breast tumor cells to tumor necrosis factor. *Mol. Cell. Biol.* **9:** 1165–1172.

10. Wong, W.M. 1999. Drug update. Trastuzumab: anti-HER2 antibody for treatment of metastatic breast cancer. *Cancer. Pract.* **7:** 48–50.

11. Huber, R., J. Deisenhofer, P.M. Colman, *et al.* 1976. Crystallographic structure studies of an IgG molecule and an Fc fragment. *Nature* **264:** 415–420.

12. Franklin, E.C. 1975. Structure and function of immunoglobulins. *Acta Endocrinol. Suppl.* **194:** 77–95.

13. Nimmerjahn, F. & J.V. Ravetch. 2006. Fcgamma receptors: old friends and new family members. *Immunity* **24:** 19–28.

14. Nimmerjahn, F. & J.V. Ravetch. 2007. Antibodies, Fc receptors and cancer. *Curr. Opin. Immunol.* **19:** 239–245.

15. Nimmerjahn, F. & J.V. Ravetch. 2008. Fcgamma receptors as regulators of immune responses. *Nat. Rev. Immunol.* **8:** 34–47.

16. Ravetch, J.V. & S. Bolland. 2001. IgG Fc receptors. *Annu. Rev. Immunol.* **19:** 275–290.

17. Ravetch, J.V. & F. Nimmerjahn. 2007. Fc receptors and their role in immune regulation and inflammation. In *Fundamental Immunology*. W.E. Paul, Ed.: 684–705. Lippincott Williams and Wilkins. Philadelphia.

18. Arnold, J.N., M.R. Wormald, R.B. Sim, *et al.* 2007. The impact of glycosylation on the biological function and structure of human immunoglobulins. *Annu. Rev. Immunol.* **25:** 21–50.

19. Kaneko, Y., F. Nimmerjahn & J.V. Ravetch. 2006. Anti-inflammatory activity of immunoglobulin G resulting from Fc sialylation. *Science* **313:** 670–673.

20. Feige, M.J., S. Nath, S.R. Catharino, *et al.* 2009. Structure of the murine unglycosylated IgG1 Fc fragment. *J. Mol. Biol.* **391:** 599–608.

21. Sazinsky, S.L., R.G. Ott, N.W. Silver, *et al.* 2008. Aglycosylated immunoglobulin G1 variants productively engage activating Fc receptors. *Proc. Natl. Acad. Sci. USA* **105:** 20167–20172.

22. Stanley, P., H. Schachter & N. Taniguchi. 2009. N-Glycans. In *Essentials of Glycobiology*, ed. Ajit Varki et al. Cold Spring Harbor Laboratory Press. New York.

23. Gornik, I., G. Maravic, J. Dumic, *et al.* 1999. Fucosylation of IgG heavy chains is increased in rheumatoid arthritis. *Clin. Biochem.* **32:** 605–608.

24. Nakagawa, H., M. Hato, Y. Takegawa, *et al.* 2007. Detection of altered N-glycan profiles in whole serum from rheumatoid arthritis patients. *J. Chromatogr. B Analyt. Technol. Biomed. Life Sci.* **853:** 133–137.

25. Guo, N., Y. Liu, Y. Masuda, *et al.* 2005. Repeated immunization induces the increase in fucose content on antigen-specific IgG N-linked oligosaccharides. *Clin. Biochem.* **38:** 149–153.

26. Ferrara, C., S. Grau, C. Jager, *et al.* 2011. Unique carbohydrate-carbohydrate interactions are required for high affinity binding between Fc{gamma}RIII and antibodies lacking core fucose. *Proc. Natl. Acad. Sci. USA* **108:** 12669–12674.

27. Ferrara, C., F. Stuart, P. Sondermann, *et al.* 2006. The carbohydrate at FcgammaRIIIa Asn-162. An element required for high affinity binding to non-fucosylated IgG glycoforms. *J. Biol. Chem.* **281:** 5032–5036.

28. Shields, R.L., J. Lai, R. Keck, *et al.* 2002. Lack of fucose on human IgG1 N-linked oligosaccharide improves binding to human Fcgamma RIII and antibody-dependent cellular toxicity. *J. Biol. Chem.* **277:** 26733–26740.

29. Shinkawa, T., K. Nakamura, N. Yamane, *et al.* 2003. The absence of fucose but not the presence of galactose or bisecting *N*-acetylglucosamine of human IgG1 complex-type oligosaccharides shows the critical role of enhancing antibody-dependent cellular cytotoxicity. *J. Biol. Chem.* **278:** 3466–3473.

30. Shoji-Hosaka, E., Y. Kobayashi, M. Wakitani, *et al.* 2006. Enhanced Fc-dependent cellular cytotoxicity of Fc fusion proteins derived from TNF receptor II and LFA-3 by fucose removal from Asn-linked oligosaccharides. *J. Biochem.* **140:** 777–783.

31. Natsume, A., M. Wakitani, N. Yamane-Ohnuki, *et al.* 2005. Fucose removal from complex-type oligosaccharide enhances the antibody-dependent cellular cytotoxicity of single-gene-encoded antibody comprising a single-chain antibody linked the antibody constant region. *J. Immunol. Methods* **306:** 93–103.

32. Davies, J., L. Jiang, L.Z. Pan, *et al.* 2001. Expression of GnTIII in a recombinant anti-CD20 CHO production cell line: expression of antibodies with altered glycoforms leads to an increase in ADCC through higher affinity for FC gamma RIII. *Biotechnol. Bioeng.* **74:** 288–294.

33. Selman, M.H., E.H. Niks, M.J. Titulaer, *et al.* 2011. IgG fc N-glycosylation changes in Lambert–Eaton myasthenic syndrome and myasthenia gravis. *J. Proteome Res.* **10:** 143–152.

34. Scallon, B.J., S.H. Tam, S.G. McCarthy, *et al.* 2007. Higher levels of sialylated Fc glycans in immunoglobulin G molecules can adversely impact functionality. *Mol. Immunol.* **44:** 1524–1534.

35. Anthony, R.M., F. Nimmerjahn, D.J. Ashline, *et al.* 2008. Recapitulation of IVIG anti-inflammatory activity with a recombinant IgG Fc. *Science* **320:** 373–376.

36. Kaneko, Y., F. Nimmerjahn, M.P. Madaio, & J.V. Ravetch. 2006. Pathology and protection in nephrotoxic nephritis is determined by selective engagement of specific Fc receptors. *J. Exp. Med.* **203:** 789–797.

37. Espy, C., W. Morelle, N. Kavian, *et al.* 2011. Sialylation levels of anti-proteinase 3 antibodies are associated with the activity of granulomatosis with polyangiitis (Wegener's). *Arthritis Rheum* **63:** 2105–2115.

38. van de Geijn, F.E., M. Wuhrer, M.H. Selman, *et al.* 2009. Immunoglobulin G galactosylation and sialylation are associated with pregnancy-induced improvement of rheumatoid arthritis and the postpartum flare: results from a large prospective cohort study. *Arthritis Res. Ther.* **11:** R193.

39. Nimmerjahn, F., R.M. Anthony & J.V. Ravetch. 2007. Agalactosylated IgG antibodies depend on cellular Fc receptors for *in vivo* activity. *Proc. Natl. Acad. Sci. USA* **104:** 8433–8437.

40. Imbach, P., S. Barandun, V. d'Apuzzo, *et al.* 1981. High-dose intravenous gammaglobulin for idiopathic thrombocytopenic purpura in childhood. *Lancet* **1:** 1228–1231.

41. Debre, M., M.C. Bonnet, W.H. Fridman, *et al.* 1993. Infusion of Fc gamma fragments for treatment of children with acute immune thrombocytopenic purpura. *Lancet* **342:** 945–949.

42. Nimmerjahn, F. & J.V. Ravetch. 2007. The antiinflammatory activity of IgG: the intravenous IgG paradox. *J. Exp. Med.* **204:** 11–15.

43. Nimmerjahn, F. & J.V. Ravetch. 2008. Anti-inflammatory actions of intravenous immunoglobulin. *Annu. Rev. Immunol.* **26:** 513–533.

44. Korganow, A.S., H. Ji, S. Mangialaio, *et al.* 1999. From systemic T cell self-reactivity to organ-specific autoimmune disease via immunoglobulins. *Immunity* **10:** 451–461.

45. Samuelsson, A., T.L. Towers, & J.V. Ravetch. 2001. Anti-inflammatory activity of IVIG mediated through the inhibitory Fc receptor. *Science* **291:** 484–486.

46. Crow, A.R., S. Song, J. Freedman, *et al.* 2003. IVIg-mediated amelioration of murine ITP via FcgammaRIIB is independent of SHIP1, SHP-1, and Btk activity. *Blood* **102:** 558–560.

47. Crow, A.R., S. Song, J.W. Semple, *et al.* 2003. IVIG induces dose-dependent amelioration of ITP in rodent models. *Blood* **101:** 1658–1659.

48. Anthony, R.M. & J.V. Ravetch. 2010. A novel role for the IgG Fc glycan: the anti-inflammatory activity of sialylated IgG Fcs. *J. Clin. Immunol.* **30** Suppl 1: S9–S14.

49. Anthony, R.M. & F. Nimmerjahn. 2011. The role of differential IgG glycosylation in the interaction of antibodies with FcgammaRs *in vivo. Curr. Opin. Organ Transplant.* **16:** 7–14.

50. Tackenberg, B., I. Jelcic, A. Baerenwaldt, *et al.* 2009. Impaired inhibitory Fcgamma receptor IIB expression on B cells in chronic inflammatory demyelinating polyneuropathy. *Proc. Natl. Acad. Sci. USA* **106:** 4788–4792.

51. Tackenberg, B., F. Nimmerjahn & J.D. Lunemann. 2010. Mechanisms of IVIG efficacy in chronic inflammatory demyelinating polyneuropathy. *J. Clin. Immunol.* **30** Suppl 1: S65–S69.

52. Bruhns, P., A. Samuelsson, J.W. Pollard & J.V. Ravetch. 2003. Colony-stimulating factor-1-dependent macrophages are responsible for IVIG protection in antibody-induced autoimmune disease. *Immunity* **18:** 573–581.

53. Anthony, R.M., F. Wermeling, M.C. Karlsson & J.V. Ravetch. 2008. Identification of a receptor required for the anti-inflammatory activity of IVIG. *Proc. Natl. Acad. Sci. USA* **105:** 19571–19578.

54. Mombaerts, P., J. Iacomini, R.S. Johnson, *et al.* 1992. RAG-1-deficient mice have no mature B and T lymphocytes. *Cell* **68:** 869–877.

55. Mebius, R.E. & G. Kraal. 2005. Structure and function of the spleen. *Nat. Rev. Immunol.* **5:** 606–616.

56. Kraal, G. & R. Mebius. 2006. New insights into the cell biology of the marginal zone of the spleen. *Int. Rev. Cytol.* **250:** 175–215.

57. Mebius, R.E., M.A. Nolte & G. Kraal. 2004. Development and function of the splenic marginal zone. *Crit. Rev. Immunol.* **24:** 449–464.

58. Crocker, P.R. 2005. Siglecs in innate immunity. *Curr. Opin. Pharmacol.* **5:** 431–437.

59. Crocker, P.R., S. Kelm, C. Dubois, *et al.* 1991. Purification and properties of sialoadhesin, a sialic acid-binding receptor of murine tissue macrophages. *Embo. J.* **10:** 1661–1669.

60. Crocker, P.R., J.C. Paulson & A. Varki. 2007. Siglecs and their roles in the immune system. *Nat. Rev. Immunol.* **7:** 255–266.

61. Crocker, P.R. & A. Varki. 2001. Siglecs, sialic acids and innate immunity. *Trends Immunol.* **22:** 337–342.

62. Crocker, P.R., Z. Werb, S. Gordon, *et al.* 1990. Ultrastructural localization of a macrophage-restricted sialic acid binding hemagglutinin, SER, in macrophage-hematopoietic cell clusters. *Blood* **76:** 1131–1138.

63. Chen, Y., T. Pikkarainen, O. Elomaa, *et al.* 2005. Defective microarchitecture of the spleen marginal zone and impaired response to a thymus-independent type 2 antigen in mice lacking scavenger receptors MARCO and SR-A. *J. Immunol.* **175:** 8173–8180.

64. Karlsson, M.C., R. Guinamard, S. Bolland, *et al.* 2003. Macrophages control the retention and trafficking of B lymphocytes in the splenic marginal zone. *J. Exp. Med.* **198:** 333–340.

65. Palecanda, A., J. Paulauskis, E. Al-Mutairi, *et al.* 1999. Role of the scavenger receptor MARCO in alveolar macrophage binding of unopsonized environmental particles. *J. Exp. Med.* **189:** 1497–1506.

66. van der Laan, L.J., E.A. Dopp, R. Haworth, *et al.* 1999. Regulation and functional involvement of macrophage scavenger receptor MARCO in clearance of bacteria *in vivo*. *J. Immunol.* **162:** 939–947.

67. Galustian, C., C.G. Park, W. Chai, *et al.* 2004. High and low affinity carbohydrate ligands revealed for murine SIGN-R1 by carbohydrate array and cell binding approaches, and differing specificities for SIGN-R3 and langerin. *Int. Immunol.* **16:** 853–866.

68. Kang, Y.S., Y. Do, H.K. Lee, *et al.* 2006. A dominant complement fixation pathway for pneumococcal polysaccharides initiated by SIGN-R1 interacting with C1q. *Cell* **125:** 47–58.

69. Kang, Y.S., J.Y. Kim, S.A. Bruening, *et al.* 2004. The C-type lectin SIGN-R1 mediates uptake of the capsular polysaccharide of Streptococcus pneumoniae in the marginal zone of mouse spleen. *Proc. Natl. Acad. Sci. USA* **101:** 215–220.

70. Kang, Y.S., S. Yamazaki, T. Iyoda, *et al.* 2003. SIGN-R1, a novel C-type lectin expressed by marginal zone macrophages in spleen, mediates uptake of the polysaccharide dextran. *Int. Immunol.* **15:** 177–186.

71. Lanoue, A., M.R. Clatworthy, P. Smith, *et al.* 2004. SIGN-R1 contributes to protection against lethal pneumococcal infection in mice. *J. Exp. Med.* **200:** 1383–1393.

72. Geijtenbeek, T.B., P.C. Groot, M.A. Nolte, *et al.* 2002. Marginal zone macrophages express a murine homologue of DC-SIGN that captures blood-borne antigens *in vivo*. *Blood* **100:** 2908–2916.

73. Park, C.G., K. Takahara, E. Umemoto, *et al.* 2001. Five mouse homologues of the human dendritic cell C-type lectin, DC-SIGN. *Int. Immunol.* **13:** 1283–1290.

74. Wieland, C.W., E.A. Koppel, J. den Dunnen, *et al.* 2007. Mice lacking SIGNR1 have stronger T helper 1 responses to mycobacterium tuberculosis. *Microbes. Infect.* **9:** 134–141.

75. Geijtenbeek, T.B., D.S. Kwon, R. Torensma, *et al.* 2000. DC-SIGN, a dendritic cell-specific HIV-1-binding protein that enhances trans-infection of T cells. *Cell* **100:** 587–597.

76. Geijtenbeek, T.B., R. Torensma, S.J. van Vliet, *et al.* 2000. Identification of DC-SIGN, a novel dendritic cell-specific ICAM-3 receptor that supports primary immune responses. *Cell* **100:** 575–585.

77. Geijtenbeek, T.B., G.C. van Duijnhoven, S.J. van Vliet, *et al.* 2002. Identification of different binding sites in the dendritic cell-specific receptor DC-SIGN for intercellular adhesion molecule 3 and HIV-1. *J. Biol. Chem.* **277:** 11314–11320.

78. Geijtenbeek, T.B., S.J. Van Vliet, E.A. Koppel, *et al.* 2003. Mycobacteria target DC-SIGN to suppress dendritic cell function. *J. Exp. Med.* **197:** 7–17.

79. Gringhuis, S.I., J. den Dunnen, M. Litjens, *et al.* 2009. Carbohydrate-specific signaling through the DC-SIGN signalosome tailors immunity to mycobacterium tuberculosis, HIV-1 and Helicobacter pylori. *Nat. Immunol.* **10:** 1081–1088.

80. Gringhuis, S.I., J. den Dunnen, M. Litjens, *et al.* 2007. C-type lectin DC-SIGN modulates Toll-like receptor signaling via Raf-1 kinase-dependent acetylation of transcription factor NF-kappaB. *Immunity* **26:** 605–616.

81. Koppel, E.A., K.P. van Gisbergen, T.B. Geijtenbeek & Y. van Kooyk. 2005. Distinct functions of DC-SIGN and its homologues L-SIGN (DC-SIGNR) and mSIGNR1 in pathogen recognition and immune regulation. *Cell. Microbiol.* **7:** 157–165.

82. Tailleux, L., O. Schwartz, J.L. Herrmann, *et al.* 2003. DC-SIGN is the major mycobacterium tuberculosis receptor on human dendritic cells. *J. Exp. Med.* **197:** 121–127.

83. van Kooyk, Y. & T.B. Geijtenbeek. 2003. DC-SIGN: escape mechanism for pathogens. *Nat. Rev. Immunol.* **3:** 697–709.

84. van Liempt, E., C.M. Bank, P. Mehta, *et al.* 2006. Specificity of DC-SIGN for mannose- and fucose-containing glycans. *FEBS Lett.* **580:** 6123–6131.

85. Anthony, R.M., T. Kobayashi, F. Wermeling & J.V. Ravetch. 2011. Intravenous gammaglobulin suppresses inflammation through a novel T(H)2 pathway. *Nature* **475:** 110–113.

86. Schaefer, M., N. Reiling, C. Fessler, *et al.* 2008. Decreased pathology and prolonged survival of human DC-SIGN transgenic mice during mycobacterial infection. *J. Immunol.* **180:** 6836–6845.

87. Neill, D.R., S.H. Wong, A. Bellosi, *et al.* 2010. Nuocytes represent a new innate effector leukocyte that mediates type-2 immunity. *Nature* **464:** 1367–1370.

88. Wang, J., C.I. Balog, K. Stavenhagen, *et al.* 2011. Fc-glycosylation of IgG1 is modulated by B-cell stimuli. *Mol. Cell. Proteomics* **10:** M110 004655-1–M110 004655-12.

89. Chintalacharuvu, S.R. & S.N. Emancipator. 1997. The glycosylation of IgA produced by murine B cells is altered by Th2 cytokines. *J. Immunol.* **159:** 2327–2333.

90. Dodel, R., H. Hampel, C. Depboylu, *et al.* 2002. Human antibodies against amyloid beta peptide: a potential treatment for Alzheimer's disease. *Ann. Neurol.* **52:** 253–256.

91. Rinne, J.O., D.J. Brooks, M.N. Rossor, *et al.* 2010. 11C-PiB PET assessment of change in fibrillar amyloid-beta load in patients with Alzheimer's disease treated with bapineuzumab: a phase 2, double-blind, placebo-controlled, ascending-dose study. *Lancet Neurol.* **9:** 363–372.

92. Fillit, H., G. Hess, J. Hill, *et al.* 2009. IV immunoglobulin is associated with a reduced risk of Alzheimer disease and related disorders. *Neurology* **73:** 180–185.

93. Relkin, N.R., P. Szabo, B. Adamiak, *et al.* 2009. 18-month study of intravenous immunoglobulin for treatment of mild Alzheimer disease. *Neurobiol. Aging* **30:** 1728–1736.

94. Fillit, H. 2004. Intravenous immunoglobulins for Alzheimer's disease. *Lancet Neurol.* **3:** 704.

95. Hack, C.E. & P. Scheltens. 2004. Intravenous immunoglobulins: a treatment for Alzheimer's disease? *J. Neurol. Neurosurg. Psychiatr.* **75:** 1374–1375.

96. Saphire, E.O., P.W. Parren, R. Pantophlet, *et al.* 2001. Crystal structure of a neutralizing human IGG against HIV-1: a template for vaccine design. *Science* **293:** 1155–1159.

Ann. N.Y. Acad. Sci. ISSN 0077-8923

ANNALS OF THE NEW YORK ACADEMY OF SCIENCES
Issue: *Glycobiology of the Immune Response*

The effect of galectins on leukocyte trafficking in inflammation: sweet or sour?

Dianne Cooper, Asif J. Iqbal, Beatrice R. Gittens, Carmela Cervone, and Mauro Perretti

William Harvey Research Institute, Barts and The London School of Medicine, Queen Mary University of London, London, United Kingdom

Address for correspondence: Dianne Cooper or Mauro Perretti, William Harvey Research Institute, Barts and The London School of Medicine and Dentistry, Queen Mary University of London, Charterhouse Square, London EC1M 6BQ, United Kingdom. d.cooper@qmul.ac.uk, m.perretti@qmul.ac.uk

The trafficking of leukocytes from the blood stream to the surrounding tissue is a fundamental feature of an inflammatory response. Although many of the adhesion molecules and chemokines that direct leukocyte trafficking have been identified, there is still much to be discovered, particularly with regard to the persistence of leukocyte infiltrates in chronic inflammation. Elucidating the molecular mechanisms involved in this process is critical to understanding and treating inflammatory pathologies. Recent studies have identified members of the galectin family as immunoregulatory proteins. Included among the actions of galectins are modulatory effects, both positive and negative, on leukocyte recruitment. The focus of this review is to summarize current knowledge on the role of galectins in leukocyte trafficking during inflammation. A better understanding of the function of this family of endogenous lectins will open new avenues for innovative drug discovery.

Keywords: galectin; leukocyte; inflammation; adhesion molecule

Introduction

Inflammation is elicited as a response to tissue injury or infection. It is generally a protective response of the host that serves to maintain tissue homeostasis. The symptoms of inflammation—pain, fever, redness, swelling, and, in chronic cases, loss of function—occur as a result of a complex sequence of events that take place at the site of inflammation and systemically. The recruitment of leukocytes from the blood stream to the surrounding tissue is the hallmark of an inflammatory response. The first cells to traverse the endothelial barrier are neutrophils, the foot soldiers of the inflammatory response. These cells are armed with powerful enzymes and oxidants with antimicrobial properties. This is followed by a wave of mononuclear leukocytes that, depending on the initiating stimulus, might be T lymphocytes or monocytes. In acute inflammation, the inflammatory response resolves by the coordinated release of pro-resolution mediators that terminate neutrophil recruitment and promote the non-phlogisitic clearance of the spent neutrophils by tissue macrophages;

the end result being a return to tissue homeostasis.[1,2] In some circumstances, restoration is not achieved, and the inflammation becomes chronic in nature. Chronic inflammation is characterized by persistent leukocytic infiltrates that differ in nature depending on the initiating stimulus of the inflammation. As leukocyte trafficking is critical to mounting an inflammatory response and persistence of chronic inflammation, the mechanisms by which leukocytes traverse the endothelium has been the focus of intense research over the past two decades. This has led to the identification of many of the molecular determinants involved in the paradigm that is the *leukocyte recruitment cascade* (Fig. 1). For excellent reviews on the mechanisms of leukocyte trafficking, the reader is directed to Ley *et al.*[3] and Luster *et al.*[4]

Although many of the adhesion molecules and chemokines that direct leukocyte trafficking have been identified, there is still much to be understood, particularly with regard to chronic inflammatory pathologies and the persistence of leukocytic infiltrates. Recent studies have identified members of the

doi: 10.1111/j.1749-6632.2011.06291.x
Ann. N.Y. Acad. Sci. 1253 (2012) 181–192 © 2012 New York Academy of Sciences.

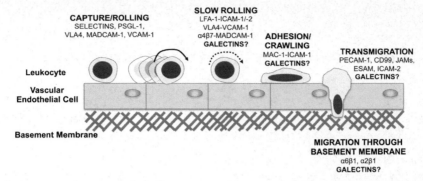

Figure 1. The leukocyte recruitment cascade. Leukocytes must undergo a series of sequential steps to migrate from the blood stream into the surrounding tissue during an inflammatory response. These stages are shown along with the key adhesion molecules involved. Stages of the cascade that may be influenced by galectins are also indicated. PSGL-1, P-selectin glycoprotein ligand-1; VLA4, very late antigen 4 ($\alpha_4\beta_1$ integrin); MAdcam-1, mucosal vascular addressin cell adhesion molecule-1; VCAM-1, vascular cell adhesion molecule-1; LFA-1, lymphocyte function-associated antigen-1 ($\alpha_L\beta_2$ integrin); ICAM-1/-2, intercellular adhesion molecule-1/-2; MAC-1, macrophage antigen-1 ($\alpha_M\beta_2$ integrin); PECAM-1, platelet/endothelial-cell adhesion molecule-1; JAM, junctional adhesion molecule; ESAM, endothelial cell selective adhesion molecule.

galectin family as exquisite modulators of the immune response; included among their many actions are effects, both negative and positive, on leukocyte recruitment in both acute and chronic inflammatory settings (Fig. 2). This review will focus on what is known to date about the galectin family members and their role in leukocyte trafficking during inflammation. For a recent review on the role of galectins in leukocyte migration across both vascular and lymphatic endothelium, the reader is referred to the recent excellent review from Thiemann and Baum.[5]

Galectins

Galectins are a family of evolutionary conserved carbohydrate-binding animal lectins. To date, 15 members of this growing family of proteins have been identified in a wide variety of tissues across species.[6–8] All galectins share close sequence homology in their carbohydrate recognition domain (CRD) but exhibit different affinities for different saccharide ligands. The affinity with which they bind saccharide ligands is altered depending on the ability of the galectin to form supramolecular arrangements or lattices.[9] In terms of structural classification, the galectins can be subdivided into three distinct groups: (1) proto-type galectins, which have one CRD and are capable of homodimerization (Gal-1, Gal-2, Gal-5, Gal-7, Gal-10, Gal-13, Gal-14, and Gal-15); (2) chimera type galectins, of which Gal-3 is the only member, that contain a single CRD

with an extended N-terminus; and (3) tandem repeat type galectins, which have two distinct CRDs joined by a short linker peptide (Gal-4, Gal-6, Gal-8, Gal-9, and Gal-12).[6,10,11] Galectins can be bi- or multi-valent in terms of their ligand-binding activity, which accounts for their ability to cross-link cell surface glycoproteins.

In contrast to the selectins, galectin binding to carbohydrates is calcium-independent.[12] The majority of galectins bind to *N*-acetyllactosamine (Galβ1,3GlcNAc or Galβ1,4GlcNAc), a common disaccharide found on many *N*- or *O*-linked glycans.[13] One exception is Gal-10, which can bind mannose-containing saccharides.[14] Binding to individual lactosamine units is of relatively low affinity (dissociation constant in the 90–100 μM range), although the arrangement of lactosamine in repeating chains (polylactosamines) increases binding affinity by 100-fold.[15] Biological responses vary quite significantly among galectins. According to structural analysis of CRD of various galectins, slight variations exist in their carbohydrate-binding specificities.[16] It is fair to speculate that these subtle variations may contribute to the distinct set of responses evoked by individual members of the galectin family; however, further analyses are necessary for complete confirmation.

Galectins are expressed both intracellularly (cytosol, nucleus, and membrane compartments) and extracellularly; however, none possess a secretion signal peptide that would direct transport through

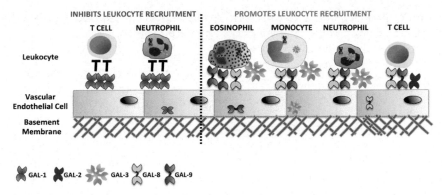

Figure 2. The effects of galectins on leukocyte recruitment. Members of the galectin family have been shown to either positively or negatively regulate leukocyte recruitment. A schematic representation of the galectin family members and the specific leukocyte subtypes they affect is shown.

the classical endoplasmic reticulum–Golgi apparatus secretory pathway.[10] Instead, some galectins are secreted via a novel/nonclassical secretory pathway similar to those used by interleukin-1β (IL-1β), annexin A1,[17–19] and fibroblast growth factor.[20] After their secretion into the extracellular environment, galectins can crosslink cell surface receptors, leading to the activation of signaling pathways involved in the modulation of a number of cellular processes, including proliferation, differentiation, apoptosis, and cytokine secretion.[21]

Over the past decade or so the volume of literature relating to the galectins has dramatically increased and continues to do so. The reason behind this growing interest relates to the broad range of functions displayed by various members of the galectin family. Essentially, galectins are key regulators in a number of physiological and pathological processes, including cell–cell adhesion, cell growth, apoptosis, and inflammatory and autoimmune responses.

Galectin expression in the inflammatory setting

To have a role in leukocyte trafficking during inflammation galectins must be present at the site of inflammation and available to interact with target cells, such as leukocytes, and the vascular endothelium. Vascular endothelial cells (ECs) express Gal-1, Gal-3, Gal-8, and Gal-9 under basal conditions *in vitro*, and these patterns of expression can be modulated by endothelial stimulation/activation with different agents.[22,23] As an example, Gal-1 expression is increased in human umbilical vein EC (HUVEC) upon culture in human serum, compared

to quiescent freshly isolated cells, and in response to pro-inflammatory cytokines,[22,23] whereas Gal-9 expression increases in response to IFN-γ and double-stranded RNA.[24,25] Endothelial Gal-3 expression is modulated at the protein level in response to IL-1β, and at the mRNA level in response to AGE-casein (in human cells) and to a high-fat diet (in murine ECs).[26–28] As well as expressing galectins intracellularly and on their surface, ECs *in vitro* secrete Gal-1, Gal-3, and Gal-9 into the extracellular environment; presumably, *in vivo* these proteins would be able to interact with leukocytes within the vessel (Ref. 29, unpublished observation). *In vivo*, endothelial expression of Gal-1 is detectable in numerous organs, with Gal-3, Gal-8, and Gal-9 being more variable, at least in healthy tissue.[23] How expression is altered during pathological states remains to be determined, but is of importance in understanding the role of galectins in pathology. Endothelial Gal-1 levels are increased in inflamed human lymph nodes,[22] while Gal-9 has been detected in the endothelium of tissues taken from patients with Sjogren's syndrome and rheumatoid synovium.[24,30] Circulating galectins are detectable in the serum from patients with rheumatoid arthritis, Behcet's disease, and heart failure, with detectable levels of Gal-3 greater than 50 ng/mL.[31–33]

Galectin expression in leukocytes is largely determined by the differentiation status of the cell. Gal-1 is not detectable in human peripheral blood leukocytes,[34] although whether leukocytes express Gal-1 once they have left the vasculature and migrated to the site of injury/infection remains to be fully determined. In T cells, Gal-1 is induced upon stimulation of peripheral blood T cells with anti-CD3 or anti-CD28 plus phorbol myristate acetate, with peak values appearing at three to five days.[35] Gal-3 is expressed in the majority of human and murine immune cells, albeit at low levels in human peripheral blood lymphocytes and murine neutrophils; in both species, Gal-3 is predominantly produced by macrophages, with expression correlating with monocyte differentiation.[36,37] Expression of Gal-3 is increased in human neutrophils upon adhesion to the endothelium,[38] which also coincides with relocalization of Gal-3 to the plasma membrane in ECs.[38] In accordance with its role in allergic inflammation, Gal-3 is also expressed by CD4[+] T cells that have infiltrated the dermis of patients with atopic dermatitis.[39] Gal-9 is detectable in peripheral blood leukocytes,[40] as well as leukocytes at the site of inflammation.[30]

One of the complexities of galectin biology relates to whether the responses attributed to galectins occur as a result of the protein acting intra- or extracellularly. Galectins clearly function extracellularly, as is evident from studies in which the recombinant protein is added to the extracellular environment and binds a cell surface–expressed receptor.[41–43] This is apparent in the effects of Gal-1 binding to receptors such as CD43 on T cells to influence migration.[29] Galectin binding to the surface of leukocytes has been demonstrated in numerous studies and is associated with a range of downstream effects.[34,44,45] The site of action is less clear in studies carried out in galectin knockout animals, in which a given galectin is absent throughout all cellular compartments, and thus the systemic effects of a global loss of the protein are likely more apparent. The issue is further complicated by the fact that some galectins may be internalized; for example, Gal-3 can bind and be internalized along with $\beta 1$ integrin.[46] Further support for an extracellular function comes from studies showing that immobilization of a galectin on the cell surface or extracellular matrix enhances or modifies activity; for example, this is evident for Gal-1 expressed on the surface of HUVEC[29] and for Gal-8.[47] It is likely that galectins can modulate leukocyte trafficking through multiple mechanisms that occur as a result of both extra- and intracellular actions of the proteins. Use of neutralizing antibodies to specific galectins can shed light on this intriguing aspect—intracellular versus extracellular site of action—of their biology.

Regulation of leukocyte trafficking by galectins

Neutrophils

Galectins affect neutrophil behavior and trafficking both *in vitro* and *in vivo*, with Gal-3, Gal-8, and Gal-9 enhancing and Gal-1 inhibiting neutrophil trafficking at various points of the leukocyte recruitment cascade.

Galectin-1. Previous data generated within our laboratory[48] demonstrated that Gal-1 inhibits PMN rolling and extravasation in inflamed postcapillary venules of wild-type mice, suggesting that a novel "anti-inflammatory loop" may exist in which Gal-1 is provided by the endothelium to target the migrating PMN, thus reducing PMN extravasation. Inhibition of neutrophil emigration by addition of exogenous Gal-1 was also shown in models of IL-1β- and zymosan-induced acute peritonitis, as well as phospholipase A$_2$- and carrageenan-induced paw edema.[48–51]

Because galectins may function both intra- and extracellularly, it is important to consider the actions of both the endogenous and the exogenous protein as a means to gain mechanistic clues. A role for endogenous Gal-1 in leukocyte trafficking has been demonstrated. Knockdown of endothelial Gal-1 resulted in increased numbers of neutrophils being captured and subsequently rolling on TNF-α–stimulated HUVEC under flow. And knockout of Gal-1 in mice resulted in significantly increased numbers of leukocytes emigrating from the microcirculation in IL-1β–inflamed cremasters.[45] A direct mechanism for these inhibitory effects of Gal-1 on neutrophil trafficking is still to be elucidated, but it may be partly due to a distinct modulation of adhesion molecule expression on the neutrophil.[45,49]

Galectin-3. Gal-3 is one of a small number of molecules that have been proposed to act as soluble cell-to-cell and cell-to-matrix adhesion proteins. Exogenous Gal-3 promotes human neutrophil adherence to EC monolayers, laminin, and fibronectin *in vitro*.[52–54] This effect was dependent on the CRD and amino terminal of Gal-3 and was temperature- and Ca^{2+}/Mg^{2+}-dependent, suggesting that Gal-3 oligomerizes at the cell surface.[52] Studies investigating the putative Gal-3 receptor on neutrophils, CD66b, found that cross-linking of antibodies binding to this protein resulted in increased adhesion of the neutrophils to ECs and release of IL-8 from intracellular stores.[55–57] This mechanism of cross-linking has also been suggested for Gal-3 interactions with neutrophils and the extracellular cell matrix protein, laminin, where FITC-labeled Gal-3 was shown to aggregate on the neutrophil cell surface.[52] Imaging studies have shown that Gal-3 clusters are concentrated at tricellular corners of the endothelium and adherent neutrophils, points of the vascular cell wall at which neutrophils are known to preferentially transmigrate;[58] such data further underscore a direct role for Gal-3 as an adhesion molecule.

In vivo, Gal-3 has been shown to accumulate in the alveolar space in a murine model of *Streptococcus* infection, and this was closely correlated with the onset of neutrophils to the area.[54] Increased levels of Gal-3 were observed in both alveolar macrophages and alveolar vascular ECs, implicating both cell types as a potential source of the increased Gal-3. Although emigrated neutrophils expressed little if any Gal-3, significant levels of protein were bound to the neutrophil surface. In a similar study, Nieminen *et al.*[59] found that there was a reduction in the number of neutrophils recruited to the alveolar space in Gal-3 knockout mice at 24 hours after infection; and this phenotype was restored by administration of exogenous Gal-3.[59] Both studies reported no effect on mice infected with *Escherichia*, suggesting that Gal-3 facilitates β_2 integrin-independent migration of neutrophils to the infected alveoli.

The effects of Gal-3 in other models of neutrophil recruitment are less clear, with Colnot *et al.*[60] reporting increased recruitment to the peritoneum of Gal-3 knockout mice at four days after thioglycollate administration, whereas Hsu *et al.*[61] observed no difference in neutrophil numbers when using a model of thioglycollate-induced peritonitis. It is clear that different inflammagens might activate distinct pathways in the peritoneal cavity[62] and thus would be variably susceptible to the modulatory properties of Gal-3 or indeed other galectins. In other words, Gal-3 functions in particular inflammatory settings to promote neutrophil recruitment either through its direct actions as an adhesion molecule or through its ability to function as a chemoattractant, as observed in the murine air pouch.[63]

Galectin-8. A role for Gal-8 in mediating neutrophil adhesion has been identified *in vitro*, although whether this function is also sustained *in vivo* has yet to be established. Gal-8 has been found to enhance neutrophil adhesion to tissue culture plates and to accelerate processing of pro-matrix metalloproteinase-9 to its active form—matrix metalloproteinase-9—an event that might be relevant for neutrophil migration.[64] Neutrophil adhesion mediated by Gal-8 was inhibited by blocking antibodies to α_M integrin (CD11b) or by abolishing the sugar-binding capabilities of the C-terminal CRD. Soluble Gal-8 has also been shown to increase neutrophil binding to HUVEC four-fold, at micromolar concentrations in a static adhesion assay.[65]

Galectin-9. Data supporting a direct effect of Gal-9 on the trafficking of cells other than eosinophils[24] is limited, and there is conflicting evidence on whether this protein may also be chemotactic for neutrophils.[40,66] The study by Tsuboi *et al.*[66] found Gal-9 to be chemotactic for murine PMN both *in vitro* and *in vivo*; of interest, the authors made the intriguing observation that PMN recruited into the peritoneal cavity by Gal-9 demonstrated an anti-inflammatory phenotype that was linked to PGE2 production. Coculture of macrophages with Gal-9–recruited PMN resulted in over a 50% reduction in LPS-induced TNF-α production, an effect that was absent when peripheral blood or casein-recruited PMN were used. The anti-inflammatory properties of Gal-9–recruited PMN were further emphasized by an enhanced inflammatory response when mice were rendered neutropenic in a murine model of the Schwartzman reaction. Studies using human leukocytes suggest that Gal-9 is not chemoattractive for neutrophils, at least *in vitro*.[40] More recently, Gal-9 has been shown to mediate adhesion of neutrophils to HUVEC in a static system, although it was tested at a single concentration of 1 μM.[65]

Eosinophils

Potential roles for galectins in allergic inflammation have been uncovered in models of atopic dermatitis and asthma in Gal-3 knockout mice[39,67] and in *in vitro* studies in which micromolar levels of soluble Gal-8 and Gal-9 increased the extent of eosinophil adhesion to HUVEC four-fold.[65]

Galectin-3. Recombinant human Gal-3 can directly support rolling and adhesion of eosinophils from allergic donors in an α_4 integrin-dependent manner, with an effect comparable to that evoked by VCAM-1.[28] Furthermore, Gal-3 also supports eosinophil rolling and adhesion on IL-1β-stimulated HUVEC, with a function identified for both endothelial- and eosinophil-derived Gal-3; in fact, preincubation of either HUVEC or eosinophils with a Gal-3–blocking antibody markedly reduced eosinophil rolling and adhesion. As is the case for neutrophils, Gal-3 also cross-links CD66b on the surface of eosinophils leading to increased adhesion, superoxide generation, and degranulation.[68]

In vivo studies in Gal-3 knockout mice have found significantly lower numbers of eosinophils recruited to the lungs and dermis in models of allergic airway inflammation and atopic dermatitis, respectively. This may indicate a direct effect for Gal-3 on eosinophil trafficking or be a result of the Th2-promoting function of Gal-3; serum IgE and IL-4 (Th2 cytokine) levels are reduced in Gal-3 knockout mice compared to their wild-type counterparts.[39,67]

Galectin-8. Gal-8 enhances eosinophil adhesion to HUVEC *in vitro*, an effect likely due to its ability to bind numerous integrins on the surface of the leukocyte.[65]

Galectin-9. Gal-9 was first identified as a potent T cell–derived chemoattractant of eosinophils.[40] Gal-9 was also shown to directly act on eosinophils, promoting aggregation and superoxide production, and was thus identified as a novel activator of eosinophils.[69] Although much of the evidence regarding Gal-9 and eospinophils suggests a predominant chemotactic effect, IFN-γ–induced Gal-9 expression in HUVEC was found to support adhesion of an eosinophil cell line;[24] similarly, incubation of eosinophils with soluble Gal-9 enhanced their adhesion to HUVEC monolayers.[65] A strong correlation between Gal-9 levels and the degree of eosinophil infiltrate has been identified in patients with nasal polyposis and in both acute and chronic eosinophilic pneumonia.[70,71] Interestingly, the response of eosinophils to Gal-9 may differ depending on the source; for instance, eosinophils from healthy donors succumbing to Gal-9 by entering the process of apoptosis, whereas those from eosinophilic pneumonia patients are resistant.[72] This further highlights the complex nature of galectin biology with not only galectin expression but also responsiveness being determined by the activation and/or differentiation status of the cell, with the associated degree of expression of specific and possibly multiple counter receptors.

Monocytes

Evidence is scant for a direct effect of galectins on the interaction between monocytes and the vascular endothelium; however, both Gal-1 and Gal-3 have been shown to promote monocyte chemotaxis,[63,73] whereas the tandem-repeat galectins (Gal-8 and Gal-9) promote monocyte adhesion to HUVEC, which may be due to their ability to directly bridge the monocyte to the endothelial surface.[65]

Galectin-1. Recent *in vitro* evidence has implicated a role for Gal-1 as a chemotactic factor for monocytes but not macrophages. The authors reported that this chemotactic effect of Gal-1 was sensitive to pertussis toxin, strongly suggestive of a G protein-coupled receptor, with engagement of an inhibitory G protein and linkage to activation of the p42/44 MAP kinase pathway.[73] Gal-1 can also function to alter the phenotype of macrophages recruited in response to inflammatory stimuli through negative regulation of MHC-II expression, in a p42/44 MAP kinase-dependent manner.[74] Recent *in vivo* evidence also highlights a role for exogenous Gal-1 in recruiting mononuclear phagocytic cells during the second phase ($>$24 hours) of leukocyte recruitment in a model of zymosan-induced peritonitis.[49]

Galectin-3. Gal-3 is also a chemoattractant for human monocytes and, more unusually, macrophages *in vitro*. In monocytes, Gal-3 exerts chemokinetic effects at low concentration while eliciting a classical chemotactic response of the monocyte at higher concentrations ($\geq$100 nM). This cellular response seemed independent from any known chemoattractant receptors; however, it was coupled to an increase in intracellular calcium in addition to being pertussis toxin sensitive, suggesting a role for G protein–coupled receptors. The same study reported that Gal-3 increased the number of monocytes that migrated to mouse dorsal air-pouches *in vivo*.[63] With regard to modulation of the leukocyte recruitment cascade, Gal-3 promoted monocyte adhesion to porcine but not human aortic ECs that express the xenoantigen galactose-α(1,3)galactose-β(1,4)GlcNAc-R;[75,76] Gal-3–induced activation of β2 integrins on the monocytes, which was in part responsible for their observed adhesion.

T cells

A wealth of literature exists describing the many functions of galectins on modulation of the immune response. In conjunction to effects on T cell proliferation, differentiation, and apoptosis, galectins have also been shown to modulate T cell trafficking from the vasculature to the site of inflammation. The effects of galectins on T cell behavior are dependent upon the developmental stage of the T cell, as not all T cells express the specific glycan ligands required to elicit the downstream effects of galectin binding.

Galectin-1. *In vitro* studies have indicated that Gal-1 inhibits T cell–adhesion to ECM glycoproteins.[77] Furthermore, the presence of Gal-1 on the surface of ECs specifically inhibited T cell transendothelial migration and reduced migration through the extracellular matrix;[29] while we reported that endogenous Gal-1 limits lymphocyte capture, adhesion, and rolling to activated ECs under flow conditions, demonstrated after siRNA knockdown of Gal-1 in the HUVEC.[78] The inhibitory effects of Gal-1 on T cell transmigration may be due to its ability to cluster CD43 on the cell surface thus preventing its movement to the uropod, a process that normally facilitates T cell transmigration.[29,79]

An inhibitory role for endogenous Gal-1 has been indicated in a model of contact hypersensitivity using Gal-1 knockout mice; absence of Gal-1 led to an increase in the recruitment of lymphocytes to the site of inflammation.[78] Collectively, these findings support the notion that in conjunction with its ability to limit inflammation through induction of T cell apoptosis, Gal-1 can also elicit its immunosuppressive/anti-inflammatory effects through inhibition of recruitment of this cell type.

Galectin-2. Gal-2 is structurally similar to Gal-1 but is preferentially localized to the gastrointestinal tract.[80,81] Gal-2 can bind carbohydrate residues on T cell surface proteins, such as β_1 integrin, in a manner that modulates their adhesion to extracellular matrix components. In contrast to Gal-1, which reduces T cell adhesion to both collagen and fibronectin, Gal-2 was shown to increase adhesion to fibronectin but reduce adhesion to collagen type I, effects that were mediated through binding to β_1 integrin.[81]

Galectin-3. Although Gal-3 has numerous effects on T cell biology, such as on cytokine release and induction of apoptosis, there is little evidence to indicate a role in T cell trafficking. A role for Gal-3 in thymocyte migration has however been identified, with migration of immature thymocytes to laminin increased 10-fold in the presence of Gal-3 *in vitro*. A role for endogenous Gal-3 in the exportation of thymocytes to the periphery during *Trypanosoma* infection was also identified.[82] These effects were proposed to be due, at least in part, to a negative regulation of thymocyte adhesion to the thymic microenvironment, as a neutralizing Gal-3 antibody significantly enhanced the interaction between thymocytes and thymic epithelial cells. With respect to peripheral lymphocytes, reduced migration into the peritoneum in response to thioglycollate broth has been reported in Gal-3 knockout mice, which is line with the largely pro-inflammatory actions ascribed to this galectin.[61]

Galectin-8. Immobilized Gal-8 supports adhesion and promotes spreading of Jurkat T cells through its interaction with $\alpha_1\beta_1$, $\alpha_3\beta_1$, and $\alpha_5\beta_1$ integrins.[47] Autoantibodies for Gal-8 detected in SLE patients were able to impede binding of Gal-8 to integrins and subsequently cell adhesion. This could be of interest because soluble Gal-8 promotes adhesion of human peripheral T cells to both tissue culture plates and HUVEC.[65]

Galectin-9. Recently, a role has been identified for Gal-9 in specifically promoting Th2 cell migration but inducing apoptosis of Th1 cells: these effects are due to differential expression of glycoprotein receptors and blockade of *N*-glycan availability to Gal-9 on Th2 cells by α2,6-linked sialic acid. Gal-9 can regulate the cell surface redox status of primary Th2 cells through an interaction with protein disulphide isomerase (PDI) expressed on the cell surface. Binding of Gal-9 to PDI increases the abundance of this enzyme at the cell surface, which leads to enhanced β_3 integrin-dependent migration of Th2 cells through matrigel.[83] In conjunction with its role as an eosinophil chemoattractant, the effects of Gal-9 on Th2 migration further suggest a positive role for Gal-9 in Th2-driven diseases such as asthma.

Cell glycosylation status and responsiveness to galectins

Figure 1 indicates many of the adhesion molecules known to be important in leukocyte trafficking during inflammation. What is not indicated is the fact that many of these molecules require posttranslational glycosylation to function. Such glycosylation is carried out by a number of enzymes termed glycosyltransferases and glycosidases. The activity of these enzymes "encode" cells with specific glycosylation signatures that allow binding of specific proteins that recognize carbohydrate residues. Efforts are being made to understand how the glycosylation profile of cells is altered during development, proliferation, and activation, as well as during different disease states. This will enable a greater understanding of the many functions of glycan-binding proteins such as galectins.

The effect of glycosylation on leukocyte trafficking has predominantly focused on selectins and their respective ligands, with studies showing that differential glycosylation of oligosaccharides on the cell surface results in altered T cell trafficking to the sites of inflammation.[84,85] Mice deficient in fucosyltransferase VII (Fuc-T VII), the rate-limiting enzyme for sialyl Lewis^x (sLe^x) synthesis, show a loss of all selectin ligands and subsequent defects in leukocyte trafficking,[86] whereas mice lacking core 2 β-1,6-N-acetylglucosaminyltransferases (C2GnT), the enzyme responsible for creating core 2 branches on O-glycans, exhibit a partial deficiency in selectin ligands and correspondingly impaired neutrophil trafficking.[87] Specifically, P-selectin–dependent leukocyte rolling is severely diminished in C2GnT null mice, whereas E-selectin–dependent rolling is only partially reliant on C2GnT.[88] Expression of these enzymes, at least in T cells, can be modified by the cytokine milieu that influences CD4$^+$ Th subset differentiation, for example, IL-12–induced STAT4 signaling is required for C2GnT expression whereas TCR activation alone is required for Fuc-T VII expression.[89] Ligands for L-selectin are unique in their requirement for sulphated oligosaccharides—which require two 6-sulfotransferases: N-acetylglucosamine-6-O-sulfotransferase-2 (GlcNAc6ST-2) and GlcNAc6ST-1.[90,91] Mice deficient in GlcNAc6ST-2 exhibit partial impairment of lymphocyte homing to peripheral lymph nodes as well as reduced lymphocyte counts in lymph nodes, whereas GlcNAc6ST-2/core 2 GlcNAcT double null mice have a marked reduction in lymphocyte homing and reduced lymphocyte counts as a result of significantly decreased 6-sulfo sLe^x on L-selectin counter-receptors.[92] In-

terestingly, transcripts for GlcNAc6ST-2 and core 2 GlcNAcT were induced in the high endothelial venules of salivary glands from non-obese diabetic mice, indicating a potential role for these enzymes in a chronic (and systemic) inflammatory status.[92]

Because galectins recognizes multiple galactose-β1–4-N-acetyl-lactosamine sequences displayed on N- and O-glycans,[93] the expression of glycosyltransferases responsible for this modification may determine susceptibility to the actions of Gal-1. Conversely, Gal-1 binding can be blocked by sialylation of ligands through the action of α2–6 sialyltransferase (ST6Gal-1).[94] Recently, susceptibility of Th1 and Th17-differentiated cells to Gal-1–induced apoptosis has been attributed to expression of a distinct set of cell surface glycans, whereas Th2 cells are afforded protection through increased expression of ST6Gal-1, leading to α2–6-linked sialylation of N- and O-glycans; such data further confirm the role of sialylation in determining immune cell responsiveness.[95] Although it is clear that particular glycosylation profiles are required for the effects of Gal-1 on T cell death,[41,95,96] their potential role(s) in T cell trafficking is less clear. The endothelial transmigration of T cell lines is inhibited by Gal-1 irrespective of whether the cells express C2GnT and, therefore, the O-linked glycan ligands required for Gal-1-induced cell death.[29,97]

It is interesting to note that glycosylation patterns in T cells are altered depending on activation and differentiation status,[98–100] whereas other leukocytes, such as neutrophils, display altered glycosylation patterns upon transmigration; this is the case for the sLex antigen, which no longer binds selectins once the neutrophil has entered the subendothelial space, a process that allows neutrophils to interact with dendritic cell–specific intercellular adhesion molecule-3–grabbing non-integrin on dendritic cells.[101] This desialylation is due to the activity of sialidases on the neutrophil cell surface.[102] Neutrophil activation, for example following exposure to phorbol myristate acetate, calcium ionophore, or fMet-Leu-Phe, has also been reported to reduce sialylation.[103] Enzymatic desialylation of resting neutrophils leads to increased binding of Gal-1, although this is not associated with increased phosphatidylserine exposure, as reported for desialylated HL-60 and Molt-4 cells.[104] In contrast, activated neutrophils are susceptible to Gal-1–induced phosphatidylserine exposure; this response was not dependent on desialylation, although a slight decrease in binding of a sialic acid–specific lectin upon neutrophil activation occurred. These results suggest that sialic acid capping of Gal-1 ligands may limit Gal-1 binding and its subsequent downstream effects in neutrophils.[104]

Modification of *N*-glycans by the glycosyltransferase β1–6 *N*-acetylglucosaminyltransferase V (Mgat V) results in generation of branched glycans with *N*-acetyllactosamine groups, which are suitable ligands for Gal-3.[53] Mgat V–modified *N*-glycans regulate T cell inflammatory responses and Mgat V knockout mice are more susceptible to autoimmune kidney disease and EAE.[105,106]

Integrins contain multiple Mgat V–modified glycosylation sites that affect clustering and adhesion,[107] whereas Gal-3 mediates endocytosis of β_1 integrins.[46] Mgat V also has a role in differentially regulating eosinophil and neutrophil recruitment during inflammation. Eosinophil recruitment to the airways of allergen-challenged Mgat V knockout mice are significantly attenuated, whereas, interestingly, neutrophil recruitment is significantly increased in response to numerous inflammatory stimuli.[108]

Conclusions

Galectins may exert a variety of effects on the process of white blood cell trafficking in acute and chronic inflammation. Focusing on a specific galectins allows defining whether it possesses positive or negative effects on the adhesion and migration of a given cell type. The majority of the studies listed and commented upon above have been conducted with human cells and in *in vitro* settings; we believe that more proof-of-concept studies with transgenic mice are required to detail the complex biology of galectins in inflammation.

Besides the clear complexity that arises when considering members of the galectin family, cellular responses would also vary in relation to the status of the cell. We have discussed the few instances where specific effects of a galectin were at odds when comparing resting cells versus activated or migrated cells; this may be linked to the presence, abundance, and type of galectin receptors on the cells used and the likely variation of galectin activation status.

Notwithstanding this complexity in galectin biology, we should pursue the goals of clarifying the functions of specific galectins (at least those found

in inflammatory exudates taken from human diseases) on the process of cell trafficking in acute and chronic inflammation and attempting to harness this knowledge for innovative drug discovery programs. If this second goal is successful, we and others may be able to capitalize on >20 years of research on this unique yet exciting family of proteins.

Acknowledgments

Funding to the authors' laboratory for the study of galectin biology in inflammation comes from Arthritis Research UK (Nonclinical Career Development Fellowship 18103 to DC), a BBSRC-Case PhD studentship (AI), the British Heart Foundation (PhD studentship FS/10/009/28166 BRG), and the William Harvey Research Foundation (MP and DC).

Conflicts of interest

The authors declare no conflicts of interest.

References

1. Gilroy, D.W. *et al.* 2004. Inflammatory resolution: new opportunities for drug discovery. *Nat. Rev. Drug Discov.* **3:** 401–416.
2. Serhan, C.N. *et al.* 2007. Resolution of inflammation: state of the art, definitions and terms. *FASEB. J.* **21:** 325–332.
3. Ley, K. *et al.* 2007. Getting to the site of inflammation: the leukocyte adhesion cascade updated. *Nat. Rev. Immunol.* **7:** 678–689.
4. Luster, A.D., R. Alon & U.H. von Andrian. 2005. Immune cell migration in inflammation: present and future therapeutic targets. *Nat. Immunol.* **6:** 1182–1190.
5. Thiemann, S. & L.G. Baum. 2011. The road less traveled: regulation of leukocyte migration across vascular and lymphatic endothelium by galectins. *J. Clin. Immunol.* **31:** 2–9.
6. Leffler, H. *et al.* 2004. Introduction to galectins. *Glycoconj. J.* **19:** 433–440.
7. Rabinovich, G.A. 1999. Galectins: an evolutionarily conserved family of animal lectins with multifunctional properties; a trip from the gene to clinical therapy. *Cell. Death Differ.* **6:** 711–721.
8. Rabinovich, G.A. *et al.* 2002. Galectins and their ligands: amplifiers, silencers or tuners of the inflammatory response? *Trends Immunol.* **23:** 313–320.
9. Di Lella, S. *et al.* 2011. When galectins recognize glycans: from biochemistry to physiology and back again. *Biochemistry* **50:** 7842–7857.
10. Cooper, D.N. & S.H. Barondes. 1999. God must love galectins; he made so many of them. *Glycobiology* **9:** 979–984.
11. Liu, F.T. 2000. Galectins: a new family of regulators of inflammation. *Clin. Immunol.* **97:** 79–88.

12. Hughes, R.C. 2001. Galectins as modulators of cell adhesion. *Biochimie.* **83:** 667–676.

13. Elola, M.T. *et al.* 2005. Galectin-1 receptors in different cell types. *J. Biomed. Sci.* **12:** 13–29.

14. Swaminathan, G.J. *et al.* 1999. Selective recognition of mannose by the human eosinophil Charcot-Leyden crystal protein (galectin-10): a crystallographic study at 1.8 A resolution. *Biochemistry* **38:** 13837–13843.

15. Cho, M. & R.D. Cummings. 1995. Galectin-1, a beta-galactoside-binding lectin in Chinese hamster ovary cells. I. Physical and chemical characterization. *J. Biol. Chem.* **270:** 5198–5206.

16. Brewer, C.F., M.C. Miceli & L.G. Baum. 2002. Clusters, bundles, arrays and lattices: novel mechanisms for lectin-saccharide-mediated cellular interactions. *Curr. Opin. Struct. Biol.* **12:** 616–623.

17. Auron, P.E. *et al.* 1984. Nucleotide sequence of human monocyte interleukin 1 precursor cDNA. *Proc. Natl. Acad. Sci. U.S.A.* **81:** 7907–7911.

18. March, C.J. *et al.* 1985. Cloning, sequence and expression of two distinct human interleukin-1 complementary DNAs. *Nature* **315:** 641–647.

19. Perretti, M. & F. D'Acquisto. 2009. Annexin A1 and glucocorticoids as effectors of the resolution of inflammation. *Nat. Rev. Immunol.* **9:** 62–70.

20. Burgess, W.H. & T. Maciag. 1989. The heparin-binding (fibroblast) growth factor family of proteins. *Annu. Rev. Biochem.* **58:** 575–606.

21. Liu, F.T. & G.A. Rabinovich. 2005. Galectins as modulators of tumour progression. *Nat. Rev. Cancer* **5:** 29–41.

22. Baum, L.G. *et al.* 1995. Synthesis of an endogeneous lectin, galectin-1, by human endothelial cells is up-regulated by endothelial cell activation. *Glycoconj J.* **12:** 63–68.

23. Thijssen, V.L., S. Hulsmans & A.W. Griffioen. 2008. The galectin profile of the endothelium: altered expression and localization in activated and tumor endothelial cells. *Am J. Pathol.* **172:** 545–553.

24. Imaizumi, T. *et al.* 2002. Interferon-gamma stimulates the expression of galectin-9 in cultured human endothelial cells. *J. Leukoc. Biol.* **72:** 486–491.

25. Ishikawa, A. *et al.* 2004. Double-stranded RNA enhances the expression of galectin-9 in vascular endothelial cells. *Immunol. Cell. Biol.* **82:** 410–414.

26. Darrow, A.L., R.V. Shohet & J.G. Maresh. 2011. Transcriptional analysis of the endothelial response to diabetes reveals a role for galectin-3. *Physiol. Genomics* **43:** 1144–1152.

27. Deo, P. *et al.* 2009. Upregulation of oxidative stress markers in human microvascular endothelial cells by complexes of serum albumin and digestion products of glycated casein. *J. Biochem. Mol. Toxicol.* **23:** 364–372.

28. Rao, S.P. *et al.* 2007. Galectin-3 functions as an adhesion molecule to support eosinophil rolling and adhesion under conditions of flow. *J. Immunol.* **179:** 7800–7807.

29. He, J. & L.G. Baum. 2006. Endothelial cell expression of galectin-1 induced by prostate cancer cells inhibits T-cell transendothelial migration. *Lab. Invest.* **86:** 578–590.

30. Seki, M. *et al.* 2007. Beneficial effect of galectin 9 on rheumatoid arthritis by induction of apoptosis of synovial fibroblasts. *Arthritis Rheum.* **56:** 3968–3976.

31. de Boer, R.A., L. Yu & D.J. van Veldhuisen. 2010. Galectin-3 in cardiac remodeling and heart failure. *Curr. Heart Fail Rep.* **7:** 1–8.

32. Lee, Y.J. *et al.* 2007. Serum galectin-3 and galectin-3 binding protein levels in Behcet's disease and their association with disease activity. *Clin. Exp. Rheumatol.* **25:** S41–S45.

33. Ohshima, S. *et al.* 2003. Galectin 3 and its binding protein in rheumatoid arthritis. *Arthritis Rheum.* **48:** 2788–2795.

34. Dias-Baruffi, M. *et al.* 2010. Differential expression of immunomodulatory galectin-1 in peripheral leukocytes and adult tissues and its cytosolic organization in striated muscle. *Glycobiology* **20:** 507–520.

35. Fuertes, M.B. *et al.* 2004. Regulated expression of galectin-1 during T-cell activation involves Lck and Fyn kinases and signaling through MEK1/ERK, p38 MAP kinase and p70S6 kinase. *Mol. Cell Biochem.* **267:** 177–185.

36. Liu, F.T. *et al.* 1995. Expression and function of galectin-3, a beta-galactoside-binding lectin, in human monocytes and macrophages. *Am. J. Pathol.* **147:** 1016–1028.

37. Sundblad, V., D.O. Croci & G.A. Rabinovich. 2011. Regulated expression of galectin-3, a multifunctional glycan-binding protein, in haematopoietic and non-haematopoietic tissues. *Histol. Histopathol.* **26:** 247–265.

38. Gil, C.D. *et al.* 2006. Interaction of human neutrophils with endothelial cells regulates the expression of endogenous proteins annexin 1, galectin-1 and galectin-3. *Cell Biol. Int.* **30:** 338–344.

39. Saegusa, J. *et al.* 2009. Galectin-3 is critical for the development of the allergic inflammatory response in a mouse model of atopic dermatitis. *Am. J. Pathol.* **174:** 922–931.

40. Matsumoto, R. *et al.* 1998. Human ecalectin, a variant of human galectin-9, is a novel eosinophil chemoattractant produced by T lymphocytes. *J. Biol. Chem.* **273:** 16976–16984.

41. Earl, L.A., S. Bi & L.G. Baum. 2010. N- and O-glycans modulate galectin-1 binding, CD45 signaling, and T cell death. *J. Biol. Chem.* **285:** 2232–2244.

42. Fulcher, J.A. *et al.* 2009. Galectin-1 co-clusters CD43/CD45 on dendritic cells and induces cell activation and migration through Syk and protein kinase C signaling. *J. Biol. Chem.* **284:** 26860–26870.

43. Stillman, B.N. *et al.* 2006. Galectin-3 and galectin-1 bind distinct cell surface glycoprotein receptors to induce T cell death. *J. Immunol.* **176:** 778–789.

44. Almkvist, J. *et al.* 2002. Activation of the neutrophil nicotinamide adenine dinucleotide phosphate oxidase by galectin-1. *J. Immunol.* **168:** 4034–4041.

45. Cooper, D., L.V. Norling & M. Perretti. 2008. Novel insights into the inhibitory effects of Galectin-1 on neutrophil recruitment under flow. *J. Leukoc. Biol.* **83:** 1459–1466.

46. Furtak, V., F. Hatcher & J. Ochieng. 2001. Galectin-3 mediates the endocytosis of beta-1 integrins by breast carcinoma cells. *Biochem. Biophys. Res. Commun.* **289:** 845–850.

47. Carcamo, C. *et al.* 2006. Galectin-8 binds specific beta1 integrins and induces polarized spreading highlighted by asymmetric lamellipodia in Jurkat T cells. *Exp. Cell. Res.* **312:** 374–386.

48. La, M. *et al.* 2003. A novel biological activity for galectin-1: inhibition of leukocyte-endothelial cell

interactions in experimental inflammation. *Am. J. Pathol.* **163**: 1505–1515.

49. Gil, C. D., C.E. Gullo & S.M. Oliani. 2010. Effect of exogenous galectin-1 on leukocyte migration: modulation of cytokine levels and adhesion molecules. *Int. J. Clin. Exp. Pathol.* **4**: 74–84.

50. Iqbal, A.J. *et al.* 2011. Endogenous galectin-1 and acute inflammation: emerging notion of a galectin-9 pro-resolving effect. *Am. J. Pathol.* **178**: 1201–1209.

51. Rabinovich, G.A. *et al.* 2000. Evidence of a role for galectin-1 in acute inflammation. *Eur. J. Immunol.* **30**: 1331–1339.

52. Kuwabara, I. & F.T. Liu. 1996. Galectin-3 promotes adhesion of human neutrophils to laminin. *J. Immunol.* **156**: 3939–3944.

53. Liu, F.T. & G.A. Rabinovich. 2010. Galectins: regulators of acute and chronic inflammation. *Ann. N. Y. Acad. Sci.* **1183**: 158–182.

54. Sato, S. *et al.* 2002. Role of galectin-3 as an adhesion molecule for neutrophil extravasation during streptococcal pneumonia. *J. Immunol.* **168**: 1813–1822.

55. Feuk-Lagerstedt, E. *et al.* 1999. Identification of CD66a and CD66b as the major galectin-3 receptor candidates in human neutrophils. *J. Immunol.* **163**: 5592–5598.

56. Schroder, A.K. *et al.* 2006. Crosslinking of CD66B on peripheral blood neutrophils mediates the release of interleukin-8 from intracellular storage. *Hum. Immunol.* **67**: 676–682.

57. Skubitz, K.M., K.D. Campbell & A.P. Skubitz. 1996. CD66a, CD66b, CD66c, and CD66d each independently stimulate neutrophils. *J. Leukoc. Biol.* **60**: 106–117.

58. Nieminen, J. *et al.* 2007. Visualization of galectin-3 oligomerization on the surface of neutrophils and endothelial cells using fluorescence resonance energy transfer. *J. Biol. Chem.* **282**: 1374–1383.

59. Nieminen, J. *et al.* 2008. Role of galectin-3 in leukocyte recruitment in a murine model of lung infection by Streptococcus pneumoniae. *J. Immunol.* **180**: 2466–2473.

60. Colnot, C. *et al.* 1998. Maintenance of granulocyte numbers during acute peritonitis is defective in galectin-3-null mutant mice. *Immunology* **94**: 290–296.

61. Hsu, D.K. *et al.* 2000. Targeted disruption of the galectin-3 gene results in attenuated peritoneal inflammatory responses. *Am. J. Pathol.* **156**: 1073–1083.

62. Ajuebor, M.N. *et al.* 1999. Role of resident peritoneal macrophages and mast cells in chemokine production and neutrophil migration in acute inflammation: evidence for an inhibitory loop involving endogenous IL-10. *J. Immunol.* **162**: 1685–1691.

63. Sano, H. *et al.* 2000. Human galectin-3 is a novel chemoattractant for monocytes and macrophages. *J. Immunol.* **165**: 2156–2164.

64. Nishi, N. *et al.* 2003. Galectin-8 modulates neutrophil function via interaction with integrin alphaM. *Glycobiology* **13**: 755–763.

65. Yamamoto, H. *et al.* 2008. Induction of cell adhesion by galectin-8 and its target molecules in Jurkat T-cells. *J. Biochem.* **143**: 311–324.

66. Tsuboi, Y. *et al.* 2007. Galectin-9 protects mice from the Shwartzman reaction by attracting prostaglandin E2-producing polymorphonuclear leukocytes. *Clin. Immunol.* **124**: 221–233.

67. Zuberi, R.I. *et al.* 2004. Critical role for galectin-3 in airway inflammation and bronchial hyperresponsiveness in a murine model of asthma. *Am. J. Pathol.* **165**: 2045–2053.

68. Yoon, J., A. Terada & H. Kita. 2007. CD66b regulates adhesion and activation of human eosinophils. *J. Immunol.* **179**: 8454–8462.

69. Matsumoto, R. *et al.* 2002. Biological activities of ecalectin: a novel eosinophil-activating factor. *J. Immunol.* **168**: 1961–1967.

70. Katoh, S. *et al.* 2010. Involvement of galectin-9 in lung eosinophilia in patients with eosinophilic pneumonia. *Int. Arch. Allergy Immunol.* **153**: 294–302.

71. Park, W.S. *et al.* 2011. Expression of galectin-9 by IFN-gamma stimulated human nasal polyp fibroblasts through MAPK, PI3K, and JAK/STAT signaling pathways. *Biochem. Biophys. Res. Commun.* **411**: 259–264.

72. Saita, N. *et al.* 2002. Association of galectin-9 with eosinophil apoptosis. *Int. Arch. Allergy Immunol.* **128**: 42–50.

73. Malik, R.K. *et al.* 2009. Galectin-1 stimulates monocyte chemotaxis via the p44/42 MAP kinase pathway and a pertussis toxin-sensitive pathway. *Glycobiology* **19**: 1402–1407.

74. Barrionuevo, P. *et al.* 2007. A novel function for galectin-1 at the crossroad of innate and adaptive immunity: galectin-1 regulates monocyte/macrophage physiology through a nonapoptotic ERK-dependent pathway. *J. Immunol.* **178**: 436–445.

75. Greenwald, A.G., R. Jin & T.K. Waddell. 2009. Galectin-3-mediated xenoactivation of human monocytes. *Transplantation* **87**: 44–51.

76. Jin, R. *et al.* 2006. Human monocytes recognize porcine endothelium via the interaction of galectin 3 and alpha-GAL. *J. Immunol.* **177**: 1289–1295.

77. Rabinovich, G.A. *et al.* 1999. Specific inhibition of T-cell adhesion to extracellular matrix and proinflammatory cytokine secretion by human recombinant galectin-1. *Immunology* **97**: 100–106.

78. Norling, L.V. *et al.* 2008. Inhibitory control of endothelial galectin-1 on in vitro and in vivo lymphocyte trafficking. *FASEB. J.* **22**: 682–690.

79. Manjunath, N. *et al.* 1995. Negative regulation of T-cell adhesion and activation by CD43. *Nature* **377**: 535–538.

80. Oka, T. *et al.* 1999. Identification and cloning of rat galectin-2: expression is predominantly in epithelial cells of the stomach. *Arch. Biochem. Biophys.* **361**: 195–201.

81. Sturm, A. *et al.* 2004. Human galectin-2: novel inducer of T cell apoptosis with distinct profile of caspase activation. *J. Immunol.* **173**: 3825–3837.

82. Silva-Monteiro, E. *et al.* 2007. Altered expression of galectin-3 induces cortical thymocyte depletion and premature exit of immature thymocytes during Trypanosoma cruzi infection. *Am. J. Pathol.* **170**: 546–556.

83. Bi, S. *et al.* 2011. Galectin-9 binding to cell surface protein disulfide isomerase regulates the redox environment to enhance T-cell migration and HIV entry. *Proc. Natl. Acad. Sci. U.S.A.* **108**: 10650–10655.

84. Chen, G. Y. *et al.* 2006. Interaction of GATA-3/T-bet transcription factors regulates expression of sialyl Lewis X homing receptors on Th1/Th2 lymphocytes. *Proc. Natl. Acad. Sci. U.S.A.* **103:** 16894–16899.

85. Mitoma, J. *et al.* 2007. Critical functions of N-glycans in L-selectin-mediated lymphocyte homing and recruitment. *Nat. Immunol.* **8:** 409–418.

86. Maly, P. *et al.* 1996. The alpha(1,3)fucosyltransferase Fuc-TVII controls leukocyte trafficking through an essential role in L-, E-, and P-selectin ligand biosynthesis. *Cell* **86:** 643–653.

87. Ellies, L.G. *et al.* 1998. Core 2 oligosaccharide biosynthesis distinguishes between selectin ligands essential for leukocyte homing and inflammation. *Immunity* **9:** 881–890.

88. Sperandio, M. *et al.* 2001. Differential requirements for core2 glucosaminyltransferase for endothelial L-selectin ligand function in vivo. *J. Immunol.* **167:** 2268–2274.

89. Lim, Y.C. *et al.* 2001. IL-12, STAT4-dependent up-regulation of CD4(+) T cell core 2 beta-1,6-n-acetylglucosaminyltransferase, an enzyme essential for biosynthesis of P-selectin ligands. *J. Immunol.* **167:** 4476–4484.

90. Kawashima, H. *et al.* 2005. N-acetylglucosamine-6-O-sulfotransferases 1 and 2 cooperatively control lymphocyte homing through L-selectin ligand biosynthesis in high endothelial venules. *Nat. Immunol.* **6:** 1096–1104.

91. Uchimura, K. *et al.* 2005. A major class of L-selectin ligands is eliminated in mice deficient in two sulfotransferases expressed in high endothelial venules. *Nat. Immunol.* **6:** 1105–1113.

92. Hiraoka, N. *et al.* 2004. Core 2 branching beta1,6-N-acetylglucosaminyltransferase and high endothelial venule-restricted sulfotransferase collaboratively control lymphocyte homing. *J. Biol. Chem.* **279:** 3058–3067.

93. Stowell, S.R. *et al.* 2004. Human galectin-1 recognition of poly-N-acetyllactosamine and chimeric polysaccharides. *Glycobiology* **14:** 157–167.

94. Amano, M. *et al.* 2003. The ST6Gal I sialyltransferase selectively modifies N-glycans on CD45 to negatively regulate galectin-1-induced CD45 clustering, phosphatase modulation, and T cell death. *J. Biol. Chem.* **278:** 7469–7475.

95. Toscano, M.A. *et al.* 2007. Differential glycosylation of TH1, TH2 and TH-17 effector cells selectively regulates susceptibility to cell death. *Nat. Immunol.* **8:** 825–834.

96. Motran, C.C. *et al.* 2008. Galectin-1 functions as a Th2 cytokine that selectively induces Th1 apoptosis and promotes Th2 function. *Eur. J. Immunol.* **38:** 3015–3027.

97. Nguyen, J.T. *et al.* 2001. CD45 modulates galectin-1-induced T cell death: regulation by expression of core 2 O-glycans. *J. Immunol.* **167:** 5697–5707.

98. Blander, J.M. *et al.* 1999. Alpha(1,3)-fucosyltransferase VII and alpha(2,3)-sialyltransferase IV are up-regulated in activated CD4 T cells and maintained after their differentiation into Th1 and migration into inflammatory sites. *J. Immunol.* **163:** 3746–3752.

99. Comelli, E.M. *et al.* 2006. Activation of murine CD4 +and CD8+ T lymphocytes leads to dramatic remodeling of N-linked glycans. *J. Immunol.* **177:** 2431–2440.

100. Daniels, M.A., K.A. Hogquist & S.C. Jameson. 2002. Sweet 'n' sour: the impact of differential glycosylation on T cell responses. *Nat. Immunol.* **3:** 903–910.

101. van Gisbergen, K.P. *et al.* 2005. Neutrophils mediate immune modulation of dendritic cells through glycosylation-dependent interactions between Mac-1 and DC-SIGN. *J. Exp. Med.* **201:** 1281–1292.

102. Gadhoum, S.Z. & R. Sackstein. 2008. CD15 expression in human myeloid cell differentiation is regulated by sialidase activity. *Nat. Chem. Biol.* **4:** 751–757.

103. Cross, A.S. & D.G. Wright. 1991. Mobilization of sialidase from intracellular stores to the surface of human neutrophils and its role in stimulated adhesion responses of these cells. *J. Clin. Invest.* **88:** 2067–2076.

104. Dias-Baruffi, M. *et al.* 2003. Dimeric galectin-1 induces surface exposure of phosphatidylserine and phagocytic recognition of leukocytes without inducing apoptosis. *J. Biol. Chem.* **278:** 41282–41293.

105. Demetriou, M. *et al.* 2001. Negative regulation of T-cell activation and autoimmunity by Mgat5 N-glycosylation. *Nature* **409:** 733–739.

106. Dennis, J.W. *et al.* 2002. UDP-N-acetylglucosamine: alpha-6-D-mannoside beta1,6 N-acetylglucosaminyltransferase V (Mgat5) deficient mice. *Biochim. Biophys. Acta.* **1573:** 414–422.

107. Demetriou, M. *et al.* 1995. Reduced contact-inhibition and substratum adhesion in epithelial cells expressing GlcNAc-transferase V. *J. Cell Biol.* **130:** 383–392.

108. Bahaie, N.S. *et al.* 2011. N-glycans differentially regulate eosinophil and neutrophil recruitment during allergic airway inflammation. *J. Biol. Chem* **286:** 38231–38241.

Ann. N.Y. Acad. Sci. ISSN 0077-8923

ANNALS OF THE NEW YORK ACADEMY OF SCIENCES
Issue: *Glycobiology of the Immune Response*

Engineering cellular trafficking via glycosyltransferase-programmed stereosubstitution

Robert Sackstein

Departments of Dermatology and of Medicine, Brigham & Women's Hospital, Department of Medical Oncology, Dana Farber Cancer Institute, Harvard Medical School, Boston, Massachusetts

Address for correspondence: Robert Sackstein, M.D., Ph.D., Harvard Institutes of Medicine, 77 Ave Louis Pasteur, Room 671, Boston, MA 02115. Rsackstein@rics.bwh.harvard.edu

The proximate hurdle for cell trafficking to any anatomic site is the initial attachment of circulating cells to target tissue endothelium with sufficient strength to overcome prevailing forces of blood flow. E-selectin, an endothelial molecule that is inducibly expressed at all sites of inflammation, is a potent effector of this primary braking process. This molecule is a member of a family of C-type lectins known as selectins that bind sialofucosylated glycans displayed on either a protein (i.e., glycoprotein) or lipid (i.e., glycolipid) scaffold. On human cells, the predominant E-selectin ligand is a specialized glycoform of CD44 known as hematopoietic cell E-/L-selectin ligand (HCELL). This review focuses on the biology of HCELL/E-selectin interactions in cell migration, and discusses the utility and applicability of glycosyltransferase-programmed stereosubstitution (GPS) for glycoengineering HCELL expression. Without compromising cell viability or native phenotype, this exoglycosylation technology literally "sweetens" CD44, licensing E-selectin–dependent vascular delivery for all cell-based therapeutics.

Keywords: HCELL; GPS; mesenchymal stem cell; hematopoietic stem cell; cell migration

Introduction

Degenerative and inflammatory diseases such as osteoporosis, chronic obstructive pulmonary disease, arthritis, atherosclerosis, inflammatory bowel disease, and multiple sclerosis affect billions of people worldwide. The clinical manifestations of these chronic diseases, and also of acute inflammatory conditions such as myocardial infarction and stroke, reflect a balance between tissue destruction and repair. Most current therapies for such diseases target the inflammatory component without acting on the repair processes. Stem cell-based therapeutics offers the real opportunity to achieve tissue regeneration, either by directly contributing to formation of new cells and/or by environmental trophic effects leading to recruitment/support of other cells necessary for tissue regrowth. Depending on stem cell type (e.g., mesenchymal stem cells), administered stem cells may concomitantly blunt inflammation via potent immunomodulatory effects.[1] Although the exact mechanism(s) by which various stem cells may exert tissue repair are unsettled at present and may differ depending on both the stem cell type and the pathologic entity, the clinical realization of this potentially curative therapeutic approach depends, at the outset, on getting the pertinent cells to the sites where they are needed. Moreover, the capability to achieve patient- and disease-specific treatments through adoptive cell therapeutics employing immune effector cells (e.g., for cancer immunotherapy) or regulatory cells (e.g., for treatment of autoimmune diseases) similarly depends on delivery of appropriate cells to sites of disease. Thus, a fundamental precondition for successful implementation of all cell-based therapies is to achieve adequate tissue colonization at requisite anatomic locations.

Delivery of cells for clinical indications can be achieved by direct (local) injection into involved tissue(s), or by intravascular (i.e., systemic) administration. At first glance, direct delivery might seem

doi: 10.1111/j.1749-6632.2011.06421.x

to be the most efficient approach, especially considering that a concentrated bolus of cells could be applied to an affected area. However, local injection may actually be counterproductive to intended tissue repair and is also limited in scope: (1) By introducing pertinent cells in media suspension under hydrostatic pressure, the injection procedure could harm the delivered cells and, furthermore, could further compromise tissue integrity and disrupt incipient repair processes, thereby exacerbating the inflammatory condition *in situ*; (2) By virtue of being an invasive method, the injection needle/device (and the suspension solution) could induce target tissue damage and/or instigate collateral tissue damage; (3) Direct injection is feasible only for organs/tissues with well-defined anatomic boundaries (e.g., the heart, but not the lung); (4) The injection procedure could be technologically demanding and labor-intensive, requiring use of sophisticated delivery systems with substantial imaging support, especially for relatively inaccessible and/or fragile organs/tissues (e.g., the central nervous system); (5) Most importantly, many degenerative and inflammatory conditions are widely distributed and multifocal in nature (e.g., osteoporosis, inflammatory bowel disease, multiple sclerosis, etc.), and thus direct injection is neither practical nor effective. The vascular route of administration is mandated for these and all generalized "systemic" disorders, as well as for any tissues with problematic access and/or anatomy not amenable to local injection (e.g., the pancreas in diabetes, the lung in chronic obstructive pulmonary disease). The capacity to administer cells repeatedly with minimal effort is another important practical advantage of systemic infusion. Therefore, creation of methodologies to optimize the expression/activity of molecular effectors directing the physiologic migration of intravascularly administered cells is key to achieving the tremendous promise of all cell-based therapeutics.

The molecular basis of cell migration

Recruitment of circulating cells to any anatomic site involves a coordinated sequence of events, conventionally described as comprising four overlapping steps. Under both steady state and pathologic conditions, extravasation typically occurs in postcapillary venules under hemodynamic shear conditions of 1–4 dynes/cm^2 (Ref. 2). This process is initiated by decelerative adhesive interactions between cells in flow and vascular endothelium at the target tissue(s), characteristically involving tethering of blood-borne cells on the endothelial surface, followed by sustained cellular rolling at velocities below that of blood flow (step 1). The molecules that mediate these adhesive contacts are called "homing receptors," because, as defined historically, these structures were thought to direct tropism of circulating cells to the respective target tissue(s). From a biophysical perspective, a homing receptor functions as a molecular brake, displaying fast on–off binding kinetics with its pertinent endothelial counterreceptor(s), which is translated into cellular torque under the action of fluid shear forces.[3] Thereafter, a cascade of events ensues, characteristically precipitated by chemokines engaging their cognate ligand(s) on the surface of the blood-borne cells, resulting in G-protein–coupled activation of integrin adhesiveness (step 2). The consequent integrin attachment to endothelial coreceptors results in firm adherence (step 3), followed by endothelial transmigration (step 4).[4] This "multistep paradigm" holds that cell migration to any tissue is regulated by a discrete combination of homing receptor and chemokine receptor expression on a given circulating cell, allowing for recognition of a pertinent "traffic signal" displayed by the relevant vascular adhesive ligands and chemokines expressed within target endothelium. Although step 1 interactions are reversible, homing receptor expression is obligatory to achieve tissue infiltration, as tethering and rolling adhesive interactions are prerequisite for elaboration of all downstream events, including extravasation. Thus, maintaining and/or enforcing expression of homing receptors is required to achieve cell delivery via the vasculature to any predetermined anatomic site.

Selectins and their ligands: the sweet homing receptors

Several molecules (including CD44 and members of the integrin family [e.g., $\alpha_4\beta_7$]) are effectors of step 1 tethering and rolling interactions, but the selectins and their ligands are the most potent mediators of these adhesive events (reviewed in Ref. 4). This family of C-type (i.e., Ca^{2+}-dependent) lectins is comprised of three glycoproteins: L-selectin expressed on mature leukocytes and hematopoietic stem cells, P-selectin expressed on platelets and endothelium, and E-selectin, which is expressed only on endothelium. By definition, "homing receptors" are displayed on circulating cells. Thus, L-selectin is a homing receptor, whereas the ligands for P- and E-selectin serve as homing receptors. L-selectin expression is regulated by cell surface proteolytic cleavage (shedding) mediated by membrane-associated metalloproteases, one of which is ADAM17.[5] Although shedding of P- and E-selectin has been described, the predominant mechanism regulating expression of these molecules is transcriptional induction. Notably, E- and P-selectin are constitutively expressed in two endothelial beds: the microvasculature of skin and marrow.[6,7] Also, P-selectin (but not E-selectin) is located in alpha granules of platelets and in Weibel–Palade bodies of endothelial cells, and its surface expression can be induced via granular translocation in response to inflammatory agents such as histamine and thrombin. However, more typically, *de novo* synthesis and surface expression of these selectins is induced by inflammatory cytokines in all microvessels (including skin and marrow).

Although P-selectin and E-selectin are prominently expressed at sites of tissue injury and inflammation, it is important to draw distinction between the cytokine induction of expression of these selectins among rodents and primates. Importantly, although interleukin-1 (IL-1) and tumor necrosis factor-α (TNF-α) each induce transcription of mRNA encoding P-selectin and E-selectin in rodents, the primate P-selectin promoter lacks the relevant response elements for these cytokines and only E-selectin is transcriptionally induced.[8] In fact, experiments in transgenic mice bearing the human P-selectin gene on a murine P-selectin knock-out

background have shown that TNF administration actually decreases human P-selectin expression (i.e., decreased human P-selectin mRNA levels are observed), whereas it increases murine P-selectin expression in wild-type animals;[9] a similar pattern of decreased human P-selectin mRNA was observed in skin of the transgenic mice undergoing contact hypersensitivity reactions. This key distinction in regulation of P-selectin expression has profound implications, indicating that results obtained from both steady state and inflammatory models in mice may overemphasize the contribution(s) of P-selectin, and underemphasize the contribution(s) of E-selectin, compared to clinical reality. Thus, in humans, optimizing the expression/activity of E-selectin ligands, not P-selectin ligands, would yield the most efficient trafficking of cells to inflammatory sites.

All three selectins display Ca^{2+}-dependent binding (i.e., are C-type lectins) to sialofucosylated glycans, the prototype of which is the tetrasaccharide known as sialyl Lewis X (sLex; see Fig. 1). This structure can be displayed on glycoproteins or glycolipids; on native cell membranes, glycoprotein E-selectin ligands generally possess greater binding activity than do glycolipids under hemodynamic flow conditions (i.e., mediate slower rolling and higher resistance to detachment under shear). Both L- and P-selectin readily bind to sulfated forms of sLex, and binding of both L- and P-selectin to the glycoprotein known as P-selectin glycoprotein ligand-1 (PSGL-1) requires display of a sulfated tyrosine adjacent to an O-linked sLex structure located within the N-terminus of the protein scaffold. However,

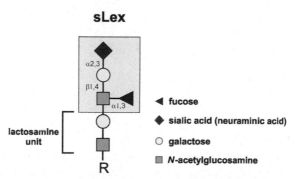

Figure 1. Structure of sialylated Lewis X (sLex). Schematic depiction of the canonical selectin binding determinant, sLex (see box), showing linkages of relevant monosaccharide units. Color key figures correspond to respective monosaccharides. The disaccharide unit consisting of galactose and N-acetylglucosamine is known as a "lactosamine" unit. In sLex, the lactosamine is capped with a sialic acid in α(2,3)-linkage to galactose; this structure is known as a "sialyllactosamine" unit. By definition, the β(1,4)-linkage between galactose and N-acetylglucosamine classifies the lactosamine unit as "type 2". The sLex structure is a terminal type 2 sialyllactosamine unit containing fucose in α(1,3)-linkage to N-acetylglucosamine.

E-selectin adherence to any of its ligands is not dependent on sulfation, and, importantly, E-selectin preferentially binds to unsulfated sLex.[2,10] Under shear conditions, binding interactions mediated by E-selectin display greater resistance to detachment and, also, slower rolling velocities than those of P-selectin and L-selectin;[11–14] these observations suggest that E-selectin receptor/ligand interactions provide optimal endothelial contact time for elaboration of steps 2 and 3 events (chemokine recognition and integrin activation, respectively).

Criteria for identification of E-selectin ligands

The identity and function of E-selectin ligands can be elucidated using both *in vivo* and *in vitro* approaches. Regarding the former, various murine models have been employed to assess the capacity of leukocyte E-selectin ligands to engage vascular E-selectin. Although these studies can offer insights under physiologic settings, rolling adhesive events attributable to E-selectin receptor/ligand interactions vary depending on numerous factors, including the animal strain employed, the inflammatory model and/or stimulus, vessel type, vessel diameter/dimension(s), and blood flow velocity/fluid shear stress. It is important to recognize that many of these "physiologic" models assess properties of nonphysiologic cells, i.e., products of genetically modified mice with either altered expression of glycosyltransferases directing synthesis of glycan determinants of E-selectin ligands or deletion of one or more scaffold proteins (e.g., PSGL-1) that may display the relevant glycan determinants. Compensatory changes frequently accompany genetic alteration(s), which could confound the identification of E-selectin ligands by unveiling and/or inducing aberrant posttranslational modifications and/or altered cell surface distribution of pertinent molecules. Thus, the data derived from these genetically altered mice do not accurately reflect the identity and function of native E-selectin ligands under authentic physiologic conditions, and, in general, these models of manipulated E-selectin ligands say more about the cellular context(s) than about the biology of cell migration. To avoid shortfalls inherent in studies of genetic mutants, one could utilize function-blocking mAb to dissect relevant contributions of (reputed) E-selectin ligands. However, to date, no such antibodies to any E-selectin lig-

and have been described. An alternative approach would be to utilize RNAi to dampen expression of relevant targets. However, gene silencing also has the potential to induce compensatory changes and, in addition, could alter cell biology in ways that could indirectly impair cellular engagement to E-selectin.

A variety of *in vitro* approaches are available to assess the ability of membrane molecules to serve as E-selectin ligands. The availability of E-selectin-immunoglobulin chimeric constructs as a probe greatly facilitates the qualitative and quantitative evaluation of E-selectin ligand activity on intact cells (by flow cytometry) or on resolved cell glycoproteins or glycolipids (by western blots and/or eastern blots, respectively).[15] However, because all selectins natively engage their cognate ligands under hemodynamic shear conditions, assays performed under non-static conditions (e.g., parallel plate flow chamber assays under fluid shear conditions[3] or Stamper–Woodruff assays[16,17]) are more reliable indicators of physiologic function than those under static conditions. In this regard, we have developed a novel assay system called the "blot rolling assay" which allows for identification of selectin ligand activity, under appropriate hydrodynamic fluid shear conditions, of glycoproteins resolved by western blot.[18–20] Although biochemical studies of isolated molecules are informative, membrane topography is a critical factor in promoting selectin receptor/ligand interactions,[21,22] and, therefore, isolation of a putative selectin ligand from the native membrane and clustering of that molecule on a support surface could either dampen or augment pertinent binding activity displayed on relevant cells. Therefore, in general, studies of ligand activity on intact cells, preferably engaging E-selectin expressed on a physiologic cell type (i.e., endothelial cells), are preferred over studies of isolated ligands displayed as nonphysiologic clusters on artificial support surfaces.

Given these considerations, the identity of a putative E-selectin ligand—or of any selectin ligand, for that matter—requires a combination of biochemical and operational criteria. First, the molecule must display the relevant glycan motif(s) that serve as binding determinants (e.g., sLex). Second, the molecule must be displayed on a native (primary) cell of interest. Third, the molecule must be capable of engaging E-selectin, in particular, under relevant physiologic fluid shear conditions. Fourth, and most importantly, expression of that ligand on a pertinent cell, ideally in absence of any other ligand(s), should endow that cell with the capacity to engage E-selectin as displayed on endothelial cells; alternatively, in settings where a cell may express additional E-selectin ligands, the contribution of the examined ligand to overall E-selectin binding should be readily apparent.

E-selectin ligands of human and murine hematopoietic stem/progenitors: gatekeepers of marrow migration

The recovery of hematopoiesis after hematopoietic stem cell transplantation depends, at the outset, on the ability of infused hematopoietic stem/progenitor cells (HSPCs) to home to bone marrow. As noted earlier, human and murine marrow microvascular endothelial cells constitutively express E-selectin,[6,23] and numerous studies over the past two decades have shown that HSPC migration to marrow involves expression of E-selectin ligands, in combination with the chemokine receptor CXCR4 and the β_1 integrin VLA-4 (for review, see Ref. 24). Intravital microscopy studies in mice have revealed that E-selectin and the chemokine that binds to CXCR4, CXCL12 (otherwise known as SDF-1), are co-expressed in a restricted fashion, expressly at those microvessels that recruit HSPCs.[23] In addition, marrow microvasculature endothelial cells constitutively express VCAM-1,[6] a ligand for VLA-4, but in a broader distribution than that of E-selectin.[23] Altogether, these studies have established that homing to marrow involves a multistep cascade, whereby E-selectin receptor/ligand interactions mediate step 1 tethering and rolling, followed by CXCL12 binding to CXCR4 yielding activation of VLA-4 (step 2), resulting in VLA-4 firm adhesion to VCAM-1 (step 3), and subsequent extravasation (step 4).

It is well known that human HSPCs express robust E-selectin ligand activity, and that the highest E-selectin ligand activity is found on primitive HSPCs (i.e., CD34$^+$ lineage$^-$ cells).[25,26] To elucidate the molecular effectors of E-selectin binding, we performed a comprehensive assessment of all glycoprotein E-selectin ligands expressed on human and mouse HSPCs.[15] These studies showed that native human HSPCs express three glycoprotein E-selectin ligands: (1) hematopoietic cell E-/L-selectin ligand (HCELL), an sLex-decorated glycoform of CD44 that binds E-selectin and L-selectin; (2) cutaneous lymphocyte antigen (CLA), a glycoform of PSGL-1

bearing abundant sLex determinants displayed on Ser/Thr-linked glycans (i.e., O-glycans) that binds E-selectin (in addition to L- and P-selectin); and (3) an sLex-bearing glycoform of CD43 that binds E-selectin (called here "CD43E"). Consistent with results of previous studies on human hematopoietic cell lines,[19,27] studies of native human HSPCs showed that HCELL is the most potent (i.e., confers the slowest rolling velocity and the highest resistance to detachment under fluid shear stress) of all the E-selectin ligands expressed on human cells, and that abrogation of HCELL expression by siRNA targeting of CD44 mRNA in hematopoietic cells resulted in a profoundly reduced ability to engage E-selectin.[15] Mouse HSPCs express only CLA and CD43E, and, consistent with absence of HCELL expression, mouse HSPCs have markedly lower E-selectin ligand activity compared to human HSPCs.[15]

Glycosyltransferase-programmed stereosubstitution (GPS) to enforce HCELL expression: glycoengineering cell migration

The CD44 molecule is best known for serving as the principal receptor for hyaluronic acid (HA; for review, see Ref. 28). Although CD44 is found on the surface of almost all human cells, expression of the HCELL glycoform in healthy individuals is restricted to HSPCs. This tight regulation of expression, coupled with this glycoprotein's high E-selectin binding activity, raised the possibility that HCELL serves an important role in directing osteotropism. To examine this issue, we developed a platform technology to custom-modify glycans of cell surface CD44 to create the HCELL glycoform. This effort involved specific formulation of glycosyltransferases and reaction buffer conditions to avoid effects on cell viability and phenotype (with exception of intended effects on CD44 structural biology; for details, see Refs. 2 and 29). Notably, cellular HCELL expression resulting from external glycosylation (i.e., "exoglycosylation") of membrane CD44 is transient, with surface turn-over to the native CD44 molecule complete within 48 hours.[30] Accordingly, after infiltration within target tissue(s), there is rapid reversion of native CD44 expression on extravasated cells.

The role of HCELL as a "bone marrow homing receptor" has been directly examined using two complementary approaches employing GPS. In the first case, we utilized a target cell devoid of step 1 effectors, human mesenchymal stem cells (MSCs). The human MSCs used in our studies expressed CD44 and VLA-4, but did not express CXCR4;[30] biochemical studies revealed that the cells expressed a CD44 glycoform decorated with N-linked terminal type 2 sialyllactosamines, indicating that the glycoprotein was missing only $\alpha(1,3)$-fucosylation at the terminal N-acetylglucosamine to complete the sLex determinant (see Fig. 1).[30] Accordingly, we treated the MSCs *in vitro* with the $\alpha(1,3)$-fucosyltransferase known as fucosyltransferase VI (FTVI) together with the nucleotide sugar donor, guanosine diphosphate- (GDP-) fucose. Follow exofucosylation, western blot showed that the only glycoprotein displaying sLex and reactive with E-selectin was CD44 (i.e., by definition, HCELL), indicating that CD44 was the predominant target of glycoengineering (see Fig. 2). When injected into immunocompromised mice, HCELL+ MSC migrated robustly to marrow within one hour of injection, whereas native (HCELL−) MSC showed minimal osteotropism. Most importantly, HCELL+ MSC infiltrated the marrow parenchyma, lodged within endosteal surfaces, and created human os-

teoid within mouse bone. Thus, enforced HCELL expression piloted colonization of human MSC within marrow, with preservation of viability and phenotype, yielding human osteoblasts that contributed to bone formation. In a complementary approach, we isolated murine HSPCs (lineage− Sca-1+ cKit+ cells; "LSK" cells) and enforced HCELL expression by GPS, again, via exofucosylation with FTVI. Notably, exofucosylation of mouse LSK cells only yielded expression of HCELL (i.e., not other E-selectin ligands), and, just as in native human HSPCs (and in MSCs expressing HCELL following exofucosylation), the relevant E-selectin binding determinants on the CD44 scaffold were expressed on N-linked glycans.[15] In short-term homing studies, enforced HCELL expression conferred ~3.5-fold increased osteotropism of LSK cells. Collectively, these findings establish that expression of HCELL itself confers high efficiency cell trafficking to marrow, supporting the designation of this structure as the human "bone marrow homing receptor."

The finding that human MSC expressing HCELL were capable of colonizing marrow in absence of CXCR4 expression prompted us to investigate the molecular basis of HCELL-mediated

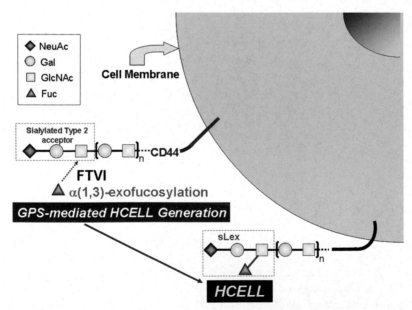

Figure 2. Application of glycosyltransferase-programmed stereosubstitution (GPS) to enforce HCELL expression. Cell surface CD44 can be converted to the HCELL glycoform by glycan engineering via GPS. In both MSC and HSPC, native CD44 displays type 2 sialyllactosaminyl glycans; color key figures correspond to respective monosaccharides (for details, see Fig. 1). *Ex vivo* treatment of cells with fucosyltransferase VI (FTVI) drives $\alpha(1,3)$-fucosylation of CD44 glycans, thereby generating sLex and, accordingly, HCELL.

transendothelial migration. Studies under both hydrodynamic flow and static conditions showed that HCELL engagement of endothelial E-selectin, or of CD44 engagement with HA, in each case triggered VLA-4 adhesion to its ligands VCAM-1 or fibronectin in absence of chemokine input.[31] This crosstalk between HCELL/CD44 and VLA-4 was mediated by signaling through engagement of a CD44-dependent G-protein coupled Rac1/Rap1-signalling cascade, resulting in transendothelial migration (without chemokine effects) on human endothelial cells expressing both E-selectin and VCAM-1.[31] These results thus refine the conventional multistep paradigm, defining the components of a novel mechanosignalling cascade that couples CD44 ligation with VLA-4 activation, yielding a "step 2-bypass" pathway of integrin activation, with commensurate firm adherence and transendothelial migration, without dependence on chemokines and other chemotactic agents. Although enhanced trafficking of HCELL+ cells to endothelial beds expressing E-selectin has been formally demonstrated thus far only for marrow (i.e., yielding robust osteotropism of both HCELL+ MSC and HCELL+ HSPC), it is likely that enforced HCELL expression would steer cell migration to other endothelial sites that express E-selectin. In all inflammatory conditions, involved tissue microvascular endothelial beds upregulate expression of both E-selectin and VCAM-1, under induction of cytokines such as TNF-α and IL-1. Because essentially all stem cells and leukocytes express CD44 as well as VLA-4, this novel regulatory pathway of cell migration has important implications for systemic cell delivery to sites of tissue injury/inflammation, for all indications of adoptive cellular therapeutics.

Regardless of whether cellular trafficking occurs through the step 2-bypass pathway or the conventional multistep cascade, glycoengineering to create HCELL will program highly efficient step 1 interactions on E-selectin, a key prerequisite for any blood-borne cell to migrate to any endothelial bed where this molecule is expressed. Notably, in all studies to date, α(1,3)-fucosylation has been the only carbohydrate substitution necessary to create the HCELL glycoform on cells that express CD44, and it is rather remarkable that a single monosaccharide substitution of this target glycoprotein can have such profound effects on cell migration. Future studies will explore how "sweetening" CD44 into HCELL via

GPS may enable not only regenerative medicine, but also cell-based immunotherapeutics employing effector lymphocytes (e.g., for cancer or infectious disease) or regulatory lymphocytes (e.g., for autoimmune conditions).

Acknowledgments

This review summarizes many years of effort to elucidate the structure and biology of HCELL, and I wish to thank all my talented and devoted coworkers for their invaluable assistance in this formidable endeavor. This work was supported by the National Institutes of Health, in particular, the National Heart Lung Blood Institute (PO1 HL107146, RO1 HL60528, and RO1 HL73714) and the National Cancer Institute (RO1 CA121335).

Conflicts of Interest

According to National Institutes of Health policies and procedures, the Brigham & Women's Hospital has assigned intellectual property rights regarding HCELL and GPS to the inventor (R.S.), who may benefit financially if the technology is licensed. R.S.'s ownership interests were reviewed and are managed by the Brigham & Women's Hospital and Partners HealthCare in accordance with their conflict of interest policy.

References

1. Bernardo, M.E., F. Locatelli & W.E. Fibbe. 2009. Mesenchymal stromal cells. *Ann. N. Y. Acad. Sci.* **1176:** 101–117.
2. Sackstein, R. 2009. Glycosyltransferase-programmed stereo-substitution (GPS) to create HCELL: engineering a roadmap for cell migration. *Immunol. Rev.* **230:** 51–74.
3. Lawrence, M.B. *et al.* 1997. Threshold levels of fluid shear promote leukocyte adhesion through selectins (CD62L,P,E). *J. Cell Biol.* **136:** 717–727.
4. Sackstein, R. 2005. The lymphocyte homing receptors: gatekeepers of the multistep paradigm. *Curr. Opin. Hematol.* **12:** 444–450.
5. Li, Y. *et al.* 2006. ADAM17 deficiency by mature neutrophils has differential effects on L-selectin shedding. *Blood* **108:** 2275–2279.
6. Schweitzer, K.M. *et al.* 1996. Constitutive expression of E-selectin and vascular cell adhesion molecule-1 on endothelial cells of hematopoietic tissues. *Am. J. Pathol.* **148:** 165–175.
7. Weninger, W. *et al.* 2000. Specialized contributions by alpha(1,3)-fucosyltransferase-IV and FucT-VII during leukocyte rolling in dermal microvessels. *Immunity* **12:** 665–676.
8. Yao, L. *et al.* 1999. Divergent inducible expression of P-selectin and E-selectin in mice and primates. *Blood* **94:** 3820–3828.

9. Liu, Z. *et al.* 2010. Differential regulation of human and murine P-selectin expression and function in vivo. *J. Exp. Med.* **207:** 2975–2987.

10. Ohmori, K. *et al.* 2000. P- and E-selectins recognize sialyl 6-sulfo Lewis X, the recently identified L-selectin ligand. *Biochem. Biophys. Res. Commun.* **278:** 90–96.

11. Lawrence, M.B. & T.A. Springer. 1993. Neutrophils roll on E-selectin. *J. Immunol.* **151:** 6338–6346.

12. Lawrence, M.B. *et al.* 1995. Rolling of lymphocytes and neutrophils on peripheral node addressin and subsequent arrest on ICAM-1 in shear flow. *Eur. J. Immunol.* **25:** 1025–1031.

13. Jung, U. *et al.* 1996. Velocity differences between L- and P-selectin-dependent neutrophil rolling in venules of mouse cremaster muscle in vivo. *Am. J. Physiol.* **271:** H2740–H2747.

14. Jung, U. & K. Ley. 1999. Mice lacking two or all three selectins demonstrate overlapping and distinct functions for each selectin. *J. Immunol.* **162:** 6755–6762.

15. Merzaban, J.S. *et al.* 2011. Analysis of glycoprotein E-selectin ligands on human and mouse marrow cells enriched for hematopoietic stem/progenitor cells. *Blood* **118:** 1774–1783.

16. Sackstein, R. & M. Borenstein. 1995. The effects of corticosteroids on lymphocyte recirculation in humans: analysis of the mechanism of impaired lymphocyte migration to lymph node following methylprednisolone administration. *J. Investig. Med.* **43:** 68–77.

17. Sackstein, R. & C.J. Dimitroff. 2000. A hematopoietic cell L-selectin ligand that is distinct from PSGL-1 and displays N-glycan-dependent binding activity. *Blood* **96:** 2765–2774.

18. Dimitroff, C.J. *et al.* 2000. A distinct glycoform of CD44 is an L-selectin ligand on human hematopoietic cells. *Proc. Natl. Acad. Sci. U. S. A.* **97:** 13841–13846.

19. Dimitroff, C.J. *et al.* 2001. CD44 is a major E-selectin ligand on human hematopoietic progenitor cells. *J. Cell. Biol.* **153:** 1277–1286.

20. Sackstein, R. & R. Fuhlbrigge. 2009. Western blot analysis of adhesive interactions under fluid shear conditions: the blot rolling assay. *Methods Mol. Biol.* **536:** 343–354.

21. Patel, K.D., M.U. Nollert & R.P. McEver. 1995. P-selectin must extend a sufficient length from the plasma membrane to mediate rolling of neutrophils. *J. Cell. Biol.* **131:** 1893–1902.

22. von Andrian, U.H. *et al.* 1995. A central role for microvillous receptor presentation in leukocyte adhesion under flow. *Cell* **82:** 989–999.

23. Sipkins, D.A. *et al.* 2005. In vivo imaging of specialized bone marrow endothelial microdomains for tumour engraftment. *Nature* **435:** 969–973.

24. Sackstein, R. 2004. The bone marrow is akin to skin: HCELL and the biology of hematopoietic stem cell homing. *J. Invest. Dermatol.* **122:** 1061–1069.

25. Greenberg, A.W., W.G. Kerr & D.A. Hammer. 2000. Relationship between selectin-mediated rolling of hematopoietic stem and progenitor cells and progression in hematopoietic development. *Blood* **95:** 478–486.

26. Dagia, N.M. *et al.* 2006. G-CSF induces E-selectin ligand expression on human myeloid cells. *Nat. Med.* **12:** 1185–1190.

27. Dimitroff, C.J. *et al.* 2001. Differential L-selectin binding activities of human hematopoietic cell L-selectin ligands, HCELL and PSGL-1. *J. Biol. Chem.* **276:** 47623–47631.

28. Sackstein, R. 2011. The biology of CD44 and HCELL in hematopoiesis: the 'step 2-bypass pathway' and other emerging perspectives. *Curr. Opin. Hematol.* **18:** 239–248.

29. Sackstein, R. 2010. Directing stem cell trafficking via GPS. *Methods Enzymol.* **479:** 93–105.

30. Sackstein, R. *et al.* 2008. Ex vivo glycan engineering of CD44 programs human multipotent mesenchymal stromal cell trafficking to bone. *Nat. Med.* **14:** 181–187.

31. Thankamony, S.P. & R. Sackstein. 2011. Enforced hematopoietic cell E- and L-selectin ligand (HCELL) expression primes transendothelial migration of human mesenchymal stem cells. *Proc. Natl. Acad. Sci. U. S. A.* **108:** 2258–2263.

Ann. N.Y. Acad. Sci. ISSN 0077-8923

ANNALS OF THE NEW YORK ACADEMY OF SCIENCES
Issue: *Glycobiology of the Immune Response*

The expanding role of α2-3 sialylation for leukocyte trafficking *in vivo*

Markus Sperandio

Walter Brendel Center of Experimental Medicine, Ludwig-Maximilians-Universität, Munich, Germany

Address for correspondence: Markus Sperandio, M.D., Walter Brendel Center of Experimental Medicine, Ludwig-Maximilians Universität, Marchioninistr.15, 81377 München, Germany. markus.sperandio@med.uni-muenchen.de

The ability of leukocytes to navigate through the different body compartments is an essential component for functioning immune defense and surveillance systems. In order to exit the blood circulation, leukocytes follow distinct recruitment steps, including capture of free-flowing leukocytes to, and rolling along, the vessel wall; firm leukocyte arrest on the endothelial lining; and postarrest modifications (spreading and crawling), which prepare the leukocyte for transmigration through the vascular wall. Post-translational glycosylation (including sialylation) has been known for many years to be functionally relevant for selectin ligands and, hence, selectin-mediated capture and rolling. Recently, sialylation by the α2-3 sialyltransferase ST3Gal-IV was identified to significantly influence chemokine-triggered firm leukocyte arrest, expanding the role of α2-3 sialylation from leukocyte rolling to subsequent chemokine-triggered leukocyte arrest. These findings make ST3Gal-IV an interesting drug target for modulating leukocyte trafficking in human disorders, including autoimmune diseases and cancer.

Keywords: sialylation; ST3Gal-IV; selectin; CXCR2; leukocyte recruitment

Introduction

Since its first descriptions, the leukocyte recruitment cascade has been considered to be a rather uniform process, consisting in the initial contact (capture) of free-flowing leukocytes to the vascular wall, leukocyte rolling along the wall, firm adhesion, and finally transmigration.[1] The early steps are mediated by selectins binding to crucial carbohydrate structures on selectin ligands.[2] Selectin-mediated rolling enables leukocytes to get into close contact with the endothelial surface layer, which allows endothelium-bound chemokines to interact with their respective chemokine receptors on the rolling leukocyte.[3] Chemokine–chemokine receptor signaling then activates leukocyte-expressed integrins, triggering firm leukocyte arrest on the endothelium and subsequent transmigration into tissue (Fig. 1).[1,3–5]

Only recently have cutting-edge molecular biology tools, combined with modern bioimaging techniques, revealed that trafficking of leukocytes between various compartments within mammalian organisms does not only allow the directed and specific navigation of single leukocyte subsets, it also allows distinct recruitment patterns of the same leukocyte subset only differing in its activation status.[6]

Leukocyte recruitment is regulated by an ever-growing number of adhesion-relevant molecules. This gains further complexity by the functional integration of posttranslational glycosylation of adhesion-relevant molecules as another regulatory mode to control leukocyte recruitment *in vivo*.[7,8] Most prominently, posttranslational glycosylation significantly contributes to the generation of functional selectin ligands and, therefore, leukocyte rolling *in vivo*. Several glycosyltransferases involved in the generation of functional selectin ligands have been described, including polypeptide *N*-acetylgalactosaminyl transferase-1 (ppGalNAcT-I),[9] core 2 β1-6-*N*-acetylglucosaminyltransferase-1 (C2GnT-I),[10,11] β1-4 galactosyltransferases (beta4GalT1 and 4),[12,13] α1-3 fucosyltransferases (FucT-VII and FucT-IV),[14,15] and α2-3 sialyltransferase IV (ST3Gal-IV).[16,17] Additional information

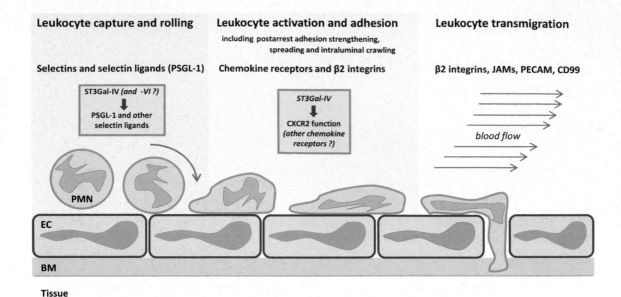

Figure 1. The role of ST3Gal-IV in regulating the multistep leukocyte recruitment cascade during inflammation. Polymorphonuclear neutrophils (PMN) get captured by the inflamed endothelium and then start to roll along the endothelium. Capture and rolling are mediated by selectins interacting with α2-3-sialylated carbohydrate determinants on selectin ligands.[7] Besides ST3Gal-IV, other ST3Gal-isoenzymes, such as ST3Gal-VI, might also contribute to selectin ligand activity. During rolling, leukocytes have ample time to screen the endothelial surface for activation signals. These activation signals are in large part provided by endothelium-bound chemokines interacting with specific chemokine receptors on the leukocyte surface. For neutrophil recruitment, endothelium-bound CXCL1 (keratinocyte-dervied chemokine KC, in mice) or CXCL8 (IL-8, in humans) bind to neutrophil-expressed CXCR2. Binding of CXCL1 or CXCL8 to CXCR2, which is most likely dependent on a putative α2-3–linked sialic acid residue on CXCR-2, triggers the activation of β_2 integrins leading to firm leukocyte arrest on and finally transmigration through the endothelium into the inflamed tissue. Whether other chemokine receptor systems are also modified by ST3Gal-IV needs to be demonstrated by future studies. BM, basal membrane; EC, endothelial cell; PMN, polymorphonuclear neutrophil; JAMs, junctional adhesion molecules; PECAM, platelet/endothelial cell adhesion molecule.

on the role of these glycosyltransferases on leukocyte rolling can be found in Table 1 and a recent review.[7]

While most of the glycosyltransferases involved in the generation of functional selectin ligands have not been reported to directly contribute to other steps of the leukocyte recruitment cascade, ST3Gal-IV has been identified to also influence chemokine receptor–triggered leukocyte arrest.[18] These findings expand the contribution of post-translational glycosylation from selectin-mediated rolling to chemokine receptor–induced leukocyte arrest (Fig. 1). Of note, two recent reports have shown that a reduction in α2-6–linked sialic acid residues on $\alpha_4\beta_1$ integrin (VLA-4) or $\alpha_M\beta_2$ integrin (Mac-1) is associated with increased binding of the respective integrins to their ligands.[19,20] These results demonstrate that altered glycosylation can also modulate integrin function, although in this case sialylation (α2-6–attached sialic acids) was

functionally linked to reduced integrin activity, with α2-6 sialylation having an anti-inflammatory effect. This review, however, will focus on the sialyltransferase ST3Gal-IV and its multifaceted proinflammatory modulation of leukocyte recruitment *in vivo*.

ST3Gal-IV and selectin ligand activity

In the mammalian genome, six genes have been described that encode the Golgi-resident sialyltransferase family (ST3Gal-I-VI), proteins that transfers sialic acid residues in α2-3 linkage to terminal galactose residues on glycoproteins and glycolipids.[21,22] Of the six sialyltransferases, ST3Gal-III, ST3Gal-IV, and ST3Gal-VI sialylate type II oligosaccharides (Galβ1-4GlcNAc) that can then be further processed to the tetrasaccharide sialyl Lewis X (sLe^x), the prototypical selectin ligand.[7] Ellies *et al.* investigated mice deficient in ST3Gal-III and ST3Gal-IV and found no difference in binding of P-selectin or E-selectin

Table 1. Glycosyltransferases involved in the generation of functional selectin ligands and respective phenotypes in knockout mouse models

Glycosyltransferase	Rolling/binding defects found in the respective knockout mouse model
ppGalNAcT-I (*Galnt1**)	• Reduced binding of Galnt1[−/−] neutrophils to P- and E-selectin in a static binding assay[9] • Reduced rolling of Galnt1[−/−] neutrophils on immobilized P- and E-selectin in a dynamic *in vitro* assay[9]
C2GnT-I (*Gcnt1**)	• Reduced binding of Gcnt1[−/−] neutrophils to E- and P-selectin in a static binding assay[10] • Reduced P- and E-selectin–dependent leukocyte rolling in inflamed cremaster muscle venules[11]
Beta4GalT1 (*B4galt1**)	• Reduced binding of B4galt1[−/−] neutrophils to P-selectin *in vitro*[12]
FucT-7 (*Fut7**)	• Severe reduction in P- and E-selectin–dependent rolling in inflamed ear vessels[15]
FucT-4 (*Fut4**)	• increased E-selectin–dependent rolling velocities in inflamed ear vessels, but otherwise no significant phenotype[15]
ST3Gal-IV (*St3gal4**)	• Reduced E-selectin–dependent rolling and increased E-selectin–mediated rolling velocity in inflamed cremaster muscle venules[16]

*Gene name

to St3gal3[−/−] neutrophils compared to wild-type neutrophils, suggesting that ST3Gal-III does not contribute to selectin ligand activity on neutrophils.[16] In contrast, neutrophils from mice lacking ST3Gal-IV showed a significant reduction in binding to both P- and E-selectin, indicating that sialylation by ST3Gal-IV contributes to selectin ligand activity on neutrophils (Table 1).[16] Additional intravital microscopic experiments in inflamed cremaster muscle venules of St3gal4[−/−] mice revealed that E-selectin–mediated rolling was modestly reduced and E-selectin–dependent rolling velocities were moderately increased, while P-selectin–mediated rolling and rolling velocity were similar to wild-type mice. This implies that under *in vivo* conditions other leukocyte-expressed sialyltransferases may compensate for the loss of ST3Gal-IV. This view is supported by the fact that P- and E-selectin binding to neutrophils from St3gal4[−/−] mice is further reduced by pretreating the cells with sialidase.[16] Potential candidate sialyltransferases in this context are ST3Gal-III and ST3Gal-VI. However, as mentioned above, St3gal3[−/−] neutrophils did not exhibit any impairment in binding to E-selectin or P-selectin.[16] ST3Gal-VI has been described to contribute to the generation of sLe[x].[23] However, its expression in hematopoietic cell lines appeared to be rather low.[23] To test the contribution of ST3Gal-VI

on selectin ligand activity, St3gal6[−/−] mice have been recently generated and are currently under investigation. Results from this study will not only elucidate how ST3Gal-VI contributes to selectin ligand activity but also uncover distinct and/or overlapping functions of ST3Gal-IV and ST3Gal-VI in the generation of functional selectin ligands *in vivo*.

Sialylation and chemokine-triggered leukocyte arrest

About a decade ago, Bannert *et al.* reported for the first time that posttranslational sialylation affects chemokine receptor binding to chemokines. They found that sialyation of chemokine receptor CCR5 is essential for its binding to CCR5-interacting chemokines CCL3 (MIP-1α) and CCL4 (MIP-1β).[24] The authors had generated several cell lines stably transfected with wild-type CCR5 or different CCR5 mutants in which specific serine or threonine residues had been exchanged with alanine to remove putative O-glycosylation sites.[24,25] Subsequent binding assays using the different CCR5-transfected cell lines demonstrated that CCR5 binding to CCL3 and CCL4 was strongly dependent on a sialic acid-carrying O-glycan linked to serine 6 at the N-terminus of CCR5 but independent of other putative O-glycosylation sites on the second and third loop of CCR5.[24] These findings indirectly

Table 2. ST3Gal-IV and its role in chemokine-triggered neutrophil recruitment[7,18]

In vivo findings in *St3gal4*[−/−] mice
- Reduced leukocyte adhesion in trauma-stimulated cremaster muscle venules
- Reduced induction of leukocyte arrest in cremaster muscle venules after systemic injection of CXCR2 chemokines CXCL1 and CXCL8
- Reduced leukocyte extravasation into inflamed cremaster muscle tissue
- Reduced leukocyte extravasation in the thioglycollate-induced peritonitis model

In vitro findings in *St3gal4*[−/−] mice
- Reduced binding of fluorescently labeled CXCL8 to isolated *St3gal4*[−/−] neutrophils
- Reduced adhesion of *St3gal4*[−/−] neutrophils on immobilized P-selectin, ICAM-1, and CXCL1 in an *ex vivo* microflow chamber assay

---

indicated that chemokine-induced leukocyte recruitment could be dependent on posttranslational sialylation of the chemokine receptor.

Stimulated by the work of Bannert *et al.*[24] and by our findings that *St3gal4*[−/−] mice exhibited reduced leukocyte adhesion during inflammation *in vivo*—a defect that cannot be explained by the mild impairment in leukocyte rolling—our group had performed intravital microscopy studies to test whether CXCR2-mediated leukocyte adhesion is dependent on ST3Gal-IV.[18] Our *in vivo* data revealed that the systemic application of CXCR2-binding chemokines CXCL1 (KC, keratinocyte-derived chemokine) or CXCL8 (IL-8) led to a strong and rapid induction of leukocyte arrest in cremaster muscle venules of wild-type mice, while in *St3gal4*[−/−] mice the increase in the number of arrested leukocytes following systemic injection of CXCL1 or CXCL8 was dramatically reduced.[18] Additional *in vitro* experiments using FITC-labeled CXCL8 demonstrated reduced binding of FITC-CXCL8 to *St3gal4*[−/−] neutrophils, WT neutrophils treated with sialidase, and *Cxcr2*[−/−] neutrophils, compared to FITC-CXCL8 binding to WT neutrophils. An overview on how ST3Gal-IV influences chemokine receptor function is provided in Table 2.

To investigate whether recruitment of other leukocyte subsets and/or chemokine receptor–chemokine systems depend on ST3Gal-IV, we also investigated CCR7-mediated migration of dendritic cells (DC). However, the absence of ST3Gal-IV did not alter CCR7-mediated migration of dendritic cells, which implies that sialylation by ST3Gal-IV only affects distinct chemokine receptor systems and/or leukocyte subsets.[18,26] Taken together, posttranslational sialylation by ST3Gal-IV is essential for CXCR-2–triggered firm leukocyte arrest, although it has to be mentioned that the sialylation-dependent binding of CXCL1 or CXCL8 to isolated CXCR2 has not been directly demonstrated thus far.

Subsequent studies will be necessary to clarify whether the influence of α2-3 sialylation on chemokine receptor function is directly related to crucial ST3Gal-IV-dependent sialylation sites on the chemokine receptor, or whether it involves sialylation sites on chemokine receptor-associated molecules. In addition, distinct expression patterns of specific sialyltransferases such as ST3Gal-IV might also influence sialylation-dependent chemokine receptor function.

Conclusion

Trafficking of immune cells is a tightly regulated and highly specific process that involves many groups of different adhesion-relevant molecules. Posttranslational sialylation by ST3Gal-IV has been shown to regulate selectin ligand activity and, perhaps more relevant under *in vivo* conditions, chemokine receptor function. In view of a potential role of ST3Gal-IV in regulating chemokine-triggered leukocyte recruitment, interfering with ST3Gal-IV activity might be an interesting way to control cell recruitment within the blood circulation in a whole variety of pathological conditions ranging from chronic inflammatory disorders and autoimmune diseases to cancer.

Acknowledgments

Original work from my laboratory cited or mentioned in this review was in part funded by Deutsche Forschungsgemeinschaft (DFG) SP621/1 and SP621/3-1, Mizutani Foundation (Grant No. 090063), and FöFoLe-LMU München (07/09).

Conflicts of interest

The authors declare no conflicts of interest.

References

1. Ley, K. 2008. The microcirculation in inflammation, In *Handbook of Physiology: Microcirculation*. R.F. Tuma, W.N. Duran & K. Ley, Eds.: 387–448. Academic Press. San Diego, USA.
2. Sperandio, M. 2006. Selectins and glycosyltransferases in leukocyte rolling in vivo. *FEBS J.* **273:** 4377–4389.
3. Ley, K, C. Laudanna, M.I. Cybulsky, *et al.* 2007. Getting to the site of inflammation: the leukocyte adhesion cascade updated. *Nat. Rev. Immunol.* **7:** 678–689.
4. Springer, T.A. 1994. Traffic signals for lymphocyte recirculation and leukocyte emigration: the multistep paradigm. *Cell* **76:** 301–314.
5. Butcher, E.C. 1994. Leukocyte-endothelial cell recognition—three (or more) steps to specificity and diversity. *Cell* **67:** 1033–1036.
6. Mora, J.R. & U.H. von Andrian. 2006. T-cell homing specificity and plasticity: new concepts and future challenges. *Trends Immunol.* **27:** 235–243.
7. Sperandio, M., C.A. Gleissner & K. Ley. 2009. Glycosylation in immune cell trafficking. *Immunol. Rev.* **230:** 97–113.
8. Marth, J.D. & P.K. Grewal. 2008. Mammalian glycosylation in immunity. *Nat. Rev. Immunol.* **8:** 874–887.
9. Tenno, M., K. Ohtsubo, F.K. Hagen, *et al.* 2007. Initiation of Protein O-Glycosylation by Polypeptide GalNAcT-1 in Vascular Biology and Humoral Immunity. *Mol. Cell. Biol.* **27:** 8783–8796.
10. Ellies, L.G., S. Tsuboi, B. Petryniak, *et al.* 1998. Core 2 oligosaccharide biosynthesis distinguishes between selectin ligands essential for leukocyte homing and inflammation. *Immunity* **9:** 881–890.
11. Sperandio, M., A. Thatte, D. Foy, *et al.* 2001. Severe impairment of leukocyte rolling in venules of core 2 glucosaminyltransferase-deficient mice. *Blood* **97:** 3812–3819.
12. Asano, M., S. Nakae, N. Kotani, *et al.* 2003. Impaired selectin-ligand biosynthesis and reduced inflammatory responses in beta-1,4-galactosyltransferase-I-deficient mice. *Blood* **102:** 1678–1685.
13. Seko, A., N. Dohmae, K. Takio, *et al.* 2003. Beta 1,4-galactosyltransferase (beta 4GalT)-IV is specific for GlcNAc 6-O-sulfate. Beta 4GalT-IV acts on keratan sulfate-related glycans and a precursor glycan of 6-sulfosialyl-Lewis X. *J. Biol. Chem.* **278:** 9150–9158.
14. Maly, P., A.D. Thall, B. Petryniak, *et al.* 1996. The a (1,3)-fucosyltransferase Fuc-TVII controls leukocyte trafficking through an essential role in L-, E-, and P-selectin ligand biosynthesis. *Cell* **86:** 643–653.
15. Weninger, W., L.H. Ulfman, G. Cheng, *et al.* 2000. Specialized contributions by a (1,3)-fucosyltransferase-IV and FucT-VII during leukocyte rolling in dermal microvessels. *Immunity* **12:** 665–676.
16. Ellies, L.G., M. Sperandio, G.H. Underhill, *et al.* 2002. Sialyltransferase specifity in selectin ligand formation. *Blood* **100:** 3618–3625.
17. Sperandio, M., D. Frommhold, I. Babushkina, *et al.* 2006. alpha2,3-sialyltransferase-IV is essential for L-selectin ligand function in inflammation. *Eur. J. Immunol.* **36:** 3207–3215.
18. Frommhold, D., A. Ludwig, M.G. Bixel, *et al.* 2008. Sialyltransferase ST3Gal-IV controls CXCR2-mediated firm leukocyte arrest during inflammation. *J. Exp. Med.* **205:** 1435–1446.
19. Woodard-Grice, A.V., A.C. McBrayer, J.K. Wakefield, *et al.* 2008. Proteolytic shedding of ST6Gal-I by BACE1 regulates the glycosylation and function of alpha4beta1 integrins. *J. Biol. Chem.* **283:** 26364–26373.
20. Feng, C., L. Zhang, L. Almulki, *et al.* 2011. Endogenous PMN sialidase activity exposes activation epitope on CD11b/CD18 which enhances its binding interaction with ICAM-1. *J. Leukoc. Biol.* **90:** 313–321.
21. Harduin-Lepers, A., R. Mollicone, P. Delannoy, *et al.* 2005. The animal sialyltransferases and sialyltransferase-related genes: a phylogenetic approach. *Glycobiology* **15:** 805–817.
22. Takashima, S. 2008. Characterization of mouse sialyltransferase genes: their evolution and diversity. *Biosci. Biotechnol. Biochem.* **72:** 1155–1167.
23. Okajima, T., S. Fukumoto, H. Miyazaki, *et al.* 1999. Molecular cloning of a novel alpha2,3-sialyltransferase (ST3Gal VI) that sialylates type II lactosamine structures on glycoproteins and glycolipids. *J. Biol. Chem.* **274:** 11479–11486.
24. Bannert, N., S. Craig, M. Farzan, *et al.* 2001. Sialylated O-glycans and sulfated tyrosines in the NH2-terminal domain of CC chemokine receptor 5 contribute to high affinity binding of chemokines. *J. Exp. Med.* **194:** 1661–1673.
25. Farzan, M., T. Mirzabekov, P. Kolchinsky, *et al.* 1999. Tyrosine sulfation of the amino terminus of CCR5 facilitates HIV-1 entry. *Cell* **96:** 667–676.
26. Videira, P.A., I.F. Amado, H.J. Crespo, *et al.* 2008. Surface alpha 2-3- and alpha 2-6-sialylation of human monocytes and derived dendritic cells and its influence on endocytosis. *Glycoconj. J.* **25:** 259–268.

Ann. N.Y. Acad. Sci. ISSN 0077-8923

ANNALS OF THE NEW YORK ACADEMY OF SCIENCES
Issue: *Glycobiology of the Immune Response*

Beyond glycoproteins as galectin counterreceptors: tumor-effector T cell growth control via ganglioside GM1

Robert W. Ledeen,[1] Gusheng Wu,[1] Sabine André,[2] David Bleich,[3] Guillemette Huet,[4] Herbert Kaltner,[2] Jürgen Kopitz,[5] and Hans-Joachim Gabius[2]

[1]Department of Neurology and Neurosciences, New Jersey, Medical School-University of Medicine and Dentistry of New Jersey, Newark, New Jersey. [2]Faculty of Veterinary Medicine, Institute of Physiological Chemistry, Ludwig-Maximilians-University Munich, Munich, Germany, gabius@lectins.de [3]Department of Medicine, New Jersey Medical School-University of Medicine and Dentistry of New Jersey, Newark, New Jersey. [4]Centre de Recherche, INSERM U837, CHRU, Lille, France. [5]Institute of Pathology, Ruprecht-Karls-University, Heidelberg, Germany.

Addresses for correspondence: Robert W. Ledeen, Department of Neurology and Neurosciences, New Jersey Medical School, University of Medicine and Dentistry of New Jersey, Newark, New Jersey, ledeenro@umdnj.edu and Hans-Joachim Gabius, Institute of Physiological Chemistry, Faculty of Veterinary Medicine, Ludwig-Maximilians-University Munich, Munich, Germany, gabius@lectins.de

Glycoprotein glycan chains, by virtue of structure, topology of presentation and connection to signal-inducing units, are functional galectin counterreceptors. As example, cross-linking of the $\alpha_5\beta_1$ integrin by galectin-1 on carcinoma cells leads to G_1 arrest or anoikis. Contact-dependent switching from proliferation to differentiation in cultured neuroblastoma cells (SK-N-MC) also utilizes galectin-1. Activity enhancement of a cell surface sialidase underlies the shift in glycan display to ganglioside GM1. Its pentasaccharide within microdomains becomes the target. Similarly, this recognition pair is upregulated upon T cell activation. Cross-linking of GM1 along with associated $\alpha_4/\alpha_5\beta_1$ integrins elicits Ca^{2+}-influx via TRPC5 channels as the relevant response for T effector cell (T_{eff}) suppression. Unlike T_{eff} cells from wild-type mice, those from genetically altered mice lacking GM1 are not suppressed by galectin-1 or regulatory T cells. Similarly, in the context of GM1 deficiency in NOD mice, T_{eff} cells are associated with resistance to regulatory T cell suppression, which is reversed by applied GM1. The broad array of glycosphingolipid structures suggests the possible existence of several novel counterreceptors targeted to endogenous lectins, with sulfatide–galectin-4 interplay within apical delivery serving as recent example.

Keywords: anoikis; carcinoma; diabetes; galectin-1; glycosphingolipid; GM1 ganglioside; immune suppression; T cells

Background

Broadly speaking, the cell surface must present a large number of signals on a limited area to ensure diverse routes of communication with the environment at high accuracy and efficiency. Given the obvious restrictions in available space, biochemical high-density coding is mandatory, along with the possibility for dynamic surface remodeling, to facilitate swift alterations if necessary, thereby avoiding complete degradation/*de novo* synthesis. The ideal chemical platform to meet these requirements is the carbohydrates. Their properties, prominently the series of chemically rather equivalent hydroxyl groups, enable them to reach an unsurpassed structural versatility in oligomer ("code word") generation. Recruiting both anomeric forms (α/β) and different hydroxyl groups as acceptors to glycosidic linkages contributes to establishing the wide scope of the sugar code.[1] Explicitly, the number of oligomers ("words") that can be formed by the given alphabet of monomers ("letters"), selecting a hexamer as a test case, is several orders of magnitude larger for sugars than for the same number of amino acids.[2] Cellular glycoconjugates thus harbor structural information encoded in their glycan chains, and these are attached to various types of scaffold.

[Correction added after online publication May 15, 2012: Article title has been corrected.]

doi: 10.1111/j.1749-6632.2012.06479.x

Table 1. Six levels of regulation of affinity of glycan binding to a lectin

1.	Mono- and disaccharides (including anomeric position and substitutions)
2.	Oligosaccharides (including branching and substitutions)
3.	Spatial parameters of oligosaccharides
	a. Shape of oligosaccharide (differential conformer selection)
	b. Conformational flexibility differences between isomers (e.g., α2,3/6-sialylation)
4.	Spatial parameters of glycans in natural glycoconjugates
	a. Shape of glycan chain (examples: modulation of conformation by substitutions not acting as a lectin ligand, such as core fucosylation or introduction of bisecting GlcNAc in *N*-glycans, influence of protein part)
	b. Cluster effect with bi- to pentaantennary *N*-glycans or branched *O*-glycans (including modulation by substitutions, please see 4a)
5.	Cluster effect with different, but neighboring, glycan chains on the same glycoprotein (e.g., in mucins) or a glycoprotein–glycolipid complex (e.g., integrin–ganglioside GM1 complexes)
6.	Cluster effect with different glycoconjugates on the cell surface in spatial vicinity forming microdomains (e.g., ganglioside GM1-rich areas)

Adapted from Ref. 122, with permission.

Common glycoproteome analysis directs immediate attention to proteins as carriers. Because glycosylation of proteins is a major event in co- and posttranslational processing along their route from the endoplasmic reticulum and Golgi stacks to the plasma membrane,[3] cell surface proteins are expected to be glycoproteins. Looking at *N*-glycosylation, its occurrence is indeed characteristic of most extracytoplasmic proteins, as revealed by comprehensive N-glycoproteomic mapping.[4,5] As reinforcement of the assumed significance of sugar coding by looking more deeply into frequency of occurrence, mucin-type *O*-glycans are known to accompany N-linked sugars on the same protein.[6] Of note, any change in the enzymatic assembly line by gene regulation (up- and downregulation, genetic deficiencies, including additional glycogenes coding for the transporters to make activated sugars available at the site of assembly) will bring about alterations in the profile of these products.[7] This marked sensitivity ensures an adequate plasticity and responsiveness of the system. Fittingly, environmental factors, such as the physiological regulator nitric oxide or manipulations at the genetic level, have been documented to reshape specific aspects of cell surface glycosylation, in the latter case disclosing that compensation of disease-associated deficiencies not directly involved in *N*-glycosylation nonetheless has a significant impact on glycan features.[8,9] Shifts in branch-end tailoring and the level of core substitutions, which can have a substantial bearing on the conformational dynamics of glycan chains, are common effects, and the adaptable presence of surface glycohydrolases/glycosyltransferases fulfills the condition for on-site tailoring.[10–12]

Experimentally, the ensuing transitions between glycoprofiles can readily be detected *in situ* by glycan-binding probes, that is, antibodies or lectins. The commercially available phytohemagglutinins are popular tools for this purpose. This cell surface glycophenotyping by a panel of lectins is not only an analytical approach. It also inspires the idea that a recognitive interplay with endogenous receptors could translate the information of a cognate glycoepitope into the respective cellular response. Six levels of affinity regulation of carbohydrate–protein recognition (summarized in Table 1) are operative to restrict functional binding to few sites, thereby setting the essential strict limits to physiological high-affinity binding of lectins.[13] In other words, only certain glycoproteins are structurally and topologically able to serve as lectin counterreceptors. An instructive example for such an interplay, yielding effective lectin-dependent growth regulation, is given by the $\alpha_5\beta_1$ integrin (fibronectin receptor) and its role as target of a galectin.

Gangliosides constitute another form of cellular glycoconjugates that in principle are constituted, structurally and topologically, to serve as lectin counterreceptors. The following review, starting from a view on galectin–glycoprotein interactions then turns to glycosphingolipids (GSLs). Among them, ganglioside GM1 is emerging as an important modulator of neural cell growth and T cell communication. These findings point to gangliosides as functionally comparable in terms of galectin counterreceptor activity to glycoproteins.

Growth regulation by glycoprotein/galectin contact

A hallmark of lectin activity, reflected by the classical hemagglutination assay, is ligand crosslinking. Association of two monovalent modules to a homodimer, as shown in Figure 1A, is an often-encountered route to bivalency. The resulting protein will be capable to build aggregates (lattices) on the cell surface with counterreceptors and thereby initiate signaling.[14–17] The galectin–glycoprotein lattice can underlie regulation of cell surface glycoprotein clustering/signaling and of endocytosis to affect cell function. Homodimeric human galectin-1, with its central contact site for a galactose moiety established by the strictly conserved Trp residue, as depicted in Figure 1B,[18] has been identified as a growth inhibitor for carcinoma cells, either attached associated to substratum or kept in suspension.[19,20] In both cases, the cellular reactivity depends on the presence of the α_5 integrin subunit and its appropriate glycosylation, especially the absence of $\alpha2,6$ sialylation, which precludes galectin-1 reactivity, from strategic sites, as also observed in T cell death.[19–22] Cross-linking of integrin units by galectin-1 is the starting point for the outward-in signaling toward growth control. The way it works depends on whether the cells are adherent or kept in suspension: it is either by G_1 cell cycle arrest via both upregulation of transcription of the genes for the cyclin-dependent kinase inhibitors p21/p27 and enhancement of p27 protein stability (adherent cells) or by induction of caspase-8–dependent anoikis (cells in suspension).[19,20] Interestingly, the latter mechanism is also operative when expression of the tumor suppressor p16[INK4a] is reconstituted *in vitro*.[23] This multifunctional protein has been identified as a master regulator of distinct glycorelated aspects driving cells into anoikis, that is, by (i) enhancing galectin-1 and α_5 integrin expression, (ii) increasing reactivity of the integrin's glycans for galectin-1 through appropriate expression

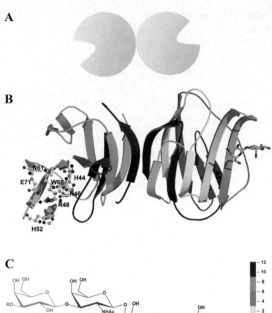

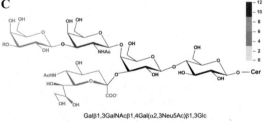

Galβ1,3GalNAcβ1,4Gal(α2,3Neu5Ac)β1,3Glc

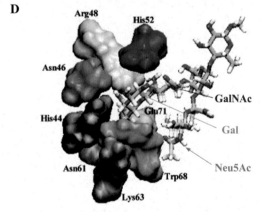

Figure 1. Schematic illustrations of homodimeric galectin-1. (A and B) The positions of the lectin sites, their key amino acids and contact to lactose as pan-galectin ligand in one of its low-energy conformers are shown along with secondary-structure elements. The oligosaccharide structures of gangliosides GM1 (R = H) and GD1a (R = sialic acid; Cer = ceramide). (C) The color coding reflects the Coulomb/van der Waals energy term (in kcal/mol) of interaction between the sugar moieties and human galectin-1 obtained by computational interaction analysis as reported previously.[58] (D) A detailed view of the topology of the interaction between human galectin-1's binding site and the terminal β-galactoside and the sialic acid of the branch, as reported previously.[58,59]

of the necessary glycogenes, and (iii) reducing antiapoptotic activity through downregulating a respective effector, that is, galectin-3.[23–25] Overall, that galectin-1 preferentially selects a distinct glycoprotein for binding, which thereby becomes spatially organized in clusters, gives the downstream signaling a clear direction.

With relevance for immunology, studies on activated T cells have taught the lesson that the cell type will determine the actual target for the galectin and the eventual response.[26–29] The same peptide core unit can be recruited differently to cell-fate decision by altering its glycosylation in terms of glycan structures and the topology of glycan presentation. Lectin responsiveness of the glycoprotein is under dynamic control. The switch for signaling by its protein part can be turned on upon demand, for example, by decreasing the degree of $\alpha 2,6$ sialylation that masks galectin-reactive sites[22,28] (analogous to anoikis induction in pancreatic carcinoma cells[23]). Of particular note, activities dependent upon the level of sialylation for regulatory purposes are not confined to glycoprotein glycans. In principle, it can also occur for another class of cellular glycoconjugates, GSLs (e.g., gangliosides), which have received much less attention than glycoproteins as lectin counterreceptors, despite a notable degree of structural sophistication.[30–33]

Growth regulation by ganglioside/galectin contact

The panoply of lipid glycan structures identified to date, based on sugar moieties alone, include 187 sialylated, 24 sulfated, and 172 neutral GSLs.[34] Additional structural variations can occur, though to a lesser degree, within the hydrophobic ceramide units (e.g., occurrence of long-chain fatty acids; see later) that serve as membrane anchors and contribute in some instances to signaling specificity. Whereas ligand properties of cell surface gangliosides such as GM1 for bacterial toxins (in this case cholera toxin or heat-labile *E. coli* enterotoxin) are well appreciated,[35] the physiological rationale for such diversity in glycan presentation, which might conceivably include novel counterreceptors for endogenous lectins, is at present an enigma.[36]

Gangliosides, by definition, are GSLs that contain one or multiple sialic acids, all of which are susceptible to sialidase removal except that of GM1 (in rare cases that too is susceptible to certain type of sialidase). The term ganglioside was proposed by Ernst Klenk in recognition of their special abundance in "Ganglionzellen" (neurons) and they indeed constitute the major sialoglycoconjugate class in the nervous system with about 75% of total bound sialic acid.[32] As with glycoproteins, they are subject to *in situ* remodeling in the process of becoming adaptable docking sites for lectins. This can occur, for example, via membrane-bound sialidase[37] or sialyltransferase,[38] as well as through a variety of other GSL-modifying enzymes located in the plasma membrane.[39] To give an example of nomenclature, ganglioside (G) GD1a (disialo = D) with four neutral sugars (chain type: gangliotetraose), the number 1 reached by the subtraction (5 minus the number of neutral sugars), is the metabolic precursor of GM1 (monosialo = M) by removal of sialic acid from the terminal galactose moiety (see Fig. 1C for the structure of the branched pentasaccharide chain of GM1 and the marking of the position of sialic acid addition to turn GM1 into GD1a; for details on nomenclature, see Refs. 31, 32, and 40). Sialidase conversion of disialoganglioside GD1a to GM1 promotes major signaling pathways in neurons and T cells (see later). Although it would appear justified to anticipate new revelations regarding GSLs functioning as counterreceptors of endogenous lectins, it is well to recall growing evidence of their ability to regulate a variety of other protein activities, as for example in GM1 interaction with TrkA NGF receptor,[41] nuclear Na^+/Ca^{2+} exchanger,[42] and δ-opioid receptor.[43]

Four lines of evidence have converged to give substance to this idea, that is, that gangliosides are functional galectin counterreceptors: (1) gangliosides, especially GM1, have neuritogenic/neuronotrophic properties and also induce differentiation in neuroblastoma cells *in vitro*,[44–47] (2) cell surface ganglioside sialidase activity is tied to differentiation and growth control in murine and human neuroblastoma cell lines,[48–51] (3) carrier-immobilized lysoganglioside GM1 binds in a specific and carbohydrate-dependent manner to human lymphocytes/monocytes and various cancer cells,[52,53] and (4) lysoganglioside GM1-exposing silica beads could react with human galectin-1 to a certain extent.[54] The human SK-N-MC neuroblastoma cells qualified as object for further study because of a conspicuous modulation of the level of surface presentation of GM1, a parameter change required for appearance of differentiation markers (acetylcholinesterase, neurofilament proteins, and neuron-specific enolase).[49–51] To collect evidence for the biochemical nature of a receptor, specific binding of lysoganglioside GM1 presented on biotinylated albumin was demonstrated, as was preferential cell adhesion to magnetic beads with ganglioside GM1, among a bead panel comprising six gangliosides and lactosylceramide tested under the same conditions.[55] As a crucial test of the hypothesis for a role of the galectin, both binding processes were impaired by the presence of a galectin-1–specific antibody, and endogenous galectin-1 could indeed be detected in the cells.[55,56] Intriguingly, its trafficking to the surface occurred, and the increase of its cell surface presentation matched the activity profile of the cell surface ganglioside sialidase.[56] This evidence builds the case for orchestrated coregulation of ligand/lectin on the level of a ganglioside, akin to the coordinated glycoprotein/galectin-1 expression described earlier. Of course, substantiation of the assumed role of ganglioside GM1 requires more experimental evidence.

Further solidifying the concept for a ganglioside GM1/galectin-1 contact, inhibition of glucosylceramide synthesis or the sialidase, the effector for the increase in GM1 presentation by desialylation of oligosialogangliosides (e.g., GD1a), as well as presence of the GM1-blocking cholera toxin B-subunit (CtxB)[55,57] significantly reduced the extent of galectin-1 association with cells (Fig. 2A and B). Glycoprotein glycans apparently cannot replace GM1 as a binding partner. Single-site mutants of galectin-1 with only minor lectin activity (W68Y/E71Q) failed to be active, excluding a carbohydrate-independent process.[57] The biochemical details of contact building were revealed by a strategic combination of NMR spectroscopic methods and molecular docking.[58] In structural terms, both the chain-end disaccharide and the sialic acid moiety of the branch (please see the details of the ganglioside's pentasaccharide chain given in Fig. 1C) are engaged in the molecular rendezvous with galectin-1.[58] Mapping the contact profile experimentally (STD NMR spectroscopy) and computationally revealed a grading in energy terms from *N*-acetylgalactosamine and terminal galactose to sialic acid, as illustrated in Figure 1C. Remarkably, this topological interplay is smoothly facilitated,

because galectin-1 selects one of the three energetically favorable conformers of the glycan chain, a feature conserved also for other galectins.[58,59] Compared to *N*-acetyllactosamine, the sialic acid moiety of the branch adds to the contact profile (Fig. 1D). Considering the thermodynamic aspects of the binding process, the inherent limits to glycan flexibility are an advantage for minimizing entropic penalties during contact formation.[13,60] The noted sharing of reactivity to the pentasaccharide among galectins prompted coincubations of lectin mixtures with cells. These experiments disclosed competitive inhibition of galectin-1 by galectin-3, a trimodular protein with one lectin site, a collagenase-sensitive tail and an N-terminal section for serine phosphorylation (Fig. 2B, inset). The affinity of cell binding in the submicromolar range depended on the integrity of cholesterol-rich microdomains,[57] being drastically lowered by a factor of 10–12 through its perturbance (Fig. 2C). This result suggests an affinity regulation on level six (Table 1). That the ganglioside sialidase has also been found associated with detergent-resistant microdomains[61] is fully in accord with the concept of an intimately coregulated recognition system. The requirement of clustered ganglioside arrangement for becoming high-affinity sites provides clues for the interpretation of results from binding assays. The apparent importance of the density of presentation may underlie the ganglioside's rather weak reactivity in the chromatographic system[54] and the reported lack of interaction in frontal affinity chromatography.[62]

These results strongly suggested the physiological relevance of ganglioside GM1/galectin-1 interplay on the cell surface. To put this reasoning to the test, inhibition of sialidase (no GM1 increase) removed one participant from this putative recognition system and released the cells from density-dependent growth inhibition (Fig. 2D). Conversely, when added in excess, galectin-1 (but not galectin-3) demostrated activity as a growth inhibitor (Fig. 2E). That galectins act not only as competitive inhibitors but also interfere with each other functionally is demonstrated by the neutralizing effect of galectin-3, the chimera-type module of this lectin family (Fig. 2F). In addition to binding, the mode of lattice formation determines the type of cellular responses, a clue to the intrafamily diversity in the galectin network.

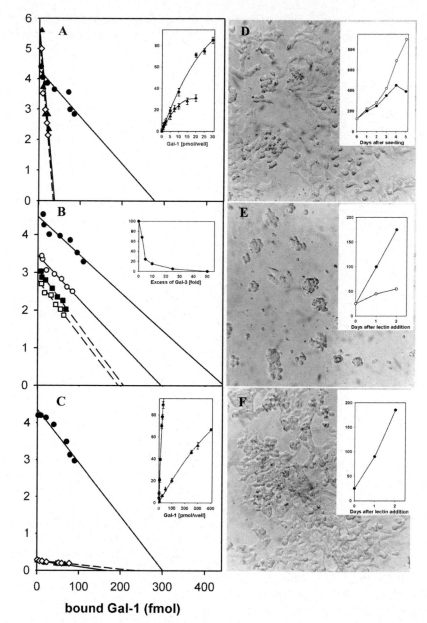

Figure 2. Analysis of binding of galectin-1 (Gal-1) to human neuroblastoma cells (Scatchard plots, binding curves as insets) exposed to different reagents (A–C), and photomicrographs of cell cultures (growth curves as insets) illustrating effects of reagents on Gal-1dependent growth inhibition (D–F). (A) Effect of the presence of two inhibitors of glucosylceramide synthesis (▲/◇) vs. control (●). (B) effect of presence of a sialidase inhibitor (○) or cholera toxin B-subunit (■) and both reagents combined (□) vs. control (●); extent of binding was also decreased by the presence of Gal-3 (inset). (C) effect of presence of two reagents for cholesterol depletion of cell membranes (▲/◇) vs. control (●). (D) Effect on cell growth of a sialidase inhibitor (○) vs. control (●), illustrating contact inhibition of cell growth. (E) Effect on cell growth of Gal-1 (○) vs. Gal-3 (●); the latter binds to cells but fails to affect growth). (F) Effect of the presence of Gal-1 together with a 10-fold excess of Gal-3. For comparison to the effects of each galectin separately, please see panel E; to Gal-3's effect on cell binding of Gal-1, please see inset to panel B). For further details, please see Refs. 51 and 55–57.

In aggregate, the coordinated cell surface presentation of galectin-1 and its cognate ganglioside in microdomains is the prerequisite to elicit growth control. Therefore, the ganglioside legitimately joins the list of galectin-1 counterreceptors. Viewed from the perspective that ganglioside GM1 is a proven modulator of growth capacity in neural systems, galectin-1 can now be counted among its receptors, for example, various growth factors and laminin-1 (here the LG4 module of the α1-chain).[46,47,63,64] At the same time, this points to the role of the ubiquitous ganglioside as a functional galectin-1 counterreceptor beyond neuroblastoma cells. Supporting evidence of this point is the colocalization of the ganglioside with galectin-1 in raft-dependent lectin endocytosis in diverse cells, including leukemic T cells.[65] Pointing to a putative role in immune regulation, regulatory T cells (T_{reg} cells) ($CD4^+$ $CD25^+$ $FoxP3^+$ T cells) contain galectin-1 and upregulate this lectin upon activation; for example, prominent staining was seen in nuclei and the inner side of the plasma membrane.[66,67] Some of the upregulated galectin-1 is secreted to the external medium upon T_{reg} cell activation.[68] This process generally occurs in synchrony with the T cell receptor (TCR) activation of T_{eff} cells ($CD4^+$ $CD25^-$ T cells), which can cause autoreactive injury unless kept at bay by activated T_{reg} cells. The apparent involvement of galectin-1 in this inter-T cell communication[66,67] suggested a possible connection between TCR activation and ganglioside GM1 on T_{eff} cells.

Suppression by inter-T cell communication has been linked to a change in the GM1 level following T_{eff} cell activation.[69,70] That cell surface sialidase activity is promoted[71] during T_{eff} cell activation supports an analogy to the neuroblastoma case described earlier. Also of interest were reports that the lectin parts of GM1-specific bacterial toxins (cholera toxin, *E.coli* heat-labile enterotoxin) are strikingly effective in the ability to ameliorate a variety of autoimmune disorders in animal models.[72–74] These observations suggested that galectin-1 secreted and/or presented by T_{reg} cells may engage the ganglioside on T_{eff} cells as counterreceptor, with the ensuing consequence to suppress their proliferation. Thus, galectin-1 exerts a paracrine pathway of communication between two cell types, as seen in epithelial–stromal cross talk in pancreatic cancer.[75]

Ganglioside–galectin contact and signaling in T cell communication

In vitro evidence for this proposed immunosuppressive mechanism was obtained with splenic T cells from wild-type mice and genetically engineered (KO) animals with disrupted GalNAcT gene (GM2/GD2 synthase, *B4galnt1*$^{-/-}$) lacking gangliotetraose gangliosides including GM1.[68] Mixed culture experiments, in which neutralization of suppression by blocking galectin-1 demonstrated the lectin's activity (Fig. 3A, left-hand side), revealed impaired communication if the tested T_{eff} cells came from KO mice (Fig. 3A, right-hand side). Fittingly, binding of labeled galectin-1 to spleen T_{eff} cells was markedly weaker for cell preparations from KO mice compared to wild-type mice (Fig. 3B). Similar to the situation in neuroblastoma cells, the loss of GM1 could not be compensated by glycoproteins. In addition, the presence of the label-free GM1-specific CtxB was shown to mask galectin-1 binding to GM1 on wild-type cells. These cells showed major elevation of whole cell and surface levels of GM1,[68] as previously reported;[69,70] GD1a, the metabolic precursor of GM1, was similarly upregulated.[68] As mentioned, activation of the T_{reg} cells resulted in pronounced increase of galectin-1 in the plasma membrane and the surrounding medium.[68] Thus, the *in vitro* evidence suggests that galectin-1 from T_{reg} cells can engage GM1 ganglioside on wild-type T_{eff} cells as the primary galectin-1 counterreceptor in T_{eff} cell suppression.

In vivo support for this mechanism was indicated in a study of murine experimental autoimmune encephalomyelitis, a widely used model for multiple sclerosis, in which wild-type animals suffered only mild disease severity when treated with galectin-1, compared to untreated mice; notably, GM1-KO animals showed significantly enhanced symptoms compared to wild-type animals.[68] A key role for GM1 cross-linking was suggested in the observation that CtxB was similarly effective in suppressing symptoms. A prior study reporting substantial benefit of galectin-1 in this autoimmune model also showed enhanced disease susceptibility in galectin-1–deficient mice.[22]

Ca^{2+}-influx as essential to T_{eff} cell suppression was suggested in earlier studies showing induction of Ca^{2+} influx by CtxB application to rat lymphocytes[76] and a leukemic T cell line;[77] galectin-1,

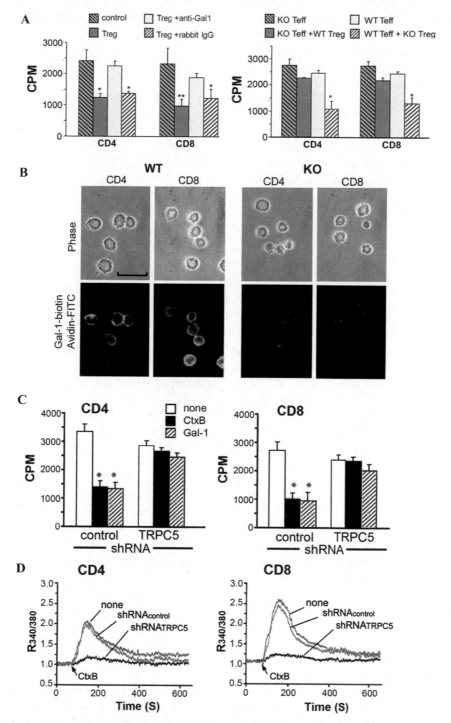

Figure 3. Involvement of ganglioside GM1, galectin-1 (Gal-1), and the TRPC5 Ca^{2+}-channel in effector T cell suppression. (A) Reduced effect of regulatory T cells on ^{3}H-thymidine incorporation by activated $CD4^{+}$ and $CD8^{+}$ T cells of wild-type mice in the presence of Gal-1-specific antibody (left-hand side) and by T cells from GM1-KO mice (right-hand side). (B) Different levels of binding of labeled Gal-1 to cells from wild-type and GM1-KO mice revealed the requirement of GM1 for substantial Gal-1 binding. (C) Knock-down of TRPC5 by shRNA blocked suppression of activated $CD4^{+}$ and $CD8^{+}$ T cells by Gal-1 and CtxB. (D) This knock-down effectively precluded Ca^{2+} influx upon CtxB binding to $CD4^{+}$ and $CD8^{+}$ T cells, as monitored by fura 2-AM. Abstracted from Ref. 68, with permission.

too, generated Ca^{2+}-influx in a nonvoltage regulated manner.[78] Such channel recordings were suggestively similar to the situation in neuroblastoma (NG108–15) cells in which the TRPC5 channel (a member of the transient receptor potential [canonical] channel subfamily belonging to the signal transduction-gated ion channels[79]) was identified as the central effector in CtxB-induced outgrowth of axon-like neurites.[80,81] Robust Ca^{2+}-flux increase was detected in activated T_{eff} cells upon cell surface GM1 cross-linking, as was TRPC5 channel upregulation on the mRNA/protein levels.[68] Viewed together, the signal-eliciting pair (GM1/galectin-1) and the putative effector after downstream signaling ("modulation at a distance"[32]) are under the control of the TCR activation process. Both the bacterial and the mammalian lectins fully depended on TRPC5 channel presence to trigger growth control, as evidenced by failure of such control following small hairpin RNA (shRNA)-mediated suppression of TRPC5 expression (Fig. 3C). The measured Ca^{2+}-influx upon GM1 cross-linking no longer occurred when cells were made devoid of TRPC5 channels (Fig. 3D). These findings raised the possibility of detecting molecular deficiencies in the delineated parameters as factors relevant for an autoimmune disease.

Type 1 diabetes was examined in that light. With focus on the insufficient control of T_{eff} cells in type 1 diabetes, the nonobese diabetic (NOD) mouse is considered to be a suitable study model.[82] Similar to the situation in human type 1 diabetes patients,[83] the murine NOD T_{eff} cells were characterized as unresponsive to regulatory T cells.[84] One reason for this might be insufficient GM1 presentation, as pinpointed in neuroblastoma (SK-N-MC) growth control impaired by the sialidase inhibitor (Fig. 2D). Accordingly, a reduction in GM1/GD1a content by ~50% or more was determined for T_{eff} cells from NOD mice compared to other tested strains (Fig. 4A, left-hand side).[85] This level of decrease was also detectable for $CD8^+$ T cells from NOD mice (Fig. 4A, right-hand side). Such cells are increasingly recognized to share center stage with $CD4^+$ T_{eff} cells in a number of autoimmune diseases.[86] Remarkably, they were reported to contain a higher level of GM1 than the $CD4^+$ cells[87] and enhanced apoptosis induction by GM1-specific enterotoxin.[87,88] Experiments with mixed cultures of T cell populations from NOD and BALB/c mice,

the latter not subject to defects in GM1, supported the notion for the molecular cause of dysfunctional communication to reside in the T_{eff} cells (Fig. 4B); this was further supported by GM1 supplementation. GM1, like gangliosides in general, is able to insert spontaneously into the plasma membrane of cells in culture and thereafter assume regular functions at that locus.[89] When testing this approach, exogenous GM1 restored responsiveness of NOD T_{eff} cells to a significant extent, both on the level of growth inhibition (Fig. 4C) and Ca^{2+} flux in the presence of galectin-1 (Fig. 4D). In line with this hypothesis, T_{eff} cell suppression was blocked by use of a noncross-linking anti-GM1 IgG antibody (probably precluding galectin-1–dependent lattice formation in the same way as galectin-3 does on neuroblastoma cells).[85] The role of TRPC5 channels was further verified in the diminished Ca^{2+} influx caused by a TRP channel inhibitor (SK&F 96365) and the above mentioned noncross-linking anti-GM1 IgG antibody. At the *in vivo* level, these results for GM1 (and indirectly GD1a) were consonant with its ability to delay and reduce the frequency of type 1 diabetes in NOD mice.[90] In a broader context, a brain ganglioside mixture, which included GM1 as a major component, proved effective in suppressing autoimmune demyelination in animal models.[91–95] Potential consideration for human therapy is thus implied if T_{eff} cells from type 1 diabetes patients prove deficient in GM1/GD1a, which is analogous to cells from NOD mice.

The signaling sequence from the initial contact to actual Ca^{2+} influx was found to involve *in situ* association of GM1 with $\alpha_4/\alpha_5\beta_1$ integrins within the plasma membrane, thus likely levels five and six in affinity regulation (Table 1).[68] Cross-linking of GM1 thus results in mutual cross-linking of the heterodimeric integrin, which, as noted earlier for the contact of galectin-1 with the fibronectin receptor in carcinoma cells, ignites an outside–in signaling cascade. Considering the processes *en route* to neurite outgrowth by opening of TRPC5 Ca^{2+} channels, induced signaling included autophosphorylation of focal adhesion kinase as well as activation of phospholipase Cγ and phosphoinositide-3 kinase.[81] Because galectin-1 could recently be demonstrated to function as an endogenous initiator of axonogenesis in cultures of murine cerebellar granule neurons (Wu *et al.*, in preparation), the contact and effector systems appear to be maintained, making similar

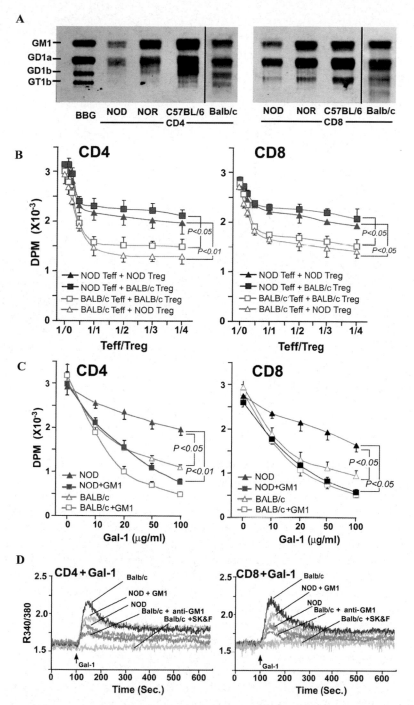

Figure 4. Detection of deficiency in GM1/GD1a in activated effector T cells (T_{effs}) of NOD mice, its consequence for inter-T cell communication and its amelioration by GM1 replacement *in vitro*. (A) High-performance thin-layer chromatography using bovine brain ganglioside (BBG) mixture as control revealed a NOD strain-specific defect. (B) Experiments with mixed cultures of cells from NOD and BALB/c mice to determine the level of T_{eff} cell proliferation revealed maintained effectiveness of T_{reg} cells but impaired responsiveness of T_{eff} cells. (C) Robust suppression of Balb/c T_{eff} cells occurred with galectin-1 (Gal-1), and failed suppression of NOD T_{eff} cells was corrected by GM1 pretreatment. (D) Involvement of TRPC5 Ca^{2+}-channels (see also Fig. 3D) was evidenced by measuring the Ca^{2+}-influx for these cell populations; TRP channel inhibitor (SK&F) blocked Ca^{2+}-influx, addition of a noncross-linking anti-GM1 antibody impaired the Gal-1-dependent effect, and pretreatment with GM1 improved the defective signal for activated T_{eff} cells from NOD mice. From Ref. 85, with permission.

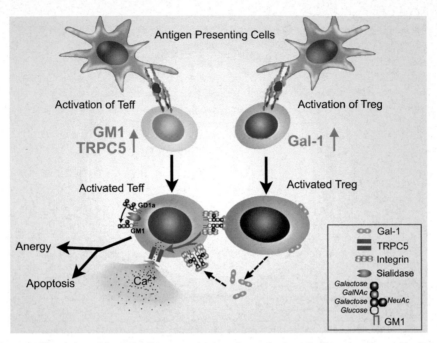

Figure 5. Schematic illustration of inter-T cell communication after activation of effector/regulatory T cells via ganglioside GM1/galectin-1 contact. T cell receptor activation of T_{reg} cells by antigen-presenting cells causes upregulation of galectin-1 (Gal-1) that is expressed on the T_{reg} cells surface and released into the surroundings. As a homodimer it cross-links GM1, which has been elevated through sialidase reaction (and possibly *de novo* synthesis) in the plasma membrane of T_{eff} cell following activation of the latter. This induces co-cross-linking of heterodimeric integrin, which is associated with GM1, and this in turn induces a signaling sequence resulting in activation of TRPC5 Ca^{2+} channels. Elevated intracellular Ca^{2+} in T_{eff} cells prevents proliferation through anergy and/or apoptosis.

signal routing likely, even including participation of a sialidase.[47] Bringing these lines of evidence together, Figure 5 depicts a flow scheme to summarize the major conclusions. The metabolic conversion of GD1a to GM1, the galectin-1 counterreceptor, can be carried out by a sialidase, for T cells, neuroblastoma cells, and neurons alike. This does not preclude enhanced *de novo* synthesis of GM1/GD1a as an additional source of GM1 elevation. The cross-linking of GM1, along with the associated integrin(s), then builds the signaling complex, the starting point for the flow of information to eventually open TRPC5 Ca^{2+} channels. The required amplitude of the signal is guaranteed by upregulating GM1, galectin-1, and TRPC5 channels in the course of T cell activation. Strategically, this orchestrated upregulation of galectin-1, its counterreceptor and the growth-control effector upon T cell activation bears remarkable resemblance to the way the tumor suppressor p16[INK4a] coordinates anoikis induction described earlier. Interestingly, a case of another Ca^{2+} channel (i.e., TRPV5, which is essential for transcel-

lular Ca^{2+} reabsorption in distal nephron) is known, where direct *N*-glycan remodeling (i.e., $\alpha 2,6$ desialylation by Klotho) prolongs cell surface retention by lattice formation with galectin-1.[96] Galectin-mediated counterreceptor cross-linking involving *N*-glycan branching (level 4b in Table 1) appears to play into basal and activation signaling through the TCR and CD45 via differential partitioning to ganglioside GM1-containing rafts, and growth arrest by cytotoxic T lymphocyte antigen-4. Deficiency in $\beta 1,6$ *N*-acetylglucosaminyltransferase V (Mgat5), an enzyme that regulates branching within attached *N*-glycans,[97,98] was found to lower T cell activation thresholds by directly enhancing TCR clustering (Boscher *et al.*[16] provides further reading and details).[99]

That sugar coding either in glycoprotein or glycolipid glycans—depending on the cell type—accounts for the cellular response convincingly illustrates the legitimate place of gangliosides as integral components in the toolbox of galectin signaling for growth control. In other words, the galectin reliably traces its cell type–associated cognate glycans in an exquisite manner, thereby giving postbinding events a clear direction and eliciting growth regulation[26] (and cell adhesion/attachment[100]). The case of galectin-7 illustrates that the same (GM1-neuroblastoma) or different counterreceptors (T cell death) can be selected in different cells *en route* to growth regulation.[101,102] Looking beyond growth regulation, recent evidence has disclosed the pertinent role of a tandem-repeat-type galectin (i.e., galectin-4, with two different lectin domains linked by a peptide) in apical membrane trafficking.

Glycosphingolipid/galectin contact in apical glycoprotein delivery

The cellular polarity of apical and basolateral surfaces in the epithelium depends on sorting signals. Segregation of glycoproteins to be routed to the apical side takes place in the trans-Golgi network by their recruitment to detergent-resistant membranes (microdomains rich in cholesterol and sphingolipids), which characteristically also contain galectin-4.[103,104] This bivalent lectin, together with GSLs particularly enriched in long-chain fatty acids (up to C26), plays a crucial role. Galectin-4 binds sulfatide (3′-sulfated galactosylceramide, SM4) bearing such 2′-hydroxylated fatty acids and is essential both for the apical machinery and the loading of apical cargo in the appropriate microdomains.[105,106] The above-average chain length may help make the compact sugar headgroup spatially accessible for contact. Cross-linking by galectin-4 favors microdomain stability and aggregation to "superrafts."[102] Moreover, the lectin is capable of singling out glycoproteins "for the ride" on these microdomains. Consequently, several glycoproteins destined to be routed to the apical side, such as the marker dipeptidylpeptidase-IV, fail to associate to the microdomains in the absence of galectin-4, and thus accumulate in the cytoplasm.[106] A high density of *N*-acetyllactosamine at branch ends is the "recognition feature" of the respective glycoproteins (a "sugar-encoded ZIP code," if you will) that is read by galectin-4.[107–110] Before it will come into contact with the microdomains and the cargo, galectin-4 needs to be secreted by the nonclassical pathway, a common process for galectins, which have no signal sequence.[111] After externaliza-

tion, galectin-4 is taken up by endocytosis and travels to the trans-Golgi network for its assignment.

In general terms, the microdomain architecture with its high density of GSLs (here sulfatide, GM1 in neuroblastoma, and T_{eff} cells) thus establishes a special area suited for ligand presentation to endogenous lectins, in growth control and transport. It should at this point be noted that other lectin classes can also exploit such a recognition platform, for example, the C-type lectins L- and P-selectin both binding sulfatide.[112–114] The same also applies to places of permanent GSL abundance such as the myelin sheath, where L-selectin is a sensor in the process cascade of sulfatide-induced generation of inflammatory mediators, and GD1a plus trisialo ganglioside GT1b are counterreceptors of siglec-4 (myelin-associated glycoprotein) to lead to axon–myelin interaction and regulation of axon outgrowth.[115]

Conclusions

The study of nerve cells holds salient lessons for our view on the standing of the glycan part of GSLs in biocoding. Their abundance, high-density presentation, and dynamic remodeling are all factors arguing in favor of their bioactivity as lectin counterreceptors. First, the study of model systems helped discern common principles of coordinated glycan/lectin expression for glycoproteins and glycolipids. Then analogies between nerve and immune cells in lectin-mediated growth control were unraveled—in terms of coregulation of GM1/galectin and the postbinding signaling. Because galectins and gangliosides (GM1/GD1a) are present in the nucleus and nuclear envelope, respectively,[116–121] it is a reasonable possibility to also consider functional interactions at these sites. The case for GSLs as genuine counterreceptors was finally strengthened by illustrating a role in intracellular glycoprotein routing. That other classes of lectins also use them as contact sites underscores their broad role in biocoding. Looking for potential medical applications, the detection of deficiencies in the glycolipid/lectin network, as illustrated for T_{eff} cells and GM1 presentation in the NOD mouse, encourages monitoring of this parameter in patients and testing glycolipid/lectin-specific supplementation with the aim of reestablishing proper inter-T cell communication.

Conflict of interest

The authors declare no conflicts of interest.

References

1. Gabius, H.-J. Ed. 2009. *The Sugar Code. Fundamentals of Glycosciences.* Wiley-VCH. Weinheim, Germany.
2. Laine, R.A. 1997. The information-storing potential of the sugar code. In *Glycosciences: Status and Perspectives.* H.-J. Gabius & S. Gabius, Eds.: 1–14. Chapman & Hall. London/Weinheim.
3. Zuber, C. & J. Roth. 2009. N-Glycosylation. In *The Sugar Code. Fundamentals of Glycosciences.* H.-J. Gabius, Ed.: 87–110. Wiley-VCH. Weinheim, Germany.
4. Gahmberg, C.G. & M. Tolvanen. 1996. Why mammalian cell surface proteins are glycoproteins. *Trends Biochem. Sci.* **21:** 308–311.
5. Zielinska, D.F., F. Gnad, J.R. Wisniewski & M. Mann. 2010. Precision mapping of an *in vivo* N-glycoproteome reveals rigid topological and sequence constraints. *Cell* **141:** 897–907.
6. Patsos, G. & A. Corfield. 2009. O-Glycosylation: structural diversity and function. In *The Sugar Code. Fundamentals of Glycosciences.* H.-J. Gabius, Ed.: 111–137. Wiley-VCH. Weinheim, Germany.
7. Haltiwanger, R.S. & J.B. Lowe. 2004. Role of glycosylation in development. *Annu. Rev. Biochem.* **73:** 491–537.
8. Patsos, G., S. André, N. Roeckel, *et al.* 2009. Compensation of loss of protein function in microsatellite-unstable colon cancer cells (HCT116): a gene-dependent effect on the cell surface glycan profile. *Glycobiology* **19:** 726–734.
9. van de Wouwer, M., S. André, H.-J. Gabius & A. Villalobo. 2011. Nitric oxide changes distinct aspects of the glycophenotype of human neuroblastoma NB69 cells. *Nitric Oxide* **24:** 91–101.
10. André, S., T. Kozár, S. Kojima, C. Unverzagt & H.-J. Gabius. 2009. From structural to functional glycomics: core substitutions as molecular switches for shape and lectin affinity of N-glycans. *Biol. Chem.* **390:** 557–565.
11. Cummings, R.D. 2009. The repertoire of glycan determinants in the human glycome. *Mol. BioSyst.* **5:** 1087–1104.
12. Parker, R.B. & J.J. Kohler. 2010. Regulation of intracellular signaling by extracellular glycan remodeling. *ACS Chem. Biol.* **5:** 35–46.
13. Gabius, H.-J., S. André, J. Jiménez-Barbero, A. Romero & D. Solís. 2011. From lectin structure to functional glycomics: principles of the sugar code. *Trends Biochem. Sci.* **36:** 298–313.
14. Brewer, C.F., M.C. Miceli & L.G. Baum. 2002. Clusters, bundles, arrays and lattices: novel mechanisms for lectin-saccharide-mediated cellular interactions. *Curr. Opin. Struct. Biol.* **12:** 616–623.
15. Rabinovich, G.A., M.A. Toscano, S.S. Jackson & G.R. Vasta. 2007. Functions of cell surface galectin-glycoprotein lattices. *Curr. Opin. Struct. Biol.* **17:** 513–520.
16. Boscher, C., J.W. Dennis, I.R. Nabi. 2011. Glycosylation, galectins and cellular signaling. *Curr. Opin. Cell Biol.* **23:** 383–392.

17. Kaltner, H. & H.-J. Gabius. 2012. A toolbox of lectins for translating the sugar code: the galectin network in phylogenesis and tumors. *Histol. Histopathol.* **27:** 397–416.
18. López-Lucendo, M.F., D. Solís, S. André, *et al.* 2004. Growth-regulatory human galectin-1: crystallographic characterisation of the structural changes induced by single-site mutations and their impact on the thermodynamics of ligand binding. *J. Mol. Biol.* **343:** 957–970.
19. Fischer, C., H. Sanchez-Ruderisch, M. Welzel, *et al.* 2005. Galectin-1 interacts with the $\alpha_5\beta_1$ fibronectin receptor to restrict carcinoma cell growth via induction of p21 and p27. *J. Biol. Chem.* **280:** 37266–37277.
20. Sanchez-Ruderisch, H., K.M. Detjen, M. Welzel, *et al.* 2011. Galectin-1 sensitizes carcinoma cells to anoikis via the fibronectin receptor $\alpha_5\beta_1$-integrin. *Cell Death Differ.* **18:** 806–816.
21. Amano, M., M. Galvan, J. He & L.G. Baum. 2003. The ST6Gal1 sialyltransferase selectively modifies N-glycans on CD45 to negatively regulate galectin-1-induced CD45 clustering, phosphatase modulation, and T cell death. *J. Biol. Chem.* **278:** 7469–7475.
22. Toscano, M.A., G.A. Bianco, J.M. Ilarregui, *et al.* 2007. Differential glycosylation of T_H1, T_H2 and T_H-17 effector cells selectively regulates susceptibility to cell death. *Nature Immunol.* **8:** 825–834.
23. André, S., H. Sanchez-Ruderisch, H. Nakagawa, *et al.* 2007. Tumor suppressor p16^{INK4a}: modulator of glycomic profile and galectin-1 expression to increase susceptibility to carbohydrate-dependent induction of anoikis in pancreatic carcinoma cells. *FEBS J.* **274:** 3233–3256.
24. Plath, T., K. Detjen, M. Welzel, *et al.* 2000. A novel function for the tumor suppressor p16^{INK4a}: induction of anoikis via upregulation of the $\alpha_5\beta_1$ fibronectin receptor. *J. Cell Biol.* **150:** 1467–1478.
25. Sanchez-Ruderisch, H., C. Fischer, K.M. Detjen, *et al.* 2010. Tumor suppressor p16^{INK4a}: downregulation of galectin-3, an endogenous competitor of the pro-anoikis effector galectin-1, in a pancreatic carcinoma model. *FEBS J.* **277:** 3552–3563.
26. Villalobo, A., A. Nogales-Gonzáles & H.-J. Gabius. 2006. A guide to signaling pathways connecting protein-glycan interaction with the emerging versatile effector functionality of mammalian lectins. *Trends Glycosci. Glycotechnol.* **18:** 1–37.
27. Garner, O.B. & L.G. Baum. 2008. Galectin-glycan lattices regulate cell-surface glycoprotein organization and signalling. *Biochem. Soc. Trans.* **36:** 1472–1477.
28. Bi, S. & L.G. Baum. 2009. Sialic acids in T cell development and function. *Biochim. Biophys. Acta* **1790:** 1599–1610.
29. Blasko, A., R. Fajka-Boja, G. Ion & E. Monostori. 2011. How does it act when soluble? Critical evaluation of mechanism of galectin-1 induced T-cell apoptosis. *Acta Biol. Hung.* **62:** 106–111.
30. Merrill, A.H., Jr., M.D. Wang, M. Park & M.C. Sullards. 2007. (Glyco)sphingolipidology: an amazing challenge and opportunity for systems biology. *Trends Biochem. Sci.* **32:** 457–468.
31. Kopitz, J. 2009. Glycolipids. In *The Sugar Code. Fundamentals of Glycosciences.* H.-J. Gabius, Ed.: 177–198. Wiley-VCH. Weinheim, Germany.

32. Ledeen, R.W. & G. Wu. 2009. Neurobiology meets glyco-sciences. In *The Sugar Code. Fundamentals of Glycosciences.* H.-J. Gabius, Ed.: 495–516. Wiley-VCH. Weinheim, Germany.

33. Lopez, P.H. & R.L. Schnaar. 2009. Gangliosides in cell recognition and membrane protein regulation. *Curr. Opin. Struct. Biol.* **19:** 549–557.

34. Yu, R.K., M. Yanagisawa & T. Ariga. 2007. Glycosph-ingolipid structures. In *Comprehensive Glycoscience.* J.P. Kamerling, Ed.: 73–122. Elsevier. Oxford, UK.

35. Svennerholm, L. 1976. Interaction of cholera toxin and ganglioside G(M1). *Adv. Exp. Med. Biol.* **71:** 191–204.

36. Ledeen, R.W. & G. Wu. 2010. In search of a solution to the Sphinx-like riddle of GM1. *Neurochem. Res.* **35:** 1867–1874.

37. Miyagi, T., T. Wada, A. Iwamatsu, *et al.* 1999. Molecu-lar cloning and characterization of a plasma membrane-associated sialidase specific for gangliosides. *J. Biol. Chem.* **274:** 5004–5011.

38. Durrie, R., M. Saito & A. Rosenberg. 1988. Endogenous glycosphingolipid acceptor specificity of sialosyltransferase systems in intact Golgi membranes, synaptosomes, and synaptic plasma membranes from rat brain. *Biochemistry* **27:** 3759–3764.

39. Prinetti, A., V. Chigorno, L. Mauri, *et al.* 2007. Mod-ulation of cell functions by glycosphingolipid metabolic remodeling in the plasma membrane. *J. Neurochem.* **103**(Suppl 1): 113–125.

40. *Nomenclature of Glycolipids.* International Union of Pure and Applied Chemistry. Available at: http://www.chem.qmul.ac.uk/iupac.

41. Mutoh, T., T. Hamano, S. Yano, *et al.* 2002. Stable trans-fection of GM1 synthase gene into GM1-deficient NG108–15 cells, CR-72 cells, rescues the responsiveness of Trk-neurotrophin receptor to its ligand, NGF. *Neurochem. Res.* **27:** 801–806.

42. Xie, X., G. Wu, Z.H. Lu & R.W. Ledeen. 2002. Potentiation of a sodium-calcium exchanger in the nuclear envelope by nuclear GM1 ganglioside. *J. Neurochem.* **81:** 1185–1195.

43. Wu, G., Z.H. Lu & R.W. Ledeen. 1997. Interaction of the δ-opioid receptor with GM1 ganglioside: conversion from inhibitory to excitatory mode. *Mol. Brain Res.* **44:** 341–346.

44. Facci, L., A. Leon, G. Toffano, *et al.* 1984. Promotion of neuritogenesis in mouse neuroblastoma cells by exogenous gangliosides. Relationship between the effect and the cell association of ganglioside GM1. *J. Neurochem.* **42:** 299–305.

45. Ledeen, R.W. 1984. Biology of gangliosides: neuritogenic and neuronotrophic properties. *J. Neurosci. Res.* **12:** 147–159.

46. Ledeen, R.W., G.S. Wu, Z.H. Lu, D. Kozireski-Chubak & Y. Fang. 1998. The role of GM1 and other gangliosides in neuronal differentiation – overview and new findings. *Ann. NY Acad. Sci.* **845:** 341–348.

47. Abad-Rodriguez, J. & A. Robotti. 2007. Regulation of ax-onal development by plasma membrane gangliosides. *J. Neurochem.* **103**(Suppl 1): 47–55.

48. Schengrund, C.L. & M.A. Repman. 1982. Density-dependent changes in gangliosides and sialidase activity of murine neuroblastoma cells. *J. Neurochem.* **39:** 940–947.

49. Kopitz, J., C. von Reitzenstein, C. Mühl & M. Cantz. 1994.

50. Kopitz, J., C. von Reitzenstein, K. Sinz & M. Cantz. 1996. Selective ganglioside desialylation in the plasma membrane of human neuroblastoma cells. *Glycobiology* **6:** 367–376.

Role of plasma membrane ganglioside sialidase of human neuroblastoma cells in growth control and differentiation. *Biochem. Biophys. Res. Comm.* **199:** 1188–1193.

51. Kopitz, J., C. Mühl, V. Ehemann, *et al.* 1997. Effects of cell surface ganglioside sialidase inhibition on growth control and differentiation of human neuroblastoma cells. *Eur. J. Cell Biol.* **73:** 1–7.

52. Gabius, S., K. Kayser, K.P. Hellmann, *et al.* 1990. Carrier-immobilized derivatized lysoganglioside GM1 is a ligand for specific binding sites in various human tumor cell types and peripheral blood lymphocytes and monocytes. *Biochem. Biophys. Res. Commun.* **169:** 239–244.

53. Hassid, S., I. Salmon, N.V. Bovin, *et al.* 1996. Histochemical expression of binding sites for labelled hyaluronic acid and carrier-immobilized synthetic (histo-blood group trisac-charides) or biochemically purified (ganglioside GM1) gly-coligands in nasal polyps and other human lesions includ-ing neoplasms. *Histol. Histopathol.* **11:** 985–992.

54. Caron, M., R. Joubert-Caron, J.R. Cartier, *et al.* 1993. Study of lectin-ganglioside interactions by high-performance liq-uid affinity chromatography. *J. Chromatography* **646:** 327–333.

55. Kopitz, J., C. von Reitzenstein, M. Burchert, *et al.* 1998. Galectin-1 is a major receptor for ganglioside GM1, a prod-uct of the growth-controlling activity of a cell surface gan-glioside sialidase, on human neuroblastoma cells in culture. *J. Biol. Chem.* **273:** 11205–11211.

56. Kopitz, J., C. von Reitzenstein, S. André, *et al.* 2001. Negative regulation of neuroblastoma cell growth by carbohydrate-dependent surface binding of galectin-1 and functional divergence from galectin-3. *J. Biol. Chem.* **276:** 35917–35923.

57. Kopitz, J., M. Bergmann & H.-J. Gabius. 2010. How adhesion/growth-regulatory galectins-1 and -3 attain cell specificity: case study defining their target on neurob-lastoma cells (SK-N-MC) and marked affinity regulation by affecting microdomain organization of the membrane. *IUBMB Life* **62:** 624–628.

58. Siebert, H.-C., S. André, S.-Y. Lu, *et al.* 2003. Unique conformer selection of human growth-regulatory lectin galectin-1 for ganglioside GM1 versus bacterial toxins. *Bio-chemistry* **42:** 14762–14773.

59. André, S., H. Kaltner, M. Lensch, *et al.* 2005. Determina-tion of structural and functional overlap/divergence of five proto-type galectins by analysis of the growth-regulatory interaction with ganglioside GM1 *in silico* and *in vitro* on human neuroblastoma cells. *Int. J. Cancer* **114:** 46–57.

60. Solís, D., A. Romero, M. Menéndez & J. Jiménez-Barbero. 2009. Protein-carbohydrate interactions: basic concepts and methods for analysis. In *The Sugar Code. Fundamentals of Glycosciences.* H.-J. Gabius, Ed.: 233–245. Wiley-VCH. Weinheim, Germany.

61. Kalka, D., C. von Reitzenstein, J. Kopitz & M. Cantz. 2001. The plasma membrane ganglioside sialidase cofractionates with markers of lipid rafts. *Biochem. Biophys. Res. Commun.* **283:** 989–993.

62. Hirabayashi, J., T. Hashidate, Y. Arata, *et al.* 2002. Oligosaccharide specificity of galectins: a search by frontal affinity chromatography. *Biochim. Biophys. Acta* **1572:** 232–254.

63. Inokuchi, J.-i. & K. Kabayama. 2008. Modulation of growth factors receptors in membrane microdomains. *Trends Glycosci. Glycotechnol.* **20:** 353–371.

64. Ichikawa, N., K. Iwabuchi, H. Kurihara, *et al.* 2009. Binding of laminin-1 to monosialoganglioside GM1 in lipid rafts is crucial for neurite outgrowth. *J. Cell Sci.* **122:** 289–299.

65. Fajka-Boja, R., A. Blasko, F. Kovacs-Solyom, *et al.* 2008. Co-localization of galectin-1 with GM1 ganglioside in the course of its clathrin- and raft-dependent endocytosis. *Cell. Mol. Life Sci.* **65:** 2586–2593.

66. Sugimoto, N., T. Oida, K. Hirota, *et al.* 2006. Foxp3-dependent and -independent molecules specific for CD25$^+$CD4$^+$ natural regulatory T cells revealed by DNA microarray analysis. *Int. Immunol.* **18:** 1197–1209.

67. Garin, M.I., C.C. Chu, D. Golshayan, *et al.* 2007. Galectin-1: a key effector of regulation mediated by CD4$^+$CD25$^+$ T cells. *Blood* **109:** 2058–2065.

68. Wang, J., Z.H. Lu, H.-J. Gabius, *et al.* 2009. Cross-linking of GM1 ganglioside by galectin-1 mediates regulatory T cell activity involving TRPC5 channel activation: possible role in suppressing experimental autoimmune encephalomyelitis. *J. Immunol.* **182:** 4036–4045.

69. Tuosto, L., I. Parolini, S. Schroder, *et al.* 2001. Organization of plasma membrane functional rafts upon T cell activation. *Eur. J. Immunol.* **31:** 345–349.

70. Brumeanu, T.D., A. Preda-Pais, C. Stoica, *et al.* 2007. Differential partitioning and trafficking of GM gangliosides and cholesterol-rich lipid rafts in thymic and splenic CD4 T cells. *Mol. Immunol.* **44:** 530–540.

71. Wang, P., J. Zhang, H. Bian, *et al.* 2004. Induction of lysosomal and plasma membrane-bound sialidases in human T-cells via T-cell receptor. *Biochem. J.* **380:** 425–433.

72. Sobel, D.O., B. Yankelevich, D. Goyal, *et al.* 1998. The B-subunit of cholera toxin induces immunoregulatory cells and prevents diabetes in the NOD mouse. *Diabetes* **47:** 186–191.

73. Williams, N.A., T.R. Hirst & T.O. Nashar. 1999. Immune modulation by the cholera-like enterotoxins: from adjuvant to therapeutic. *Immunol. Today* **20:** 95–101.

74. Salmond, R.J., J.A. Luross & N.A. Williams. 2002. Immune modulation by the cholera-like enterotoxins. *Expert Rev. Mol. Med.* **4:** 1–16.

75. Roda, O., E. Ortiz-Zapater, N. Martín-Bosch, *et al.* 2009. Galectin-1 is a novel functional receptor for tissue plasminogen activator in pancreatic cancer. *Gastroenterology* **136:** 1379–1390.

76. Dixon, S.J., D. Stewart, S. Grinstein & S. Spiegel. 1987. Transmembrane signaling by the B subunit of cholera toxin: increased cytoplasmic free calcium in rat lymphocytes. *J. Cell Biol.* **105:** 1153–1161.

77. Gouy, H., P. Deterre, P. Debre & G. Bismuth. 1994. Cell calcium signaling via GM1 cell surface gangliosides in the human Jurkat T cell line. *J. Immunol.* **152:** 3271–3281.

78. Walzel, H., M. Blach, J. Hirabayashi, *et al.* 2000. Involvement of CD2 and CD3 in galectin-1 induced signaling in human Jurkat T-cells. *Glycobiology* **10:** 131–140.

79. Clapham, D.E., L.W. Runnels & C. Strubing. 2001. The TRP ion channel family. *Nat. Rev. Neurosci.* **2:** 387–396.

80. Fang, Y., X. Xie, R.W. Ledeen & G. Wu. 2002. Characterization of cholera toxin B subunit-induced Ca^{2+} influx in neuroblastoma cells: evidence for a voltage-independent GM1 ganglioside-associated Ca^{2+} channel. *J. Neurosci. Res.* **69:** 669–680.

81. Wu, G., Z.H. Lu, A.G. Obukhov, *et al.* 2007. Induction of calcium influx through TRPC5 channels by cross-linking of GM1 ganglioside associated with $\alpha_5\beta_1$ integrin initiates neurite outgrowth. *J. Neurosci.* **27:** 7447–7458.

82. Anderson, M.S. & J.A. Bluestone. 2005. The NOD mouse: a model of immune dysregulation. *Annu. Rev. Immunol.* **23:** 447–485.

83. Schneider, A., M. Rieck, S. Sanda, *et al.* 2008. The effector T cells of diabetic subjects are resistant to regulation via CD4$^+$ FOXP3$^+$ regulatory T cells. *J. Immunol.* **181:** 7350–7355.

84. D'Alise, A.M., V. Auyeung, M. Feuerer, *et al.* 2008. The defect in T-cell regulation in NOD mice is an effect on the T-cell effectors. *Proc. Natl. Acad. Sci. USA* **105:** 19857–19862.

85. Wu, G., Z.H. Lu, H.-J. Gabius, *et al.* 2011. Ganglioside GM1 deficiency in effector T cells from NOD mice induces resistance to regulatory T cell suppression. *Diabetes* **60:** 2341–2349.

86. Walter, U. & P. Santamaria. 2005. CD8$^+$ T cells in autoimmunity. *Curr. Opin. Immunol.* **17:** 624–631.

87. de Mello Coelho, V., D. Nguyen, B. Giri, *et al.* 2004. Quantitative differences in lipid raft components between murine CD4$^+$ and CD8$^+$ T cells. *BMC Immunol.* **5:** 2.

88. Nashar, T.O., N.A. Williams & T.R. Hirst. 1996. Cross-linking of cell surface ganglioside GM1 induces the selective apoptosis of mature CD8$^+$ T lymphocytes. *Int. Immunol.* **15:** 731–736.

89. Wu, G. & R.W. Ledeen. 1994. Gangliosides as modulators of neuronal calcium. *Progr. Brain Res.* **101:** 101–112.

90. Vieira, K.P., A.R. de Almeida e Silva Lima Zollner, C. Malaguti, *et al.* 2008. Ganglioside GM1 effects on the expression of nerve growth factor (NGF), Trk-A receptor, proinflammatory cytokines and on autoimmune diabetes onset in non-obese diabetic (NOD) mice. *Cytokine* **42:** 92–104.

91. Shimada, K., C.S. Koh, K. Uemura & N. Yanagisawa. 1994. Suppression of experimental allergic encephalomyelitis in Lewis rats by administration of gangliosides. *Cell. Immunol.* **154:** 231–239.

92. Sekiguchi, Y., M. Ichikawa, A. Inoue, *et al.* 2001. Brain-derived gangliosides suppress the chronic relapsing-remitting experimental autoimmune encephalomyelitis in NOD mice induced with myelin oligodendrocyte glycoprotein peptide. *J. Neuroimmunol.* **116:** 196–205.

93. Monteiro de Castro, G., M. Eduarda Zanin, D. Ventura-Oliveira, *et al.* 2004. Th1 and Th2 cytokine immunomodulation by gangliosides in experimental autoimmune encephalomyelitis. *Cytokine* **26:** 155–163.

94. Ponzin, D., A.M. Menegus, G. Kirschner, *et al.* 1991. Effects of gangliosides on the expression of autoimmune

demyelination in the peripheral nervous system. *Ann. Neurol.* **30:** 678–685.

95. Ledeen, R.W., B. Oderfeld-Nowak, C.F. Brosnan & A. Cervone. 1990. Gangliosides offer partial protection in experimental allergic neuritis. *Ann. Neurol.* **27**(Suppl): S69–S74.

96. Cha, S.K., B. Ortega, H. Kurosu, *et al.* 2008. Removal of sialic acid involving Klotho causes cell-surface retention of TRPV5 channel via binding to galectin-1. *Proc. Natl. Acad. Sci. USA* **105:** 9805–9810.

97. Chen, I.-J., H.-L. Chen & M. Demetriou. 2007. Lateral compartmentalization of T cell receptor versus CD45 by galectin-N-glycan binding and microfilaments coordinate basal and activation signaling. *J. Biol. Chem.* **282:** 35361–35372.

98. Lau, K.S., E.A. Partridge, A. Grigorian, *et al.* 2007. Complex N-glycan number and degree of branching cooperate to regulate cell proliferation and differentiation. *Cell* **129:** 123–134.

99. Demetriou, M., M. Granovsky, S. Quaggin & J.W. Dennis. 2001. Negative regulation of T-cell activation and autoimmunity by Mgat5 N-glycosylation. *Nature* **409:** 733–739.

100. Gabius, H.-J. 2006. Cell surface glycans: the why and how of their functionality as biochemical signals in lectin-mediated information transfer. *Crit. Rev. Immunol.* **26:** 43–79.

101. Sturm, A., M. Lensch, S. André, *et al.* 2004. Human galectin-2: novel inducer of T cell apoptosis with distinct profile of caspase activation. *J. Immunol.* **173:** 3825–3837.

102. Kopitz, J., S. André, C. von Reitzenstein, *et al.* 2003. Homodimeric galectin-7 (p53-induced gene 1) is a negative growth regulator for human neuroblastoma cells. *Oncogene* **22:** 6277–6288.

103. Danielsen, E.M. & B. van Deurs. 1997. Galectin-4 and small intestinal brush border enzymes form clusters. *Mol. Biol. Cell* **8:** 2241–2251.

104. Danielsen, E.M. & G.H. Hansen. 2006. Lipid raft organization and function in brush borders of epithelial cells. *Mol. Membr. Biol.* **23:** 71–79.

105. Delacour, D., V. Gouyer, J.-P. Zanetta, *et al.* 2005. Galectin-4 and sulfatides in apical membrane trafficking in enterocytelike cells. *J. Cell Biol.* **169:** 491–501.

106. Stechly, L., W. Morelle, A.F. Dessein, *et al.* 2009. Galectin-4-regulated delivery of glycoproteins to the brush border membrane of enterocyte-like cells. *Traffic* **10:** 438–450.

107. Wu, A.M., J.H. Wu, J.-H. Liu, *et al.* 2004. Effects of polyvalency of glycotopes and natural modifications of human blood group ABH/Lewis sugars at the β-terminated core saccharides on the binding of domain-I of recombinant tandem-repeat-type galectin-4 from rat gastrointestinal tract (G4-N). *Biochimie* **86:** 317–326.

108. Dam, T.K., H.-J. Gabius, S. André, *et al.* 2005. Galectins bind to the multivalent glycoprotein asialofetuin with enhanced affinities and a gradient of decreasing binding constants. *Biochemistry* **44:** 12564–12571.

109. Ideo, H., A. Seko & K. Yamashita. 2005. Galectin-4 binds to sulfated glycosphingolipids and carcinoembryonic antigen in patches on the cell surface of human colon adenocarcinoma cells. *J. Biol. Chem.* **280:** 4730–4737.

110. Morelle, W., L. Stechly, S. André, *et al.* 2009. Glycosylation pattern of brush border-associated glycoproteins in enterocyte-like cells: involvement of complex-type N-glycans in apical trafficking. *Biol. Chem.* **390:** 529–544.

111. Hughes, R.C. 1999. Secretion of the galectin family of mammalian carbohydrate-binding proteins. *Biochim. Biophys. Acta* **1473:** 172–185.

112. Aruffo, A., W. Kolanus, G. Walz, *et al.* 1991. CD62/P-selectin recognition of myeloid and tumor cell sulfatides. *Cell* **67:** 35–44.

113. Suzuki, Y., Y. Toda, T. Tamatani, *et al.* 1993. Sulfated glycolipids are ligands for a lymphocyte homing receptor, L-selectin (LECAM-1), binding epitope in sulfated sugar chain. *Biochem. Biophys. Res. Commun.* **190:** 426–434.

114. Garcia, J., N. Callewaert & L. Borsig. 2007. P-selectin mediates metastatic progression through binding to sulfatides on tumor cells. *Glycobiology* **17:** 185–196.

115. Jeon, S.-B., H.J. Yoon, S.-H. Park, *et al.* 2008. Sulfatide, a major lipid component of myelin sheath, activates inflammatory responses as an endogenous stimulator in brain-resident immune cells. *J. Immunol.* **181:** 8077–8087.

116. Purkrábková, T., K. Smetana Jr., B. Dvoránková, *et al.* 2003. New aspects of galectin functionality in nuclei of cultured bone marrow stromal and epidermal cells: biotinylated galectins as tool to detect specific binding sites. *Biol. Cell* **95:** 535–545.

117. Smetana Jr., K., B. Dvoránková, M. Chovanec, *et al.* 2006. Nuclear presence of adhesion/growth-regulatory galectins in normal/malignant cells of squamous epithelial origin. *Histochem. Cell Biol.* **125:** 171–182.

118. Ledeen, R.W. & G. Wu. 2007. GM1 in the nuclear envelope regulates nuclear calcium through association with a nuclear sodium-calcium exchanger. *J. Neurochem.* **103**(Suppl 1): 126–134.

119. Haudek, K.C., K.J. Spronk, P.G. Voss, *et al.* 2010. Dynamics of galectin-3 in the nucleus and cytoplasm. *Biochim. Biophys. Acta* **1800:** 181–189.

120. Kodet, O., B. Dvoránková, L. Lacina, *et al.* 2011. Comparative analysis of nuclear presence of adhesion/growth-regulatory galectins and galectin reactivity in interphasic and mitotic cells. *Folia Biol. (Praha)* **57:** 125–132.

121. Ledeen, R.W. & G. Wu. 2011. New findings on nuclear gangliosides: overview on metabolism and function. *J. Neurochem.* **116:** 714–720.

122. Gabius, H.-J. 2009. Animal and human lectins. In *The Sugar Code. Fundamentals of Glycosciences.* H.-J. Gabius, Ed.: 317–328. Wiley-VCH. Weinheim, Germany.